Lecture Notes in Mathematics 1617

Springer
Berlin
Heidelberg
New York
Barcelona
Budapest
Hong Kong
London
Milan
Paris
Santa Clara
Singapore
Tokyo

Vadim Yurinsky

Sums and Gaussian Vectors

Springer

Author

Vadim Vladimirovich Yurinsky
Sobolev Institute of Mathematics
Siberian Division of the Russian Academy of Sciences
Universitetsky prospekt, 4
Novosibirsk, 630090, Russia

Library of Congress Cataloging-in-Publication Data

Yurinsky, Vadim, 1945-
Sums and Gaussian vectors / Vadim Yurinsky.
p. cm. -- (Lecture notes in mathematics ; 1617)
Includes bibliographical references (p. -) and index.
ISBN 3-540-60311-5 (soft cover)
1. Gaussian sums. 2. Limit theorems (Probability theory)
I. Title. II. Series: Lecture notes in mathematics (Springer-Verlag) ; 1617.
QA3.L28 no. 1617
[QA246.8.G38]
510 s--dc20
[519.2'6] 95-20482
CIP

Mathematics Subject Classification (1991): 60B12, 60F05, 60F10

ISBN 3-540-60311-5 Springer-Verlag Berlin Heidelberg New York

Typesetting: Camera-ready TEX output by the author
SPIN: 10479502 46/3142-543210 - Printed on acid-free paper

Preface

In a traditional course of probability theory, the main result is the central limit theorem — the assertion that Gaussian distributions approximate those of sums of independent random variables. Since its discovery, probabilists have had to work hard, both learning calculus and inventing their own tools, before the view of the effects that appear in summing real random variables became reasonably clear.

The more recent shift to study of distributions in "very high" or infinite dimensions brought new discoveries and disappointments.

A fact that can safely be included into either category is that many results for sums of infinite-dimensional random vectors are quite similar to their real-line counterparts, and can be obtained by extremely simple means.

This book is an exposition of this homely part of probability in infinite dimensions — inequalities and asymptotic expressions for large deviation probabilities, normal and related approximations for distributions of sums — all of them considered as preparations for the feat of getting a very general CLT with an estimate of convergence rate.

Naturally, something is lost upon the passage from one to infinitely many dimensions — distributions in high dimensions can develop all real-line pathologies as well as some unthinkable in the classical context. Thus, most results concerning convergence rates in the CLT become fairly bulky if attention is paid to details. This makes the choice between transparency and completeness of exposition even less easy.

The compromise attempted here is to provide a reasonably detailed view of the ideas that have already gained a firm hold, to make the treatment as unified as possible, and sacrifice some of the details that do not fit into the scheme or tend to inflate the text beyond reasonable limits. The price is that such a selection inevitably results biased, and one of the sacrifices was the refined CLT itself. Bibliographic commentary is intended as a partial compensation of bias.

Most of the text of this book was written in Novosibirsk at the *Institute of Mathematics* of Siberian Division of the Russian Academy of Sciences. At the final stage, the author stayed for some time at Departamento de Matemáticas de *Universidad de Oviedo* as a scholarship holder of the *FICYT* (Fundación por el Fomento en Asturias de Investigación Científica Aplicada Y Tecnología). I am sincerely grateful to these two institutions for the opportunity to work without haste they gave me.

I am sincerely grateful to my colleagues, in Novosibirsk, Moscow, Vilnius, Kiev, and many other places, who shared their ideas and created the intellectual stimuli indispensable for any mathematical research, and I profit of the opportunity to thank specially the librarians of the Institute of Mathematics who manage to keep our Novosibirsk library fairly complete despite all the difficulties.

I express my gratitude to Professors Yu.V.Prokhorov and V.V.Sazonov for advice and encouragement.

Contents

Chapter 1

Gaussian Measures in Euclidean Space

This chapter exposes some theorems about the distributions of convex functions with a Gaussian random vector for argument. The results included are mainly inequalities for specific characteristics of such laws. Some of them will later serve to derive similar estimates in infinite dimensions.

The terms "normal" and "Gaussian" (distribution, random variable, etc.) are used as synonyms. The former usually refers to the real line, and the latter to the case of several dimensions.

1.1 Preliminaries

This section is a reminder of basic facts concerning Gaussian distributions in Euclidean spaces.

1.1.1 Standard Normal Distribution

On the real line, the density and distribution function (DF) of the standard normal distribution are

$$\varphi(x) = \exp\left\{-\tfrac{1}{2}x^2\right\}/\sqrt{2\pi}, \quad \Phi(x) = \int_{-\infty}^{x} \varphi(y)dy. \tag{1.1.1}$$

Its characteristic function (CF) is $\int_{-\infty}^{\infty} e^{itx}\varphi(x)dx = \exp\{-t^2/2\}$.

For $x \to \infty$, the behavior of the standard normal DF is described by the asymptotic relation

$$1 - \Phi(x) = [1 + O(1/x^2)]\varphi(x)/x. \tag{1.1.2}$$

There are similar expressions for its derivatives:

$$\Phi'(x) = x\,[1 - \Phi(x)]\,(1 + o(1)), \quad \Phi''(x) = -x^2\,[1 - \Phi(x)]\,(1 + o(1)). \tag{1.1.3}$$

The following estimates for integrals with respect to the standard normal distribution (1.1.1) are used in some calculations below.

Lemma 1.1.1 *For $t \in \mathbf{R}$ and $l = 1, 2$,*

$$I_l(t) \equiv \int_t^\infty (x-t)^l d\Phi(x) \le (1+t_-)^l .$$

Proof of Lemma 1.1.1. The inequality $I_2(t) \le I_2(0) = \frac{1}{2}$ is evident for $t \ge 0$ since the derivative $I_2'(t)$ is nonpositiive.

If $t < 0$, then

$$I_2(t) \le \int_{-\infty}^\infty \left[x^2 + 2|x|t_- + t_-^2\right] d\Phi(x) \le (1+t_-)^2 .$$

The estimate for $I_1(t)$ is derived similarly. ○

Lemma 1.1.2 *If $\tau > t \ge 0$, then*

$$\int_\tau^\infty (x-t)^l d\Phi(x) \le l!(\tau - t)^l [1-\Phi(\tau)] \sum_{m=0}^{l} \frac{\alpha^m}{(l-m)!}, \quad \alpha = \frac{1}{\tau(\tau-t)}.$$

Proof of Lemma 1.1.2. If $\tau > 0$, then the functions

$$p_1(x) = \tau \exp\{-\tau(x-\tau)\}, \quad p_2(x) = \varphi(x)/[1-\Phi(\tau)]$$

are both densities of distributions concentrated on the half-line $[\tau, \infty)$.

The ratio $p_1(x)/p_2(x)$ is increasing and the integrals of both functions over $[\tau, \infty)$ equal 1, so there is a number $A \ge \tau$ such that $p_1(x) \ge p_2(x)$ for $x > A$ and $p_1(x) \le p_2(x)$ for $x < A$. Consequently, for each increasing function f

$$\int_\tau^\infty f(x)[p_1(x) - p_2(x)]\, dx = \int_\tau^\infty [f(x) - f(A)][p_1(x) - p_2(x)]\, dx \ge 0,$$

and there is inequality

$$\begin{aligned} \int_\tau^\infty f(x) d\Phi(x) &\le [1-\Phi(\tau)] \int_\tau^\infty f(x) \tau e^{-\tau(x-\tau)} dx \\ &= [1-\Phi(\tau)] \int_0^\infty f(\tau + x/\tau)\, e^{-x}\, dx. \end{aligned}$$

It is easy to calculate the integral if $f(x) = (x-t)^l$. The calculation yields the inequality of the lemma. ○

1.1.2 Gaussian Distributions: Basic Definitions

The *standard Gaussian distribution* in $\mathbf{R}^k$, $N_{k,0,I}$, is one with the density

$$n_{k,0,I}(x) = (2\pi)^{-k/2} \exp\left\{-\frac{1}{2}|x|^2\right\}. \tag{1.1.4}$$

A random vector (RV) with distribution (1.1.4) is also called standard Gaussian. Its coordinates are independent random variables (rv's) with distribution (1.1.1).

If the distribution of the RV $\eta \in \mathbf{R}^l$ is $N_{l,0,I}$, then, whatever the choice of a constant vector $a \in \mathbf{R}^k$ and $k \times l$ matrix A, the RV

$$\xi = a + A\eta \in \mathbf{R}^k \tag{1.1.5}$$

has the CF

$$g_{k,a,V}(t) \equiv \mathbf{E}\exp\{i(t,\xi)\} = \exp\left\{i(a,t) - \frac{1}{2}(Vt,t)\right\}, \tag{1.1.6}$$

where $t \in \mathbf{R}^k$ and $V = AA^T$. The covariance matrix of ξ and its expectation are

$$\left\|\mathbf{cov}\left(\xi^{(i)},\xi^{(j)}\right)\right\| = V, \quad \mathbf{E}\xi = a, \tag{1.1.7}$$

where $\mathbf{cov}\left(\xi^{(i)},\xi^{(j)}\right) = \mathbf{E}\xi^{(i)}\xi^{(j)} - \mathbf{E}\xi^{(i)}\mathbf{E}\xi^{(j)}$. Each nonnegative matrix admits some factorization $V = AA^T$ with $l \geq k$. Consequently, the right-hand side of (1.1.6) defines the CF of a probability law in $\mathbf{R}^k$ for any a and $V \geq 0$.

A ***Gaussian distribution*** in $\mathbf{R}^k$ is a distribution with CF (1.1.6). It is denoted by $N_{k,a,V}$. Formula (1.1.5) shows that a RV with an arbitrary Gaussian distribution can be obtained from a standard Gaussian one with same or greater number of coordinates by an affine transform. This proves, in particular, that for each Gaussian RV

$$\mathbf{E}\exp\left\{h\,|\xi|^2\right\} < \infty \text{ if } |h| < h_0(V).$$

The parameters a and V in (1.1.6) are the expectation and covariance matrix of the distribution. The Gaussian distribution $N_{k,a,V}$ is *nondegenerate* if $V > 0$.

Theorem 1.1.1 *A nondegenerate Gaussian distribution $N_{k,a,V}$ has density*

$$n_{k,a,V}(x) = \frac{1}{\sqrt{\det(2\pi V)}}\exp\left\{-\frac{1}{2}\left(V^{-1}[x-a],[x-a]\right)\right\}. \tag{1.1.8}$$

If the matrix V is degenerate, the Gaussian distribution $N_{k,a,V}$ is concentrated in the affine manifold $L = a + V^{1/2}\mathbf{R}^k$. It is absolutely continuous with respect to the Lebesgue measure in this Euclidean space (whose metric is inherited from the original space). This is easily verified, e.g., using (1.1.5).

Proof of Theorem 1.1.1. The right-hand side of (1.1.6) is summable if $V > 0$, so the distribution corresponding to this CF does indeed have the continuous density

$$n_{k,a,V}(x) = (2\pi)^{-k}\int\exp\left\{-i(x-a,t) - \tfrac{1}{2}(Vt,t)\right\}\,dt.$$

By changing variables to $y = V^{1/2}t$ and $z = V^{-1/2}(x-a)$, the integral is transformed into

$$n_{k,a,V}(x) = n_{k,0,I}(z)/\sqrt{\det(V)} = g_{k,0,I}(z)/\sqrt{\det(2\pi V)}.$$

This proves the theorem. ○

1.1.3 Characterization Theorems

Gaussian distributions are infinitely divisible and stable. It follows from the definition of a multidimensional Gaussian distribution that a RV is Gaussian if and only if its projection onto each one-dimensional subspace is Gaussian. This circumstance allows one to extend known theorems concerning characterization of Gaussian distributions on the real line to multidimensional Euclidean spaces.

Theorem 1.1.2 *Let $\xi,\ \eta \in \mathbf{R}^k$ be independent RV's.*
If $\xi + \eta$ is Gaussian, then both ξ and η are Gaussian as well.

Theorem 1.1.3 *Let ξ and η be i.i.d. RV's. Put*

$$\xi(a) = \cos(a)\xi + \sin(a)\eta,\ \eta(a) = -\sin(a)\xi + \cos(a)\eta.$$

a) *If, for some $a \in (0, \pi/2)$, the RV's $\xi(a)$ and $\eta(a)$ are independent and have identical distributions, then ξ and η are centered Gaussian RV's: $\mathcal{L}(\xi) = \mathcal{L}(\eta) = N_{k,0,V}$ with some matrix $V \geq 0$.*

b) *If ξ and η are independent and have distribution $N_{k,0,V}$, then $\xi(a)$ and $\eta(a)$ are, for each value of a, also independent with the same distribution: $\mathcal{L}(\xi) = \mathcal{L}(\eta) = N_{k,0,V}$.*

1.1.4 Monotonicity in Covariances

The next theorem is a special case of the so-called Slepian inequality. It shows that for a Gaussian RV the dependence of the *joint DF function of coordinates*

$$\mathbf{P}\{\xi < x\} = \mathbf{P}\left(\bigcap_{j=1}^{k}\{\xi^{(j)} < x^{(j)}\}\right)$$

on its covariance matrix is, in a sense, monotone.

The CF inversion formula yields the following expression for the DF of a Gaussian RV: if $\mathcal{L}(\xi) = N_{k,a,V}$, then

$$N_{k,a,V}(x) = \mathbf{P}\{\xi < x\} = \int_{y<x}\left[(2\pi)^{-k}\int e^{i(t,y)}g_{k,a,V}(t)dt\right]dy. \quad (1.1.9)$$

It follows from this equality that, for x fixed, the DF is a smooth function in its parameters a and V on the set $\{\langle a, V\rangle \in \mathbf{R}^k \times (\mathbf{R}^k \otimes \mathbf{R}^k) : V > 0\}$.

Lemma 1.1.3 *Consider the Gaussian distribution $N_{k,a,\widetilde{V}}$ with the covariance matrix*

$$\widetilde{V} = \widetilde{V}(u) = V + uC,\ u \in \mathbf{R}.$$

where $V > 0$ and the real square matrix $C = (C_{ij})$ is symmetrical with zero diagonal elements: $c_{ii} = 0,\ i = \overline{1,k}$. The DF of this distribution satisfies the relations

$$\frac{d}{du}N_{k,a,\widetilde{V}}(x)\bigg|_{u=0} = \frac{1}{2}\sum_{l\neq m} c_{lm}\int_{A_{lm}(x)} n_{k,a,V}(y)\,\mathrm{mes}_{lm}(dy) \geq 0,$$

where mes_{lm} *is the Lebesgue measure in the* $(k-2)$*-dimensional orthant*

$$A_{lm}(x) = \left\{ y : y^{(j)} < x^{(j)},\ j \neq l, m;\ y^{(l)} = x^{(l)},\ y^{(m)} = x^{(m)} \right\}.$$

Proof of Lemma 1.1.3. CF (1.1.6) decays rapidly as $|t| \to \infty$. Hence for small values of u it is possible to interchange the order of differentiation and integration. It follows that in a neighborhood of zero

$$\frac{d}{du} N_{k,a,\widetilde{V}}(x) = \frac{-1}{2(2\pi)^k} \int_{y<x} \left(\sum_{l \neq m} c_{lm} \int t^{(l)} t^{(m)}\, g_{k,a,\widetilde{V}}(t)\, e^{-i(t,y)}\, dt \right) dy.$$

The inner integral in the right-hand side can be represented as a derivative with respect to $y^{(l)}$ and $y^{(m)}$: indeed,

$$\begin{aligned} \int t^{(l)} t^{(m)}\, g_{k,a,\widetilde{V}}(t)\, e^{-i(t,y)}\, dt &= -\frac{\partial^2}{\partial y^{(l)} \partial y^{(m)}} \int g_{k,a,\widetilde{V}}(t)\, e^{-i(t,y)}\, dt \\ &= -(2\pi)^k \frac{\partial^2 n_{k,a,\widetilde{V}}}{\partial y^{(l)} \partial y^{(m)}}(y). \end{aligned}$$

To obtain the assertion of the lemma, it suffices to integrate the last identity in variables $y^{(l)}$, $y^{(m)}$ and put $u = 0$. ○

Theorem 1.1.4 *Assume that* $\mathcal{L}(\xi) = N_{k,a,V}$ *and* $\mathcal{L}(\eta) = N_{k,a,W}$. *If the covariance matrices of these Gaussian distributions satisfy the relations*

$$V_{jj} = W_{jj}, \quad V_{ij} \leq W_{ij},\ i \neq j,$$

then for all $x \in \mathbf{R}^k$

$$\mathbf{P}\{\xi < x\} \leq \mathbf{P}\{\eta < x\}.$$

Proof of Theorem 1.1.4. If both covariance matrices are nondegenerate, the inequality of the theorem follows from Lemma 1.1.3. Indeed, in this case the matrix $\widetilde{V} = uW + (1-u)V$ is also nondegenerate for all $u \in [0,1]$, and $C = W - V$ satisfies the conditions of this lemma, so $(d/du) N_{k,a,\widetilde{V}(u)}(x) \geq 0$.

If at least one of the covariance matrices degenerates, one can first derive the inequality of the theorem for the distributions with covariance matrices $V + \varepsilon I$ and $W + \varepsilon I$ assuming that $\varepsilon > 0$.

As $\varepsilon \to 0$, the distributions corresponding to these latter matrices weakly converge to the original ones. This convergence yields the inequality of the theorem at all continuity points of the DF's considered. The set of these points is dense in $\mathbf{R}^k$. Thus, one more passage to the limit shows that the inequality holds everywhere. ○

1.1.5 Conditional Distributions and Projections

For Gaussian rv's, independence and absence of correlations are equivalent.

Theorem 1.1.5 *Let the joint distribution of the real rv's ξ and η be a Gaussian one. Then ξ and η are independent if and only if they are uncorrelated, i.e., if* $\mathbf{cov}(\xi,\eta) = 0$.

Proof of Theorem 1.1.5 reduces to calculating the joint CF:

$$\mathbf{E}\exp\{is\xi + it\eta\} = \exp\left\{is\mathbf{E}\xi + it\mathbf{E}\eta - \tfrac{1}{2}s^2\mathbf{D}\xi - \tfrac{1}{2}t^2\mathbf{D}\eta - st\,\mathbf{cov}(\xi,\eta)\right\}.$$

The right-hand side of this equality splits into the product of CF's of ξ and η only if the covariance equals zero. ○

One more result of this kind is

Theorem 1.1.6 *Let $\xi \in \mathbf{R}^k$ be a Gaussian RV with the distribution $N_{k,a,V}$ and $b \in \mathbf{R}^k$ a constant vector. Assume, moreover, that $(Vb, b) > 0$, and put $\eta = (\xi, b)$.*

The conditional distribution of ξ given $\eta = y$ is Gaussian with expectation and covariance matrix

$$m(y) = a - (Vb,b)^{-1}[y - (a,b)]\,Vb, \quad \Sigma = V - (Vb,b)^{-1}Vb(Vb)^T,$$

i.e., $\mathbf{P}\{\xi \in A\,|\eta = y\} = N_{k,m(y),\Sigma}(A)$ *for each measurable subset* $A \subset \mathbf{R}^k$.

Proof of Theorem 1.1.6. The RV ξ can be represented in the form $\xi = \eta B + X$, where the coordinates of X are uncorrelated with the real rv η. To obtain a decomposition of this kind it suffices to set $B = (Vb,b)^{-1}\,Vb$ and $X = \xi - \eta B$.

Apply (1.1.6) to compute $\mathbf{E}\exp\{i(\lambda, X) + i(\mu, \eta B)\}$, the CF of the $2k$-dimensional joint distribution. It splits into the product of marginal CF's, those of X and ηB. Hence these RV's are independent. The same calculation shows they are also Gaussian, and a little more work yields the covariance matrix and expectation: thus, given $\eta = y$ the conditional expectation of ξ equals $\mathbf{E}X + yB$, etc. ○

The following special case will be used later on. Let $\xi = \left(\xi^{(j)},\ j = \overline{1,k}\right)$ be a Gaussian RV. For the conditional distribution of $\left(\xi^{(j)},\ j = \overline{2,k}\right) \in \mathbf{R}^{k-1}$ given $\xi^{(1)} = y$, the coordinates of the expectation are, by Theorem 1.1.6,

$$m^{(j)} = \mathbf{E}\xi^{(j)} + \frac{\mathbf{cov}\left(\xi^{(1)},\xi^{(j)}\right)}{\mathbf{cov}\left(\xi^{(1)},\xi^{(1)}\right)}\left(y - \mathbf{E}\xi^{(1)}\right), \tag{1.1.10}$$

and the covariance matrix has the elements

$$v'_{ij} = \mathbf{cov}\left(\xi^{(i)},\xi^{(j)}\right) - \frac{\mathbf{cov}\left(\xi^{(1)},\xi^{(i)}\right)\mathbf{cov}\left(\xi^{(1)},\xi^{(j)}\right)}{\mathbf{cov}\left(\xi^{(1)},\xi^{(1)}\right)}. \tag{1.1.11}$$

1.1.6 Laplace Transform for DF of Squared Norm

Let $\xi \in \mathbf{R}^k$ be a Gaussian RV with the distribution $N_{k,a,V}$ whose covariance matrix V has the eeigenvalues

$$\sigma_1^2 = \cdots = \sigma_\nu^2 > \sigma_{\nu+1}^2 \geq \cdots \geq \sigma_k^2,$$

where ν is the multiplicity of the principal eigenvalue.

Theorem 1.1.7 *For each complex number z such that $2\sigma_1^2 \operatorname{Re} z < 1$*

$$\mathbf{E}\exp\left\{z\,|\xi|^2\right\} = \exp\left\{z\left(\left[I - 2zV\right]^{-1} a, a\right)\right\} \prod_{j=1}^{k} 1/\sqrt{1 - 2z\sigma_j^2}.$$

The branch of the complex-valued square root in the right-hand side is selected so as to satisfy the condition $\sqrt{1 - 2z\sigma_j^2} > 0$ for $\mathbf{Im} z = 0$.

Proof of Theorem 1.1.7. Let Q be an orthogonal matrix which transforms V to its diagonal form: $Q^T V Q = \left(\sigma_j^2 \delta_{ij}\right)$. According to (1.1.5)–(1.1.6), the distribution $N_{k,a,V}$ is that of the RV $a + QD\eta$, where $D = (\sigma_j \delta_{ij})$ is a diagonal matrix and η a standard Gaussian RV. Hence

$$\mathbf{E}\exp\left\{z\,|\xi|^2\right\} = \mathbf{E}\exp\left\{z\,|\bar{a} + D\eta|^2\right\} = \prod_{j=1}^{k} I(\bar{a}^{(j)}, \sigma_j, z),$$

where $\bar{a} = Q^T a = \left(\bar{a}^{(j)}\right) \in \mathbf{R}^k$, $|\bar{a}| = |a|$, and

$$I(\bar{a}, \sigma, z) = \frac{1}{\sqrt{2\pi}} \exp\left\{z\,|\bar{a}^2|\right\} \int_{-\infty}^{\infty} \exp\left\{-\tfrac{1}{2}\left(1 - 2z\sigma^2\right) u^2 + 2z\sigma\bar{a}\right\} du.$$

The integrand in I is analytical, so the choice of integration path in the complex plane does not influence the value of the integral. If $\mathbf{Re}\left(1 - 2z\sigma^2\right) > 0$, one arrives at the equality

$$\begin{aligned} I(\bar{a}, \sigma, z) &= \frac{1}{\sqrt{2\pi}} \exp\left\{\frac{z\bar{a}^2}{\left(1 - 2z\sigma^2\right)}\right\} \int_{\operatorname{Im} u = 0} \exp\left\{-\frac{1 - 2z\sigma^2}{2}(u - c)^2\right\} du \\ &= \exp\left\{\frac{z\bar{a}^2}{1 - 2z\sigma^2}\right\} / \sqrt{1 - 2z\sigma_j^2}. \end{aligned}$$

The equality of the theorem now results from the relation

$$\sum_{j=1}^{k} \frac{\left(\bar{a}^{(j)}\right)^2}{1 - 2z\sigma_j^2} = \left(\left[I - 2zD^2\right]^{-1} \bar{a}, \bar{a}\right) = \left(\left[I - 2zV\right]^{-1} a, a\right). \bigcirc$$

1.2 Extremal Properties of Half-Spaces

1.2.1 Isoperimetric Property of Sphere

The ***uniform distribution*** on the sphere $S_r \subset \mathbf{R}^k$ is defined by the equality

$$\sigma_r(A) = |S_r|^{-1} \int_{A \bigcap \{|x| = r\}} ds(x), \tag{1.2.1}$$

where $|S_r|$ is the surface area of the sphere (see (A.1.7) and (A.1.8)).

Distribution (1.2.1) can be considered both as a measure on Borel subsets of $\mathbf{R}^k$ and as one on subsets of the surface of the sphere. Whenever this does not lead to misunderstanding, no special effort is spent on discerning between the two:

$$\sigma_r(A) = \sigma_r\left(A\bigcap S_r\right), \quad A \in \mathcal{B}\left(\mathbf{R}^k\right); \quad \sigma_r\left(\mathbf{R}^k\right) = \sigma_r(S_r) = 1. \qquad (1.2.2)$$

For a half-space

$$H_{e,\rho} = \{x : (x,e) \leq \rho\}, \; H^c_{e,\rho}\{x : (x,e) > \rho\}, \qquad (1.2.3)$$

the value of $\sigma_r(H_{e,\rho})$ does not depend on the direction of the unit vector e, $|e| = 1$. For instance, $\sigma_r(H_{e,0}) = \sigma_r\left(H^c_{e,0}\right) = \frac{1}{2}$, whatever the choice of e. If $\rho > 0$, then $\sigma_r(H_{e,\rho}) > \frac{1}{2}$ and $\sigma_r\left(H^c_{e,\rho}\right) < \frac{1}{2}$.

Half-spaces possess an important extremal property with respect to the uniform law on sphere. It is expressed by the so-called *isoperimetric inequality* for sphere which is described below in Proposition 1.2.1. Some new notation is necessary to state it.

On the sphere S_r, topology can be defined using the *geodesic distance* $\rho(x,y)$ equal to the length of the shortest geodesic connecting x and y. A geodesic neighborhood of a set $A \subset S_r$ is (cf. (A.1.1))

$$\mathcal{U}_r[A,\delta] = \{x \in S_r : \rho(x,A) < \delta\}, \; \rho(x,A) = \inf_{y\in A} \rho(x,y). \qquad (1.2.4)$$

Proposition 1.2.1 *Let A be an arbitrary measurable subset of $\mathbf{R}^k$. If*

$$\sigma_r(A) \geq \sigma_r(H)$$

for some subspace $H = H_{e,\rho}$, $e \in \mathbf{R}^k$, $\rho > 0$, then the values of the uniform distribution on geodesic neighborhoods of these sets are related through the inequality

$$\forall\,\delta > 0 \;\; \sigma_r\left(\mathcal{U}_r\left[A\bigcap S_r, \delta\right]\right) \geq \sigma_r\left(\mathcal{U}_r\left[H\bigcap S_r, \delta\right]\right). \qquad (1.2.5)$$

This proposition will not be proved here (see Appendix B for references).

1.2.2 Constructing Gaussian Law from Uniform Ones

The standard Gaussian distribution can be approximated by marginal distributions of uniform laws (1.2.1) on spheres in very high dimensions.

Below, the space $\mathbf{R}^k$ is considered as a k-dimensional subspace of the Euclidean space $\mathbf{R}^D$ with large (and later infinitely growing) number of dimensions. The corresponding orthogonal projection is denoted by $\mathcal{P} : \mathbf{R}^D \to \mathbf{R}^k$, and $\sigma_{r,D}$ is the uniform distribution (1.2.1) on the sphere of the larger space, $S_r^{(D)} = \left\{x \in \mathbf{R}^D : |x| = r\right\}$.

Lemma 1.2.1 *The distribution in $\mathbf{R}^k$ defined by the formula*

$$\nu_{r,D}(A) \equiv \sigma_{r,D}\left(\left\{z \in S_r^{(D)} : \mathcal{P}z \in A\right\}\right)$$

is absolutely continuous with respect to the Lebesgue measure in $\mathbf{R}^k$ *and has density*

$$\tilde{n}_{r,D}(x) = c\left(1 - |x|^2/r^2\right)_+^{(D-k-2)/2},$$

where c is a normalizing constant.

If k is fixed and $r = r(D) = \sqrt{D}$, *then the distribution* $\nu_{r(D),D}$ *converges in variation to the standard Gaussian law* (1.1.4) *as* $D \to \infty$.

Proof of Lemma 1.2.1. Choose for coordinates of a point z on the D-dimensional sphere $S_r^{(D)}$ the vector $x = \mathcal{P}z$ and some $(D-k-1)$-dimensional projection of $y = z - \mathcal{P}z$. Represent $\nu_{r,D}(A)$ by a "volume" integral in these coordinates. If x is fixed, the other coordinate runs over a solid ball of squared radius $r^2 - |x|^2$ in its range as z runs over the original sphere. This yields the above expression for the density.

To establish convergence in variation, evaluate the difference of the measures considered in as straightforward a fashion. ○

1.2.3 An Isoperimetric Inequality for Gaussian Laws

The isoperimetric property and connection between uniform and Gaussian laws combine to yield the following

Theorem 1.2.1 *Let* $\eta \in \mathbf{R}^k$ *be a RV with distribution* $N_{k,0,I}$, $A \subset \mathbf{R}^k$ *an arbitrary measurable set, and* $H = H_{e,\rho}$ *a half-space.*

If $\mathbf{P}\{\eta \in A\} \geq \mathbf{P}\{\eta \in H\}$, *then a similar inequality is fulfilled for the neighborhoods of A and H :*

$$\forall\, \delta > 0 \ \ \mathbf{P}\left\{\eta \in A^\delta\right\} \geq \mathbf{P}\left\{\eta \in H^\delta\right\}. \tag{1.2.6}$$

Corollary 1.2.1 *Assume that* $\mathbf{P}\{\eta \in A\} \geq \Phi(t) > 0$ *for some real t, and the assumptions of Theorem* 1.2.1 *are fulfilled. Then for each* $\delta > 0$

$$\mathbf{P}\left\{\eta \in A^\delta\right\} \geq \Phi(t+\delta), \quad \Phi^{-1}\left(\mathbf{P}\left\{\eta \in A^\delta\right\}\right) \geq \Phi^{-1}(\mathbf{P}\{\eta \in A\}) + \delta,$$

where Φ^{-1} *is the inverse of the standard normal DF.*

If the function $U : [0, \infty) \to \mathbf{R}$ *is nondecreasing, then*

$$\mathbf{E}\{U(\,\mathrm{dist}(\eta, A)\,)\,;\ \eta \notin A\} \leq \int_t^\infty U(u-t)\,d\Phi(u). \tag{1.2.7}$$

It is essential that the number of dimensions is not explicit in the statement of Theorem 1.2.1 and Corollary 1.2.1. This circumstance makes it possible to apply them in the study of infinite-dimensional Gaussian distributions.

Proof of Corollary 1.2.1. The first inequality of the corollary is, because of (1.2.10), just another form of the assertion of Theorem 1.2.1. To derive from it the second one, it is sufficient to apply the inverse function Φ^{-1} to both sides of the inequality.

The inequality for integrals follows from the identity

$$\mathbf{E}\left\{U\left(\operatorname{dist}(\eta, A)\right); \, \eta \notin A\right\} = \int_0^\infty U(u) d\mathbf{P}\left\{\operatorname{dist}(\eta, A) < u\right\}$$

after integration by parts. ○

Proof of Theorem 1.2.1. It is easier to obtain the inequality of the theorem first for the case of $N(A) > N(H)$, where $N(A) = N_{k,0,I}(A) = \mathbf{P}\{\eta \in A\}$. The case of equality can be treated through a passage to the limit using a sequence of growing half-spaces H_n which approximate H and are contained in it.

Assume that $D > k$, and that $\rho(x, y)$ is the geodesic metric (see (1.2.4)) on the sphere $S_r^{(D)}$ with $r = \sqrt{D}$.

The geodesic distance between two arbitrary points of the sphere is greater than the Euclidean one in the space containing it: $\rho(x, y) \geq |x - y|$. For this reason geodesic neighborhoods (1.2.4) $\mathcal{U}_\delta[B] = \mathcal{U}_r(B, \delta)$ admit the following inclusion:

$$S_r^{(D)} \bigcap \mathcal{P}^{-1}\left(A^\delta\right) \supset \mathcal{U}_\delta\left[S_r^{(D)} \bigcap \mathcal{P}^{-1}(A)\right], \qquad (1.2.8)$$

where $\delta > 0$ is arbitrary, A^δ is the Euclidean neighborhood in $\mathbf{R}^k$, and $\mathcal{P}^{-1}(\cdot)$ is the pre-image of a set under the projection $\mathcal{P}$.

Given $\delta > 0$, one can choose, for each half-space H, such a positive number $\bar{\delta}(D)$ that the geodesic neighborhood of the pre-image of H under the projection $\mathcal{P}$ would coincide with the pre-image of the Euclidean $\bar{\delta}$-neighborhood of the same half-space:

$$\mathcal{U}_\delta\left[S_r^{(D)} \bigcap \mathcal{P}^{-1}(H)\right] = \mathcal{P}^{-1}\left(H^{\bar{\delta}}\right). \qquad (1.2.9)$$

It is easy to see that as $D \to \infty$, the curvature of the sphere of radius $\sqrt{D}$ goes to zero and $\bar{\delta} \to \delta$, i.e., the geodesic distances on the sphere converge to the Euclidean ones.

If the Gaussian measures of A and H satisfy the inequality $N(A) > N(H)$, the same inequality would be true for the values of the uniform distribution on pre-images of these sets (provided that the number of dimensions D is large):

$$\sigma_{r,D}\left(\mathcal{P}^{-1}(A)\right) > \sigma_{r,D}\left(\mathcal{P}^{-1}(H)\right).$$

This permits one to apply the isoperimetric inequality of Proposition 1.2.1 to compare the measures of the neighborhoods: by (1.2.8) and (1.2.9), for large values of D

$$\begin{aligned}\sigma_{r,D}\left(\mathcal{P}^{-1}\left(A^\delta\right)\right) &\geq \sigma_{r,D}\left(\mathcal{U}_\delta\left[S_r^{(D)} \bigcap \mathcal{P}^{-1}(A)\right]\right) \\ &\geq \sigma_{r,D}\left(\mathcal{U}_\delta\left[S_r^{(D)} \bigcap \mathcal{P}^{-1}(H)\right]\right) \geq \sigma_{r,D}\left(\mathcal{P}^{-1}(H^{\bar{\delta}})\right) \\ &\geq \sigma_{r,D}\left(\mathcal{P}^{-1}(H^{\delta-\gamma})\right),\end{aligned}$$

where $\gamma > 0$ is arbitrarily small.

By Lemma 1.2.1, the above yields in the limit (at $D \to \infty$) the inequality

$$\forall \delta > 0 \;\; N\left(A^\delta\right) > N\left(H^{\delta-\gamma}\right), \;\; \gamma \in (0, \delta),$$

where $\lim_{\gamma \searrow 0} N\left(H^{\delta-\gamma}\right) = N\left(H^{\delta}\right)$ by continuity of Gaussian DF's on the real line. The inequality of the theorem follows from these relations. ○

1.2.4 Application to Large Deviations

The probability that a Gaussian RV hits a half-space can be expressed in terms of the standard normal DF (1.1.1). As the absolute value of its argument goes to infinity, the normal DF approaches its limit values 0 or 1 most rapidly (1.1.2). Theorem 1.2.1 shows that multidimensional Gaussian laws are as highly concentrated around their expectations.

For instance, let $|\cdot|_*$ be a seminorm in $\mathbf{R}^k$ satisfying the condition $|x|_* \leq B|x|$ and η a Gaussian RV with distribution (1.1.4). Define $H_{e,\rho}$ by (1.2.3) using some unit vector $e \in \mathbf{R}^k$. The distribution of (η, e) is (1.1.1), so

$$\mathbf{P}\{\eta \in H_{e,\rho}\} = \mathbf{P}\{(\eta, e) < \rho\} = \Phi(\rho). \tag{1.2.10}$$

The rv $|\eta|_*$ is finite. Hence there are positive numbers r_0, ρ_0, and $\alpha_0 > \frac{1}{2}$ such that

$$\mathbf{P}\{|\eta|_* \leq r_0\} = \mathbf{P}\{(\eta, e) \leq \rho_0\} = \alpha_0.$$

If $r > r_0$, then the set $\{|x|_* < r\}$ contains the $(r - r_0)/B$-neighborhood of $\{|x|_* \leq r_0\}$, and for the half-space the corresponding neigborhood is $\{(x, e) < \rho_0 + (r - r_0)/B\}$. Hence by Theorem 1.2.1

$$\mathbf{P}\{|\eta|_* > r\} \leq \mathbf{P}\{(\eta, e) > \rho_0 + (r - r_0)/B\} = 1 - \Phi\left(\rho_0 + (r - r_0)/B\right).$$

1.3 Dilations of Convex Sets

1.3.1 Some Special Symmetric Densities

For an arbitrary spherically-symmetric distribution F, its value on a measurable set $A \subset \mathbf{R}^k$ can be expressed in terms of the "radial" distribution $F_+(r) = F(B_r)$, $B_r = \{|x| \leq r\}$, and the "angle size" of the set calculated using the uniform distribution (1.2.1):

$$F(A) = \int_0^\infty \sigma_r\left(A \bigcap S_r\right) dF_+(r). \tag{1.3.1}$$

The distributions considered below are assumed to be absolutely continuous with densities of the form

$$f(x) = \exp\left\{-h\left(\ln|x|\right)\right\}, \tag{1.3.2}$$

where h is a continuously differentiable convex function. The standard Gaussian density (1.1.4) is in this class, with

$$h(u) = \tfrac{1}{2}k\ln(2\pi) + \tfrac{1}{2}\exp\{2u\}. \tag{1.3.3}$$

Theorem 1.3.1 *If F is a distribution with density* (1.3.2) *and $A \subset \mathbf{R}^k$ an* *arbitrary measurable set, then $F(\lambda A)$ is differentiable and the derivative*

$$\frac{d}{d\lambda}F(\lambda A) = \frac{k}{\lambda}F(\lambda A) - \frac{1}{\lambda}\int_{\lambda A} h'(\ln|x|)\exp\{-h(\ln|x|)\}\,dx$$

is continuous for $\lambda > 0$.

The proof of the theorem is based on the following auxiliary propositions.

Lemma 1.3.1 *Under the assumptions of Theorem 1.3.1, the function*

$$J_A(a) = \int_A \exp\{-h(a+\ln|y|)\}\,dy, \quad a \in \mathbf{R},$$

is continuously differentiable, whatever the choice of $A \subset \mathbf{R}^k$. Its derivative is

$$J'_A(a) = -\int_A h'(a+\ln|y|)\exp\{-h(a+\ln|y|)\}\,dy.$$

Lemma 1.3.2 *If the function h defines a density (1.3.2), then there exist positive constants C, U, and γ such that for $u > U$*

$$0 < h'(u)\exp\{ku - h(u)\} \le Ch'(u)\exp\{-\gamma h(u)\},$$

and for all $u < -U$ either $h'(u) \ge 0$ and

$$0 \le h'(u)\exp\{ku - h(u)\} \le h(0)\exp\{-\gamma|u|\},$$

or $h'(u) < 0$ and

$$0 < -h'(u)\exp\{ku - h(u)\} \le -Ch'(u)\exp\{-\gamma h(u)\}.$$

Proof of Lemma 1.3.2. While calculating J_A, pass to polar coordinates and then change the radial distance to the variable $u = \ln|y|$. This reduces $J_A(a)$ to the form

$$J_A(a) = \int_{-\infty}^{\infty} \chi_A(u)\exp\{ku - h(a+u)\}\,du, \tag{1.3.4}$$

where the bounded function $\chi_A(u) = \sigma_r(A \bigcap S_r)$, $r = e^u$, is the "angle size" of (1.3.1).

For $A = \mathbf{R}^k$ and $a = 0$, the integral equals one by the normalization of densities, and the angle size is constant, so conditions of the theorem include an implicit assumption that the exponential in the integrand above is summable on the real line. This is a fairly strong restriction upon the convex function h.

The exponent cannot stay bounded when $u \to +\infty$. Thus, there is some $U > 0$ such that $h(U) - kU > h(0)$ and for $u \ge U$ by convexity

$$h'(u) \ge h'(U) \ge \frac{1}{U}[h(U) - h(0)] > k.$$

Consequently, there are constants $\gamma, \varepsilon > 0$ and c such that for $u \ge U$

$$\begin{aligned} h(u) &\ge h(U) - h'(U)U + (k+\varepsilon)u, \\ h(u) - ku &= \frac{\varepsilon}{k+\varepsilon}h(u) + \frac{k}{k+\varepsilon}[h(u) - (k+\varepsilon)u] \ge \gamma h(u) + c. \end{aligned}$$

For the same values of the argument,

$$0 < h'(u) \exp\{ku - h(u)\} \le h'(u) \exp\{-c - \gamma h(u)\},$$

which is the estimate of the lemma for large positive u.

To obtain the estimates for the negative half-line, suppose first that $h'(-V) < 0$ for some $V > 0$, so that $h'(u) < 0$ for $u < -V < 0$. In this case the third inequality of the lemma is obvious:

$$0 < -h'(u) \exp\{ku - h(u)\} \le -h'(u) \exp\{-h(u)\}, \quad u \le -V < 0.$$

If $h'(u) \ge 0$ for all $u < 0$, consider the convex function $\widetilde{h}(v) = h(-v) + kv$, $v > 0$. It cannot stay bounded from above as $v \to \infty$, so there is some $V > 0$ such that $\widetilde{h}(v) > \widetilde{h}(0)$ and by convexity $\widetilde{h}'(v) > 0$ for $v \ge V$. As before,

$$\widetilde{h}(v) \ge \widetilde{h}(V) + (v - V)\widetilde{h}'(V) \ge \widetilde{c} + \widetilde{\gamma} v, \quad v > V.$$

Since the derivative of a convex function is nondecreasing,

$$0 \le h'(-v) \le h'(-V) \le h'(0),$$

and by the above

$$0 \le h'(u) \exp\{-h(u) + ku\} \le h'(0) \exp\{-\widetilde{c} - \widetilde{\gamma}\,|u|\} \text{ if } u \le -V.$$

This is the second inequality of the lemma. ○

Proof of Lemma 1.3.1. The bounds of Lemma 1.3.2 suffice to prove the existence of the derivative of $J_A(a)$. To do so, use (1.3.4) to obtain the representation

$$\frac{J_A(a+\delta) - J_A(a)}{\delta} = \int_{-\infty}^{\infty} \chi_A(u) e^{ku} \left(\int_0^1 h'(u + a + \delta w) e^{-h(u+a+\delta w)} dw \right) du.$$

The integrand stays bounded while u and δ change within a bounded set. For this reason, differentiability of the integral over the segment $[-T, T]$ follows, for each $T > 0$, from the Lebesgue theorem on majorated convergence. The integrals over the complement of the segment $[-T, T]$ are small, uniformly in $|\delta| \le 1$, by the estimates of Lemma 1.3.2 which apply to large absolute values of the argument. ○

Proof of Theorem 1.3.1. If $\lambda > 0$, the change of variable $y = \lambda^{-1} x$ leads to the equality

$$F(\lambda A) = \lambda^k \int_A \exp\{-h(\ln \lambda + \ln |y|)\}\, dy.$$

By Lemma 1.3.1, the integral in the right-hand side is continuously differentiable in parameter, and the differentiation can be performed under the sign of integration. After the variable is changed to the original one, this yields the expression for the derivative included in the statement of the theorem. ○

1.3.2 Majoration by Measures of Half-Spaces

For convex sets, Theorem 1.3.1 and the isoperimetric property (Proposition 1.2.1) can be applied to obtain some fairly general bounds in the spirit of Theorem 1.2.1. The main tool in these calculations is the following theorem.

Consider in $\mathbf{R}^k$ a closed convex set C and a half-space $H = H_{e,\rho}$.

Theorem 1.3.2 *If the number of dimensions satisfies the condition*

$$k \geq 3 \tag{1.3.5}$$

and $\sigma_{r'}(C) \geq \sigma_{r'}(H)$ *for some* $r' > 0$, *then* $\sigma_r(C) \geq \sigma_r(H)$ *for each* $r \in (0, r')$.

The proof of Theorem 1.3.2 needs some preparations.

A *central projection* with center $z \in \mathbf{R}^k$ from the sphere S_1 onto S_r is the mapping

$$\mathcal{P}_{z,r} : \mathbf{R}^k \supset S_1 \ni x \mapsto z + a(x)(x - z) \in \mathbf{R}^k, \tag{1.3.6}$$

where $a(x) > 0$ is determined from the equality $|\mathcal{P}_{z,r}(x)| = r$. Below it is supposed that $0 < |z| < r < 1$. In this case the projection $\mathcal{P}_{z,r}$ is a one-to-one mapping of S_1 onto S_r.

The points z, $x \in S_1$, $\mathcal{P}_{z,r}(x) \in S_r$, and $\frac{1}{r}\mathcal{P}_{z,r}(x) \in S_1$ lie in a two-dimensional subspace of $\mathbf{R}^k$. Their positions with respect to each other are completely determined by either of the two "lengths" $L = |x - z|$, $l = |\mathcal{P}_{z,r}(x) - z|$, or each of the "angles" α, φ, and ψ satisfying the relations

$$(x, z) = |z| \cos\varphi, \; (\mathcal{P}_{z,r}(x), z) = r\,|z| \cos\psi, \; (x - z, z) = L\,|z| \cos\alpha. \tag{1.3.7}$$

In the above,

$$r \sin\psi = l \sin\alpha, \; \sin\varphi = L \sin\alpha, \tag{1.3.8}$$

so by the cosine theorem of elementary trigonometry

$$L^2 + 2L\,|z| \cos\alpha = 1 - |z|^2, \; l^2 + 2l\,|z| \cos\alpha = r^2 - |z|^2. \tag{1.3.9}$$

In the calculations to follow the quantities above are considered as functions of the angle $\alpha \in [0, \pi]$. Some calculations show that the derivatives in α, viz., $\psi' = (d/d\alpha)\psi$, etc., are

$$l' = \left(l/\tilde{l}\right) |z| \sin\alpha, \; L' = \left(L/\tilde{L}\right) |z| \sin\alpha, \; \psi' = l/\tilde{l}, \; \varphi' = L/\tilde{L},$$

where $\tilde{l} = l + |z| \cos\alpha$ and $\tilde{L} = L + |z| \cos\alpha$.

Equations (1.3.8) and (1.3.9) can also be interpreted as definitions of the functional dependencies $\psi = \psi(\varphi)$, $L = L(\varphi)$, and $l = l(\varphi)$ in a parametric form. It follows from the above formulae for derivatives that

$$\frac{d\psi}{d\varphi} = \frac{\psi'}{\varphi'} = \frac{l}{L}\frac{\tilde{L}}{\tilde{l}} > 0. \tag{1.3.10}$$

The derivative $d\psi/d\varphi$ is strictly increasing as a function of α (or, which is the same, is increasing with the corresponding "length" $L = |x - z|$). The ratio

$l/L = r \sin\psi / \sin\varphi$ enjoys the same monotonicity property. Indeed, a calculation similar to one above leads to the relations

$$\begin{aligned} \frac{d}{d\alpha}\ln(l/L) &= \left(1/\tilde{l} - 1/\tilde{L}\right)|z|\sin\alpha > 0, \qquad (1.3.11)\\ \frac{d}{d\alpha}\ln(\psi'/\varphi') &= \left(\frac{\tilde{l}+|z|\cos\alpha}{\tilde{l}^2} - \frac{\tilde{L}+|z|\cos\alpha}{\tilde{L}^2}\right)|z|\sin\alpha. \end{aligned}$$

Central projection (1.3.6) can be used to define a new distribution on the unit sphere:

$$\tau_{z,r}(A) = \sigma_1\left(\tfrac{1}{r}\mathcal{P}_{z,r}(A)\right), \quad A \in \mathcal{B}(S_1). \qquad (1.3.12)$$

Lemma 1.3.3 *If $0 < |z| < r < 1$ and the number of dimensions satisfies the condition $k \geq 3$, then distribution (1.3.12) is absolutely continuous with respect to the uniform distribution σ_1. Its density has the form*

$$\frac{d\tau_{z,r}}{d\sigma_1}(x) = \mathbf{p}(|x-z|), \; x \in S_1, \qquad (1.3.13)$$

where the function $\mathbf{p}(u)$ is strictly increasing in $u > 0$.

Proof of Lemma 1.3.3. Using (1.3.7)–(1.3.9) one can verify that the mapping $Q: x \mapsto \frac{1}{r}\mathcal{P}_{z,r}(x)$ is one-to-one and differentiable on the unit sphere. This ensures the absolute continuity $\tau_{z,r} \ll \sigma_1$. Because of the symmetry, the density of $\tau_{z,r}$ depends only on $|x-z|$, so (1.3.13) is valid.

To calculate the density, one should compute the ratio of the infinitesimal surface area of a "spherical ring"

$$\left\{x \in S_1 : \left(x, |z|^{-1}z\right) \in [\cos(\varphi), \cos(\varphi + d\varphi)]\right\}$$

and that of its image under the mapping Q

$$\left\{x \in S_1 : \left(x, |z|^{-1}z\right) \in [\cos(\psi), \cos(\psi + d\psi)]\right\},$$

where the angles φ and ψ are related to x through equalities (1.3.7). Combined with (1.3.10), the calculation (omitted as it is quite elementary) yields for the density the expression

$$\mathbf{p}(|x-z|) = \left(\frac{\sin\psi}{\sin\varphi}\right)^{k-2}\frac{d\psi}{d\varphi} = \frac{1}{r^{k-2}}\left(\frac{l}{L}\right)^{k-1}\frac{\tilde{L}}{\tilde{l}}.$$

It remains to verify that the density is indeed monotone. It follows from (1.3.11) that

$$\begin{aligned} \frac{d}{du}\ln\mathbf{p}(u)\bigg|_{u=L} &= \left(\frac{dL}{d\alpha}\right)^{-1}\left[1/\tilde{l} - 1/\tilde{L}\right]|z|\sin\alpha\,[(k-1)+g],\\ g(\alpha) &= \left(1/\tilde{l} + 1/\tilde{L}\right)|z|\cos\alpha. \end{aligned}$$

The quantity g in the last factor decreases for $\alpha \in [0, \pi]$:

$$g'(\alpha) = -|z|\sin\alpha\left[\frac{l(l+2|z|\cos\alpha)}{\tilde{l}^3} + \frac{L(L+2|z|\cos\alpha)}{\tilde{L}^3}\right] < 0.$$

Besides, there are equalities $l = r + |z|$, $\tilde{l} = r$, $\tilde{L} = 1$, and $L = 1 + |z|$ for $\alpha = \pi$, so

$$g(\alpha) \geq g(\pi) = -|z|(1 + 1/r) > -2.$$

Consequently, the restriction $k \geq 3$ on the dimension number permits one to establish the relations

$$\frac{d}{du}\ln \mathbf{p}(u)\bigg|_{u=L} > (k-3)\left(\frac{dL}{d\alpha}\right)^{-1}\left[\frac{1}{\tilde{l}} - \frac{1}{\tilde{L}}\right]|z|\sin\alpha > 0.$$

This completes the proof. ○

Lemma 1.3.4 *Consider in* $\mathbf{R}^k$, $k \geq 3$, *a half-space* (1.2.3) $H_{e,\rho}$, $|e| = 1$, $\rho \in \mathbf{R}$, *and a central projection* (1.3.6) $\mathcal{P}_{z,r}$ *with parameters* $z = \zeta e$, r *such that* $0 \leq \zeta < r < 1$.

If a measurable set $A \subset S_1$ *and the half-space* $H_{e,\rho}$ *are related through the inequality*

$$\sigma_1(A) \geq \sigma_1\left(H^c_{e,\rho} \bigcap S_1\right),$$

then the images of these sets under $\mathcal{P}_{z,r}$ *satisfy the inequality*

$$\sigma_r\left(\mathcal{P}_{z,r}(A)\right) \geq \sigma_1\left(\mathcal{P}_{z,r}\left(H^c_{e,\rho}\bigcap S_1\right)\right).$$

Proof of Lemma 1.3.4. For $\zeta = 0$ the assertion of the lemma is evident: in this case $\mathcal{P}_{z,r}(A) = rA$ and $\sigma_r(\mathcal{P}_{z,r}(A)) = \sigma_r(rA) = \sigma_1(A)$.

Suppose that $\zeta > 0$. The closure of the "cap" $H^c_{e,\rho}\bigcap S_1$ contains those (and those only) points x for which $|x - z| \leq u$, $u = u(\rho, z)$. Hence by Lemma 1.3.3

$$H^c_{e,\rho}\bigcap S_1 = \{x \in S_1 : \mathbf{p}(|x - z|) \leq a\}$$

for some number $a > 0$. Consequently,

$$\begin{aligned}
&\sigma_r(\mathcal{P}_{z,r}(A)) - \sigma_r\left(\mathcal{P}_{z,r}\left(H^c_{e,\rho}\bigcap S_1\right)\right) \\
= \ &\sigma_1\left(\tfrac{1}{r}\mathcal{P}_{z,r}(A)\right) - \sigma_1\left(\tfrac{1}{r}\mathcal{P}_{z,r}\left(H^c_{e,\rho}\bigcap S_1\right)\right) \\
\geq \ &\sigma_1\left(\tfrac{1}{r}\mathcal{P}_{z,r}(A)\right) \\
- \ &\sigma_1\left(\tfrac{1}{r}\mathcal{P}_{z,r}\left(H^c_{e,\rho}\bigcap S_1\right)\right) - a\left[\sigma_1(A) - \sigma_1\left(H^c_{e,\rho}\bigcap S_1\right)\right] \\
= \ &\int_{S_1}\left[\mathbf{p}(|x - z|) - a\right]\left[\mathbf{1}_A(x) - \mathbf{1}_{H^c_{e,\rho}}(x)\right]\sigma_1(dx) \geq 0,
\end{aligned}$$

as both factors in the last integrand have same signs. ○

Proof of Theorem 1.3.2. Assume that $r' = 1$ (this can always be achieved by a change of scale).

a) If $\sigma_1(H) = \frac{1}{2}$, the assertion of the theorem is obvious. Indeed, under this restriction the boundary of the half-space contains the origin, so $\sigma_r(H) = \frac{1}{2}$

whatever the value of r. At the same time, the convex set C also contains the origin — otherwise it would be possible to include it into some half-space separated from the origin, and the measure of such a half-space is strictly less than $\frac{1}{2}$. Hence by convexity $r(C\bigcap S_1) \subset C\bigcap S_r$, and

$$\sigma_r\left(C\bigcap S_r\right) \ge \sigma_r\left(r\left[C\bigcap S_1\right]\right) = \sigma_1(C) \ge \tfrac{1}{2}.$$

b) If $\sigma_1(H) > \frac{1}{2}$, then $\rho > 0$. Arguing as before, it is easy to see that both the set C and the half-space contain the origin, and one can choose a family of half-spaces $\{H_{e',\rho'}\}$ indexed by pairs $\langle e',\rho'\rangle$ from some set E so that

$$C = \bigcap_{\langle e',\rho'\rangle\in E} H_{e',\rho'}. \tag{1.3.14}$$

In this representation, each pair $\langle e',\rho'\rangle$ satisfies the inequality $\rho' \ge \rho > 0$, since

$$\sigma_1(H_{e',\rho'}) \ge \sigma_1(C) \ge \sigma_1(H) \ge \tfrac{1}{2}.$$

Choose $\delta > 0$ so that $H\bigcap S_r = \mathcal{U}[r(H\bigcap S_1)]$, where $\mathcal{U}[\cdot] \equiv \mathcal{U}(\cdot,\delta)$ is the geodesic neighborhood (1.2.4) of a subset of S_r.

Under this choice of δ, for each half-space in representation (1.3.14) there is inclusion

$$H_{e',\rho'}\bigcap S_r \supseteq \mathcal{U}\left[r\left(H_{e',\rho'}\bigcap S_1\right)\right].$$

Consequently,

$$\begin{aligned} C\bigcap S_r &= \bigcap(H_{e',\rho'}\bigcap S_r) &&\supseteq \bigcap\mathcal{U}[r(H_{e',\rho'}\bigcap S_1)] \\ &\supseteq \mathcal{U}[r([\bigcap H_{e',\rho'}]\bigcap S_1)] &&= \mathcal{U}[r(C\bigcap S_1)]. \end{aligned}$$

The inequality included in the statement of the theorem now follows from the isoperimetric property of spheres (Proposition 1.2.1):

$$\sigma_r\left(C\bigcap S_r\right) \ge \sigma_r\left(\mathcal{U}\left[r\left(C\bigcap S_1\right)\right]\right) \ge \sigma_r\left(\mathcal{U}\left[r\left(H\bigcap S_1\right)\right]\right) = \sigma_r\left(H\bigcap S_r\right),$$

since

$$\sigma_r\left(r\left[C\bigcap S_1\right]\right) = \sigma_1\left(C\bigcap S_1\right) \ge \sigma_1\left(H\bigcap S_1\right) = \sigma_r\left(r\left[H\bigcap S_1\right]\right).$$

c) If $\sigma_1(H) < \frac{1}{2}$, the distance from the half-space H to the origin equals $|\rho| = -\rho > 0$.

The assertion of the theorem is trivial if $r \le |\rho|$: for these values of r the measure of the half-space is zero, $\sigma_r(H) = 0$.

Assume that $0 < |\rho| < r < 1$. The set C cannot be strictly separated from the ball $B_{|\rho|}$ by any hyperplane — otherwise the measure $\sigma_1(C\bigcap S_1)$ would be too small. Hence there exists a point z, $|z| = |\rho|$, which belongs to the closed set C.

In the following calculations $e = |z|^{-1} z$, and $H_u^c = \bar{H}_{e,u}^c$ is the closure of the "complementary" half-space (1.2.3). Evidently,

$$\sigma_1\left(H^c_{|z|}\right) = \sigma_1(H), \quad \sigma_r\left(H^c_{|z|}\right) = \sigma_r(H).$$

One can choose $b \in (0, |\rho|)$ in such a way that

$$\sigma_1(H^c_b) = \sigma_1(C).$$

Consider central projection (1.3.6) $\mathcal{P} = \mathcal{P}_{z,r}$ with the parameters $z \in \mathbf{R}^k$ and $r > |z|$. It is easy to see that the inclusion

$$S_r \bigcap H^c_{|z|} \subset \mathcal{P}(H^c_b)$$

is true if $b < |z|$. Moreover, since $z \in C$ and C is convex,

$$\mathcal{P}\left(C \bigcap S_1\right) \subset C \bigcap S_1.$$

It follows by Lemma 1.3.4 that

$$\sigma_r(H) = \sigma_r\left(H^c_{|z|} \bigcap S_r\right) \le \sigma_r(\mathcal{P}[H^c_b]) \le \sigma_r\left(\mathcal{P}\left[C \bigcap S_1\right]\right) \le \sigma_r(C).$$

This proves the theorem. ○

1.3.3 Isoperimetric Inequalities for Convex Sets

Theorem 1.3.2 can be used to gain additional information in the setting of Theorem 1.3.1 when it is applied to convex sets.

Lemma 1.3.5 *Assume that the distribution F is spherically symmetric with $F(\{0\}) = 0$, the function $\varphi : [0, \infty) \to \mathbf{R}$ is a nondecreasing one, and the "radial distribution" of* (1.3.1) *satisfies the condition*

$$\int_0^\infty |\varphi(u)| \, dF_+(u) < \infty.$$

If a closed convex set $C \subset \mathbf{R}^k$ and a half-space $H = H_{e,\rho}$ obey the condition

$$0 < F(C) = F(H) < 1, \tag{1.3.15}$$

then

$$\int_C \varphi(|x|) \, F(dx) \le \int_H \varphi(|x|) \, F(dx). \tag{1.3.16}$$

This inequality holds also under the condition $F(C) < F(H)$ if φ is nonnegative.

Proof of Lemma 1.3.5. By Theorem 1.3.2, the function

$$\delta(r) = \sigma_r(H) - \sigma_r(C), \quad r \in (0, \infty),$$

is either nonnegative or changes sign only once (from "–" to "+").

Suppose that $\delta(r) \leq 0$ for $r \leq a$ and $\delta(r) > 0$ for $r > a$. Then

$$\int_H \varphi(|x|)\ F(dx) \quad - \quad \int_C \varphi(|x|)\ F(dx) = \int_0^\infty \varphi(r)\delta(r)dF_+(r)$$
$$= \int_0^\infty [\varphi(r) - \varphi(a)]\,\delta(r)\,dF_+(r) \geq 0.$$

This proves the lemma. ○

Theorem 1.3.3 *Assume satisfied condition* (1.3.5) *and let F be a spherically-symmetrical distribution with density* (1.3.2).

a) *If $C \subset \mathbf{R}^k$ is a closed convex set and $H = H_{e,\rho}$ a half-space satisfying condition* (1.3.15), *then*

$$\frac{d}{d\lambda}F(\lambda C)\bigg|_{\lambda=1} - \frac{d}{d\lambda}F(\lambda H)\bigg|_{\lambda=1} \geq 0.$$

b) *If the derivative $h'(u)$ is nonnegative for all values of its argument, C is a convex set, and $H = H_{e,\rho}$ is a half-space, then the inequality $F(\lambda C) \geq F(\lambda H)$ implies the following inequality for the derivatives:*

$$\frac{d}{d\lambda}F(\lambda H) - \frac{d}{d\lambda}F(\lambda C) \leq \frac{k}{\lambda}[F(\lambda H) - F(\lambda C)].$$

Proof of Theorem 1.3.3 is an application of the representation for the derivative given by Theorem 1.3.1. Inequalities a) and b) follow from Lemma 1.3.5. ○

For Gaussian measures of convex sets, Theorem 1.3.3 yields estimates similar to those of Theorem 1.2.1.

Theorem 1.3.4 *Let $\eta \in \mathbf{R}^k$ be a RV with the standard Gaussian distribution $N_{k,0,I}$.*

If $C \subset \mathbf{R}^k$ is a closed convex set and $H = H_{e,\rho}$ some half-space (1.2.3) *such that*

$$\mathbf{P}\{\eta \in C\} \geq \mathbf{P}\{\eta \in H\},$$

then

$$\forall \lambda > 1 \;\; \mathbf{P}\{\eta \in \lambda C\} \geq \mathbf{P}\{\eta \in \lambda H\}. \tag{1.3.17}$$

Proof of Theorem 1.3.4. Restriction (1.3.5) is circumvented by embedding the space in one with a "superfluous" dimension.

Suppose that the inequality of Theorem 1.3.4 is violated for some $\lambda_+ > 1$. By Theorem 1.3.1, the function $\lambda \mapsto N(\lambda A) = \mathbf{P}\{\eta \in \lambda A\}$ is continuous. Hence

$$N(\lambda_0 C) - N(\lambda_0 H) = 0$$

if $\lambda_0 = \sup\{\lambda : 1 \leq \lambda \leq \lambda_+,\ N(\lambda C) \geq N(\lambda H)\}$.

Inequality $N(\lambda C) < N(\lambda H)$ holds for $\lambda_0 < \lambda < \lambda_+$ by the choice of λ_0. But it follows from Theorem 1.3.3 that

$$\frac{d}{d\lambda}[N(\lambda H) - N(\lambda C)] \leq \frac{k}{\lambda}[N(\lambda H) - N(\lambda C)]$$

for each $\lambda \in [\lambda_0, \lambda_+]$. Consequently, by the Bellman–Gronwall inequality

$$[N(\lambda H) - N(\lambda C)] = 0.$$

This contradiction proves the theorem. ○

1.4 Some Formulae for Convex Functions

The material exposed below is used in later sections to study distributions of convex functions of standard Gaussian RV's.

1.4.1 Special Coordinates

In what follows, the function $q : \mathbf{R}^k \to \mathbf{R}$ is *strictly convex*:

$$\forall\, x, y \in \mathbf{R}^k,\ \lambda \in (0,\ 1)\ \ q(\lambda x + [1-\lambda]\, y) \ <\ \lambda q(x) + [1-\lambda] q(y).$$

Besides, q has derivatives of all orders in the whole space, is bounded from below, and has a unique minimum:

$$\exists\, x_0 \in \mathbf{R}^k :\ \ x \neq x_0 \implies q(x) > q(x_0) = q_0 \stackrel{def}{=} \min_{x \in \mathbf{R}^k} q(x) \ >\ \infty.$$

Its *level sets* $Q_u = \{x :\ q(x) \le u\}$, $u > q_0$, are compact and contain interior points.

The following estimate is valid for the gradient of an arbitrary convex function in R^k.

Lemma 1.4.1 *Let the function* $\widehat{q} : \mathbf{R}^k \to \mathbf{R}$ *be convex in* $\mathbf{R}^k$ *and continuously differentiable in a neighborhood of* x.

If the inequality $\widehat{q}(x) > \widehat{q}(y)$ *holds for some point* $y \in \mathbf{R}^k$, *then*

$$|\nabla \widehat{q}(x)| \ge \frac{\widehat{q}(x) - \widehat{q}(y)}{|x-y|}.$$

Proof of Lemma 1.4.1. The function $f(s) = \widehat{q}(sx + [1-s]y)$ is convex for $s \ge 0$ and increases for s close to one. Hence $f'(1) \ge f(1) - f(0)$, and

$$|\nabla \widehat{q}(x)|\,|x-y| \ \ge\ (\nabla \widehat{q}(x), x-y) \ =\ f'(1) \ \ge\ f(1) - f(0) = \widehat{q}(x) - \widehat{q}(y). \ ○$$

By Lemma 1.4.1, the gradient ∇q does not vanish for $x \neq x_0$. At each point on the "level surface" ∂Q_u with $u > q_0$, the supporting hyperplane of Q_u is unique and orthogonal to the unit vector of outward normal to ∂Q_u.

The mapping

$$\partial Q_u \ni x \mapsto n(x) \ =\ |\nabla q(x)|^{-1}\, \nabla q(x) \in S_1$$

establishes a one-to-one correspondence between the points of the boundary ∂Q_u and those of the unit sphere.

In a number of calculations it is convenient to use a special coordinate system in the exterior of a level set Q_u.

Consider the mapping

$$\hat{\pi} : \langle t, x \rangle \mapsto y = x + tn(x), \quad t \geq 0, \; x \in \partial Q_u, \tag{1.4.1}$$

which maps a nonnegative number t and a boundary point $x \in \partial Q_u$ to a point from the exterior of ∂Q_u.

This mapping is one-to-one. The parameter t is the distance from y to Q_u:

$$t = \operatorname{dist}(y, Q_u). \tag{1.4.2}$$

Indeed, if $t = 0$ and $x \in \partial Q_u$, the derivative $(\partial/\partial t)q(x + tn(x)) = |\nabla q(x)|$ is strictly positive. Hence it follows from convexity of q that $q(y) > u$ for all points (1.4.1) with $t > 0$. Conversely, if $y \notin Q_u$, then there exists a point x which belongs to the compactum ∂Q_u and is nearest to y. The segment $[x, y]$ is normal to the support hyperplane of Q_u at x. Consequently $y - x = tn(x)$ with $t > 0$.

Further on, mapping (1.4.1) is used to parameterize the set $Q_u^c = \mathbf{R}^k \setminus Q_u$ by pairs $x \in \partial Q_u$, $t \in \mathbf{R}_+$.

Fix the point $x \in \partial Q_u$. The function q is, by convexity, strictly increasing along the ray (1.4.1) which corresponds to values $t > 0$. For this reason, the quantity $T_u(x, v)$ is unambiguously defined by the equation

$$q\left[x + T_u(x, v)\, n(x)\right] = v, \quad v \geq u. \tag{1.4.3}$$

This function can be used to discern between points of the set Q_v, $v > u$, and its complement: for $x \in \partial Q_u$ and $t > 0$

$$y = \hat{\pi} \stackrel{def}{=} x + tn(x) \in \begin{cases} Q_v, & t \leq T_u(x, v), \\ Q_v^c, & t > T_u(x, v). \end{cases} \tag{1.4.4}$$

The implicit function theorem shows that for $y_0 = x + T_u(x, v)n(x)$

$$(\partial/\partial v)T_u(x, v) = 1/\left(\, n(x), \nabla q(y_0)\, \right) \geq 0. \tag{1.4.5}$$

Given u and y in (1.4.4), both x and y_0 are unambiguously determined.

1.4.2 Volume Integrals over Complements to Level Sets

Let x_0 be a point on the smooth hypersurface ∂Q_u. Suppose that a neighborhood of this point is smoothly parameterized by some mapping $\xi \mapsto X(\xi)$, $\xi = \left(\xi^{(j)}, j = \overline{1, k-1}\right) \in \mathbf{R}^{k-1}$. One can use these coordinates on the hypersurface to define new coordinates in a "solid" k-dimensional neighborhood of x using (1.4.1):

$$\langle t, \xi \rangle \mapsto y(t, \xi) = X(\xi) + tn\left(X(\xi)\right).$$

Whatever the choice of parameterization on ∂Q_u, the Jacobian of this change of coordinates in $\mathbf{R}^k$ is easy to calculate in terms of $X_{\cdot j} = [\partial/\partial \xi^{(j)}]X \in \mathbf{R}^k$:

$$\hat{J}(t, \xi) = \sqrt{\det(G^T G)}, \quad G = \|n, X_{\cdot 1} + t\,(X_{\cdot 1}, \nabla)\, n, \dots\|,$$

where G is the $k \times k$-matrix with columns n, $X_{\cdot j}$.

For $t = 0$ the Jacobian depends only on the metric coefficients of the surface ∂Q_u, and $ds(X(\xi)) = \hat{J}(0,\xi)d\xi$ is the surface element surrounding $x(\xi)$ in integration with respect to surface area over ∂Q_u.

For $t > 0$, the Jacobian depends also on the principal curvatures of ∂Q_u at the appropriate point. Choosing the coordinates on the surface in such a way that $X_{\cdot j}$ points in the jth principal direction that corresponds to the principal curvature $K_j \geq 0$, it is easy to see that in an integral over a neighborhood of $y(t,x) = x + tn(x)$, $t > 0$, the volume element is

$$dy = J(t,x)\, dtds(x), \quad J(t,x) = \prod_{j=1}^{k-1} [1 + tK_j(x)], \tag{1.4.6}$$

and on the surface $J(0,x) = 1$.

The derivative

$$\frac{\partial}{\partial t} \log J(t,x) = \sum_{j=1}^{k-1} K_j(x) \geq 0 \tag{1.4.7}$$

equals the Gauss curvature of the hypersurface and does not depend upon the distance t from the boundary to y.

The above calculations yield the following identity for integrals over the complement to Q_u:

$$\int_A \phi(y)\, dy = \int_{A'} \phi(y(t,x))\ J(t,x)dtds(x), \quad A \subset Q_u^c = \mathbf{R}^k \setminus Q_u, \tag{1.4.8}$$

where $ds(x)$ is the surface element of ∂Q_u, the set $A' = \hat{\pi}^{-1}(A)$ is the pre-image of A in the product $\mathbf{R}^+ \times \partial Q_u$ under mapping (1.4.1).

1.4.3 Derivatives of Integrals over Level Sets

Coordinates (1.4.1) on the set Q_u^c can be used to obtain a manegeable expression for the derivative of the function

$$G(u) = \int_{\{q(y) \leq u\}} g(y)dy = G_+ - \int_{\{q(y) > u\}} g(y)dy, \quad G_+ = \int_{\mathbf{R}^k} g(y)dy. \tag{1.4.9}$$

Later on, g is a probability density.

For $v > u$, it follows from definition (1.4.3) of $T_u(x,v)$ and expression (1.4.8) for integrals over the exterior of $Q_u = \{q \leq u\}$ that

$$G(v) = G_+ - \int_{\partial Q_u} ds(x) \int_0^\infty g(y)J(t+T,x)dt, \tag{1.4.10}$$

where $y = y(t,x;v) = x + [T+t]\,n(x)$ with $T = T_u(x,v)$, and $J(T+t,x)$ is defined in (1.4.6).

Lemma 1.4.2 *Let $g : \mathbf{R}^k \to [0,\infty)$ be a smooth nonnegative function with compact support.*

If $u > q_0 = \inf_x q(x)$, then function (1.4.9) *is differentiable and*

$$G'(u) \le -\int_{\{q>u\}} \frac{(n(x), \nabla g(y))}{|\nabla q(x)|} dy,$$

where $x = x(y) \in \partial Q_u$ is found from the equation $y = x + tn(x)$.

Relation between y and x in the integrand is as in (1.4.1).

Proof of Lemma 1.4.2. Assume that $v > u > q_0$. The only element in the integrand of (1.4.10) which depends on v is $T = T_u(x, v)$, and the conditions of the lemma allow one to differentiate in v under the integration sign. Thus, the derivative does exist and equals, for $v > u$,

$$G'(v) = -\int_{\partial Q_u} ds(x) \int_0^\infty A(x,t)dt, \tag{1.4.11}$$

where

$$A(x,t) = \frac{J(t+T,x)}{(n(x), \nabla q(y_0))} \left[(n(x), \nabla g(y)) + g(y) \frac{\partial}{\partial t} \ln J(t+T,x) \right],$$

and the point $y_0 = x + Tn(x)$ lies on the boundary of the larger set ∂Q_v. Notation is that of (1.4.10), and the derivation of the last formula makes use of (1.4.5).

By (1.4.7) and (1.4.5), the second summand in the square brackets on the right-hand side of the definition of A is nonnegative by convexity of Q_u since $g \ge 0$. This yields the bound

$$\begin{aligned} G'(v) &\le -\int_{\partial Q_u} ds(x) \int_0^\infty \frac{(n(x), \nabla g(y))}{(n(x), \nabla q(y_0))} J(t+T,x)\, dt \\ &= -\int_{Q_v^c} \frac{(n(x), \nabla g(y))}{(n(x), \nabla q(y_0))} dy. \end{aligned}$$

The inequality of the lemma is derived from this relation by a passage to the limit in $v \searrow u$: in this case $T \searrow 0$ and $(n(x), \nabla q(y_0)) \to |\nabla q(x)|$. ○

Remark 1.4.1 Integrating by parts the summand in (1.4.11) which contains the derivative $\partial J/\partial t$, it is easy to establish the equality

$$G'(v) = -\int_{\partial Q_u} \frac{J(T,x)\, g(y_0)}{(n(x), \nabla q(y_0))} ds(x), \quad v > u.$$

As $v \searrow u$, this yields the relation

$$G'(u) = -\int_{\partial Q_u} \frac{g(x)}{(n(x), \nabla q(x))} ds(x) = -\int_{\partial Q_u} \frac{g(x)}{|\nabla q(x)|} ds(x). \tag{1.4.12}$$

This latter expression is easily derived using a different coordinate system, where a point $y \in \mathbf{R}^k$ is located using $Q = q(y)$ and $n(y) = |\nabla q(y)|^{-1} \nabla q(y)$.

1.4.4 Piecewise-Linear Functions and Polytopes

Some of the functions considered in later sections are not strictly convex and infinitely differentiable, but piecewise linear.

A *convex piecewise-linear function* (CPLF) in $\mathbf{R}^k$ is one of the form

$$q(x) = \max_{j=1,\dots,m}\left[\alpha_j + (b_j, x)\right], \quad m \in \mathbf{N},\ \alpha_j \in \mathbf{R},\ b_j \in \mathbf{R}^k. \tag{1.4.13}$$

Its convexity is obvious, as well as convexity of the *level sets*

$$Q_u = \{x : q(x) \le u\} = \bigcap_{j=1}^{M} \{x : \alpha_j + (b_j, x) \le u\}, \ u \in \mathbf{R}. \tag{1.4.14}$$

Sets (1.4.14) are *convex polytopes* , i.e., intersections of finite families of half-spaces. The standard representation of a polytope is

$$P = \bigcap_{j=1}^{m} H_{e_j,\rho_j}, \ |e_j| = 1, \ \rho_j \in \mathbf{R}, \tag{1.4.15}$$

with notation (1.2.3) for half-spaces. In this special case the CPLF is defined using unit vectors $b_j = e_j$. The convex hull of a finite set of points in a Euclidean space is always a convex polytope.

In what follows it is assumed that at least one of the vectors b_j in (1.4.13) is nonzero, and each closed convex set

$$L_j = \{x : q(x) = (x, b_j) + \alpha_j\} \tag{1.4.16}$$

has interior points. Otherwise one could, without changing the function, use a smaller collection of pairs $\langle \alpha_j, b_j \rangle$ in (1.4.13). It is assumed also that there are no identical pairs $\langle \alpha_j, b_j \rangle$ in representation (1.4.13).

The only value that function (1.4.13) may assume on a set of positive Lebesgue measure is

$$q_0 = \inf\left\{q(x) : x \in \mathbf{R}^k\right\} \tag{1.4.17}$$

if this lower bound is finite. In this latter case there is a unique number j_0 for which $b_{j_0} = 0$ and the equality $q(x) = q_0$ holds on the set L_{j_0} of positive measure. The boundaries of all sets (1.4.16) have zero Lebesque measure.

In the interior of each set (1.4.16), a CPLF is smooth and its gradient $\nabla q(x)$ assumes one of the values b_j.

To calculate integrals over complements to the level sets (1.4.14), it is sometimes convenient to employ constructions similar to those of subsection 1.4.1.

If $b_j \ne 0$, the unit vector $n_j = |b_j|^{-1} b_j$ is well defined, as well as the quantities (possibly infinite)

$$T_j^+(x) = \sup\{t : x + tn_j \in L_j\}, \ T_j^-(x) = \inf\{t : x + tn_j \in L_j\}. \tag{1.4.18}$$

(as usual, $\sup \emptyset = -\infty$ and $\inf \emptyset = +\infty$). For each point $\max_{j=\overline{1,m}} T_j^+(x) = \infty$. Indeed, no two vectors b_j are equal. It follows easily that $x + tb_k \in L_k$ for all large enough values of t if $|b_k| = \max_j |b_j|$.

If the interval

$$I_j = \{q(x + tn_j) : T_j^-(x) < t < T_j^+(x)\}$$

is nonempty, one can define on it the function

$$t_j(u, x) = \sup\{t : q(x + tn_j) \le u\}. \tag{1.4.19}$$

It follows immediately from definition (1.4.19) that, for a fixed point x, the function t_j is linear in its argument u inside the interval $I_j(x)$ and

$$\frac{\partial}{\partial u} t_j(u, x) = \frac{1}{|\nabla q(x + t_j(u, x)n_j)|} = \frac{1}{|b_j|}. \tag{1.4.20}$$

It is convenient to extend this function outside $I_j(x)$ by the equalities

$$t_j(u, x) = \begin{cases} T_j^-(x), & \text{if } u \le \inf I_j, \\ T_j^+(x), & \text{if } u \ge \sup I_j. \end{cases}$$

If this agreement is used, one-sided derivatives $\partial_\pm t_j(u, x)$ of (1.4.19) exist and coincide on all real line with possible exception of the ends of the interval I_j, so almost everywhere $\partial_\pm t_j(u, x) = (\partial/\partial u)t_j(u, , x)$.

If the interval $(u,\ v)$ is a subset of I_j, then

$$t_j(v, x) - t_j(u, x) = \int_u^v dw / |\nabla q[x + t(w, x)n_j]| \,. \tag{1.4.21}$$

Obviously, $q(x + t_j(u, x)n_j) = u$ for $u \in I_j$.

It is natural to choose as coordinates of a point x from (1.4.16) its projection $y = x - (x, n_j)n_j$ onto the hyperplane $E_j = \{x : (b_j, x) = 0\}$ and $t = (n_j, x)$, the value of its projection onto the normal to this hyperplane.

In these coordinates, the points which belong both to L_j and to the complement of the level set (1.4.14) are selected by the condition $t > t_j(u, y)$. If $u > q_0$ (see (1.4.17)), then the boundary of set (1.4.14) is covered by the "faces"

$$F_j(u) = \{y + t_j(u, y)n_j : y \in L_j^0,\ t_j(u, y) < T_j^+(y)\}, \tag{1.4.22}$$

where $b_j \ne 0$ and the $(k-1)$-dimensional set

$$L_j^0 = E_j \bigcap \{x : \exists t \in \mathbf{R}\ x + tn_j \in L_j\} = E_j \bigcap (L_j + \mathbf{R}n_j). \tag{1.4.23}$$

is the projection of L_j onto the hyperplane E_j.

Lemma 1.4.3 a) *There are no common interior points in the sets*

$$F_j^+(u) = \{x + tn_j : t \ge 0,\ x \in F_j(u)\}. \tag{1.4.24}$$

b) *Consider arbitrary points $x \in F_j(u)$ and $y = x + tn_j \in F_j^+(u)$ with $t \ge 0$. If $u > b > q_0$, then*

$$\operatorname{dist}(y, Q_b) \ge \operatorname{dist}(x, Q_b). \tag{1.4.25}$$

Proof of Lemma 1.4.3. a) The unit normal vectors n_j and n_l are not parallel for the sets F_j^+ and F_l^+ adjacent to different faces (1.4.22) of the polytope, viz., $(n_j, n_l) < 1$.

Indeed, assume the converse: $(n_j, n_l) = 1$.

If the point x_j belongs to $F_j(u)$ together with its $(k-1)$-dimensional neighborhood U_j, then the points from the set $U_j + t\,|b_j|^{-1}\, n_j$ lie on the boundary of Q_{u+t} for small values of $t > 0$. The same applies to points of the set $U_l + t\,|b_l|^{-1}\, n_l$, where U_l is a small neighborhood of some interior point x_l of $F_l(u)$.

Hence the vectors b_j and b_l, which are assumed parallel, have equal norms — otherwise the intersection of Q_{u+t} with the plane spanned by $n_j = n_l$ and $x_j - x_l$ would not be convex — and are, consequently, equal.

But in this case also the quantities α_j and α_l in (1.4.13) should coincide, which leads to a contradiction with the assumption that there is no redundancy in (1.4.13).

One arrives at a contradiction as well if it is assumed that $x_j + t_j n_j = x_l + t_l n_l$ for some points $x_j \in F_j(u)$ and $x_l \in F_l(u)$ with $t_j,\ t_l > 0$.

Indeed, the half-spaces $\{x : (x - x_m, n_m) \le 0\}$ for $m = j,\ l$ both contain the set Q_u. Hence the relations

$$t_j \;=\; (x_j + t_j n_j - x_j, n_j) \;=\; (x_l + t_l n_l - x_j, n_j) \;\le\; t_l\,(n_j, n_l)$$

should be valid simultaneously with the converse inequality $t_l \;\le\; t_l\,(n_j, n_l)$, which is impossible if $t_j,\ t_l > 0$. This proves the first assertion of the lemma.

b) Suppose that the point $z_0 \in Q_b$ is such that $|z - y| = \operatorname{dist}(y, Q_b)$. The half-space $\{z : (z - x, n_j) \le 0\}$ contains Q_b. Hence $(z_0 - x, n_j) \le 0$ and, consequently,

$$|z_0 - y|^2 \;=\; |z_0 - x|^2 + t^2 - 2t\,(z_0 - x, n_j) \;\ge\; |z_0 - x|^2 \;\ge\; [\operatorname{dist}(x, Q_b)]^2\,.$$

This proves the second assertion of the lemma. ○

For a CPLF, Lemma 1.4.1 has the following corollary.

Corollary 1.4.1 *If q is CPLF* (1.4.13), *and x is an interior point of the face $F_j(u)$, then*

$$|b_j| \ge \frac{u - b}{\operatorname{dist}(x, Q_b)}. \tag{1.4.26}$$

Proof of Corollary 1.4.1. For an interior point x of the face $F_j(u)$ of the boundary of a level set, the point $x' = x + t n_j$ is an interior point in L_j provided that $t > 0$ is small enough. For this reason the function q is differentiable in a neighborhood of this point and $|b_j| = |\nabla q(x')|$.

Let z be the point of Q_b nearest to x'. Then by Lemma 1.4.1

$$|b_j| \;=\; |\nabla q(x')| \;\ge\; \frac{q(x') - q(z)}{|x' - z|} \;=\; \frac{q(x') - b}{\operatorname{dist}(x', Q_b)}.$$

The inequality of the corollary follows from an obvious passage to the limit as $t \searrow 0$ since in this case one can achieve that $x' \to x$. ○

For a CPLF whose representation (1.4.13) has special properties, the lower bound of Lemma 1.4.1 for the gradient of a convex function can be complemented by the folllowing upper one.

Lemma 1.4.4 *Let q be a CPLF* (1.4.13) *with $\alpha_j > 0$ for all j.*
If $q(x) \geq 0$ and q is continuously differentiable in a neighborhood of x, then

$$1/\left|\nabla q(x)\right| \geq (n(x), x)/q(x), \quad n(x) = \left|\nabla q(x)\right|^{-1} \nabla q(x).$$

Proof of Lemma 1.4.4. Under the assumptions of the lemma, the inequality

$$[1+\delta]\, q(x) \geq q([1+\delta]x)$$

holds for each $\delta > 0$. The estimate of the lemma is follows from it after the passage to the limit in $\delta \searrow 0$:

$$q(x) \geq (x, \nabla q(x)) = \left|\nabla q(x)\right| (n(x), x)\,. \bigcirc$$

1.4.5 Integral Representation for Distribution of CPLF

For levels u higher than the minimal value q_0 of (1.4.17), the complement to the level set (1.4.14) can be represented in the form

$$Q_u^c = \mathbf{R}^k \setminus Q_u = \bigcup_{j \leq m} L_j \bigcap Q_u^c,$$

where all the sets in the union have no common interior points and correspond to those pairs $\langle \alpha_j, b_j \rangle$ in (1.4.13) which include nonzero vectors b_j.

For a CPLF q and a function $g(x)$ summable function over $\mathbf{R}^k$, consider the integral of (1.4.9)

Lemma 1.4.5 *If $u > q_0$, then*

$$\begin{aligned} G_+ - G(u) &= \textstyle\sum_{j=1}^m \int_{L_j \bigcap Q_u^c} g(x)dx \\ &= \textstyle\sum_{j=1}^m \int_{L_j^0} \mathrm{mes}_j(dy) \int_{t_j(y,u)}^{T_j^+} g(y + tn_j)\, dt, \end{aligned} \tag{1.4.27}$$

where the summands in the right-hand side correspond to nonzero vectors b_j in representation (1.4.13), *$n_j = \left|b_j\right|^{-1} b_j$ are unit outer normal vectors for faces* (1.4.22) *of the boundary ∂Q_u, and* mes_j *is the* $(k-1)$*-dimensional Lebesgue measure in $E_j = \{x : (x, b_j) = 0\}$ with the metric of* $\mathbf{R}^k$.

Proof of Lemma 1.4.5 consists in changing the integration variables to

$$y = x - (x, n_j)\, n_j \in L_j, \quad t = (x - y, n_j) \in \mathbf{R}$$

while integrating over the set $L_j \bigcap Q_u^c$. In the sum, the summands with $L_j \subset Q_u$ can be omitted: for these, $t_j = T_j^+$. $\bigcirc$

Below, a function g is said to have a ***uniform integrable majorant*** in a neighborhood of each hyperplane if for each $e \in \mathbf{R}^k$, $|e| = 1$, and $\rho \in \mathbf{R}$ there exists a real number $h = h(e, \rho) > 0$ such that

$$\forall \; |(x, e) - \rho| < h \quad |g(x)| \leq f_{e,\rho}(y), \quad y = x - (x, e)e,$$

and the function $f_{e,\rho}(y)$ is summable with respect to the $(k-1)$-dimensional Lebesgue measure in the subspace $E_{0,e} = \{x : (x, e) = 0\}$.

Lemma 1.4.6 *Let the function g be continuous and summable in $\mathbf{R}^k$. Assume, moreover, that it has a uniform integrable majorant in a neighborhood of each hyperplane.*

For values of u greater than the minimal one q_0, function (1.4.9) *is continuous and has one-sided derivatives*

$$\partial_{\pm}G(u) = \sum_{j=1}^{m} \int_{L_j^0} g\left(y + t_j(u,y)n_j\right) \partial_{\pm}t_j(u,y)\,\mathrm{mes}_j(dx), \qquad (1.4.28)$$

where the sum in the right-hand side includes only summands which correspond to nonempty faces of the boundary of Q_u.

The one-sided derivatives $\partial_{\pm}G(u)$ coincide everywhere with the possible exception of a finite set of discontinuities, and for $v > u > q_0$

$$G(v) - G(u) = \int_u^v \partial_{\pm}G(u')du'.$$

Notation of Lemma 1.4.6 is that of (1.4.27) and (1.4.20). When one-sided derivatives coincide, there is a *bona fide* derivative. If the CPLF is defined using only unit vectors, formula (1.4.28) at continuity points becomes an expression of G' by a surface integral:

$$\frac{dG}{du}(u) = \partial_{\pm}G(u) = \int_{\partial Q_u} g(x)\,ds(x) \ \text{ if } \forall j \ \ |b_j| = 1,\ G(u+0) = G(u-0).$$

Proof of Lemma 1.4.6. If g is continuous and has compact support, the assertion of the lemma is evident: it follows from (1.4.27) and the formula of differentiation of an integral with respect to its upper integration limit.

To substantiate differentiability of G and formula (1.4.28) in the general case, one can have recourse to approximation of g by continuous functions with compact support.

The right- and left-sided derivatives of G may be different only if there is a $(k-1)$-dimensional face of the boundary ∂Q_u with nonempty interior: the one-sided derivatives of $t_j(u,x)$ may differ only if $t_j(u,x) = T^{\pm}(x)$. Consequently, the set of points where G in not differentiable is, if nonempty, finite.

Continuity of G is a consequence of the fact that the CPLF q can assume on a set of positive Lebesgue measure only its minimal value q_0. ○

Lemma 1.4.7 *If the conditions of Lemma* 1.4.6 *are satisfied, and the function g has continuous derivatives which are summable, then the right-hand side of* (1.4.28) *can be represented as a sum of "volume" integrals:*

$$\partial_{+}G(u) = -\sum_{j=1}^{m} \frac{1}{|b_j|} \int_{F_j^+(u)} (\nabla g(x), n_j)\,dx.$$

In the formula of Lemma 1.4.7, the summation index runs only over the numbers of nonempty faces of the boundary of Q_u, so all the sets (1.4.24) which it uses lie in the closure of Q_u^c and have no common interior points (see Lemma 1.4.3).

Proof of Lemma 1.4.7 follows from (1.4.28), (1.4.20), and the obvious identity

$$g\left(y+t_j(u,y)n_j\right)=-\int_0^{\infty}\left(\nabla g\left(y+[t+t_j(u,y)]\,n_j\right),n_j\right)dt. \bigcirc$$

1.4.6 Approximation of Piecewise-Linear Functions

It is well known (see Appendix B) that each compact convex set can be approximated, to an arbitrary degree of accuracy, by the level set of an analytical convex function possessing the properties stipulated in the beginning of this section. In some calculations, it is convenient to approximate by smooth strictly convex functions the piecewise linear ones.

Assume that $\widehat{q}$ is a CPLF (1.4.13) which is bounded from below and has compact level sets $\widehat{Q}_u=\{\widehat{q}(x)\le u\}$. There is a sequence of smooth strictly convex functions which approximates $\widehat{q}$ and its derivatives uniformly on each compact subset of $\mathbf{R}^k$. Approximation of this kind is provided, e.g., by the sequence of convolutions $q_n=w_n*\widetilde{q}_n$, where $w_n(x)=n^k w(nx)$, $w(x)$ is an infinitely differentiable probability density in $\mathbf{R}^k$ with compact support, and $\widetilde{q}_n(x)=\widehat{q}(x)+\frac{1}{n}|x|^2$. By convexity of both $\widehat{q}$ and $\widetilde{q}_n$, there is inequality $q_n\ge\widehat{q}$. As $n\to\infty$, convergence $q_n\to\widehat{q}$ is uniform on each compact set.

Moreover, for each $x\in\mathbf{R}^k\setminus\widehat{Q}_u$

$$\operatorname{dist}\left(x,\widehat{Q}_u^{(n)}\right)\to\operatorname{dist}\left(x,\widehat{Q}_u\right),\quad \widehat{Q}_u^{(n)}=\{q_n\le u\},$$

if $u>\widehat{q}_0$. Indeed, if $x\in\widehat{Q}_u^c$ and $z\in\partial\widehat{Q}_u$ is the point of $\widehat{Q}_u$ nearest to x, then each neighborhood of z contains at least one point of $\widehat{Q}_u^{(n)}$ for large n. Consequently,

$$\limsup_{n\to\infty}\operatorname{dist}\left(x,\widehat{Q}_u^{(n)}\right)\le\operatorname{dist}\left(x,\widehat{Q}_u\right).$$

The converse inequality, $\operatorname{dist}\left(x,\widehat{Q}_u^{(n)}\right)\ge\operatorname{dist}\left(x,\widehat{Q}_u\right)$, is obvious.

A similar reasoning shows that $\operatorname{dist}\left(x,\widehat{Q}_u^{(n)}\right)\to 0$ if $x\in\widehat{Q}_u$.

For an approximating sequence of the above described type one can always find a common majoration: as the growth of $\widehat{q}$ is linear, it is possible to find some $c>0$ such that $|q_n|\le c\left(1+|x|^2\right)$ for all n and x.

For a CPLF q, the boundary $\{x:q(x)=v\}$ of the level set $\{x:q(x)\le v\}$ splits into the "faces" $F_j(v)=\{x:q(x)=v\}\bigcap L_j$. If $q_\varepsilon(x)$ is a smooth approximation of the kind described above, and no face of $\{q(x)=v\}$ that has nonzero surface area is a part of the boundary separating some sets $L_{j'}$ and $L_{j''}$ with $b_{j'}\ne b_{j''}$, then for a smooth function g with compact support

$$\int_{\{q_\varepsilon=v\}}\frac{g(x)ds(x)}{|\nabla q_\varepsilon(x)|}\to\int_{\{q=v\}}\frac{g(x)ds(x)}{|\nabla q(x)|},$$

where the right-hand side is the sum of $(k-1)$-dimensional integrals over the boundary's faces divided by the corresponding quantities $|b_j|$. The set of exceptional levels is, at most, finite.

One may, use smooth approximations of CPLF'se.g., to prove a version of Lemma 1.4.6 using formulae of subsection 1.4.3.

Let the function g be continuous and summable in $\mathbf{R}^k$. Assume, moreover, that it has a uniform integrable majorant in a neighborhood of each hyperplane (cf. Lemma 1.4.6).

Lemma 1.4.8 *For a CPLF q, function* (1.4.9) *is continuous.*

It is differentiable on (q_0, ∞) with the possible exception of a finite set. Outside this exceptional set, its derivative is continuous and admits representation (1.4.12). *Moreover, for $v > u > q_0$*

$$G(v) - G(u) = \int_u^v G'(u')du'.$$

The density G' satisfies the inequality of Lemma 1.4.2 *almost everywhere.*

Proof of Lemma 1.4.8. If the function q were replaced by an infinitely differentiable and strictly convex one, and g were continuous with compact support, the assertions of the lemma would follow from (1.4.12) and the formula of differentiation of an integral with respect to its upper limit.

For g continuous with compact support and piecewise-linear q, the expression for the derivative and the equality of the lemma are verified using smooth strictly convex approximations for q. The equality of the lemma serves to identify expression (1.4.12) with the derivative. Thus, formula (1.4.12) holds for a CPLF on (q_0, ∞) with the possible exception of a finite set.

The final step in the general case is approximation of g by continuous functions with compact support, justified using the majoration condition.

The last assertion of the lemma is proved in a similar manner. ○

1.5 Smoothness of Gaussian CPLF "Tails"

1.5.1 "Tail" Density of CPLF: Existence and Lower Bound

Let $\eta \in \mathbf{R}^k$ be a standard Gaussian RV with distribution (1.1.4) and q a convex function. Put

$$\zeta = q(\eta), \quad G(u) = \mathbf{P}\{q(\eta) \le u\}. \tag{1.5.1}$$

The right-continuous DF above is positive to the right of the minimal value that ζ can attain,

$$u_0 = \inf\{u : G(u) > 0\}. \tag{1.5.2}$$

If q is homogeneous, $q(\lambda x) = \lambda q(x)$, Theorem 1.3.1 shows that G is differentiable, and Theorem 1.3.3 provides a bound on its derivative.

Distributions of convex piecewise-linear functions (1.4.13) also have an absolutely continuous component.

Theorem 1.5.1 *For a CPLF* 1.4.13), *the DF G is absolutely continuous on the half-line (u_0, ∞). Its derivative $G'(x)$ is defined on this half-line with a possible* *exception of a finite set of points. It is continuous outside the exceptional set and strictly positive if* $\max_j |b_j| > 0$ *in* (1.4.13). *For* $u_0 < u_1 < u_2$

$$G(u_2) - G(u_1) = \int_{u_1}^{u_2} G'(u)du.$$

Proof of Theorem 1.5.1 follows immediately from Lemma 1.4.6 or Lemma 1.4.8. Expression (1.4.12) for the derivative shows that it is bounded. It is also strictly positive because the Gaussian density is positive everywhere. ○

The additional restriction

$$\alpha_j \geq 0, \quad j = \overline{1, m}, \tag{1.5.3}$$

permits one to obtain for the DF of (1.5.1) and its derivative some estimates in terms of Φ and Φ', the standard normal DF and density (1.1.1). These estimates do not depend explicitly on the number of dimensions (cf. Sections 1.2 and 1.3).

Under restriction (1.5.3), $u > 0$, and the origin lies in the interior of the level set $Q_u = \{q(x) \leq u\}$ if $G(u) > \frac{1}{2}$. Indeed, otherwise it would be possible to include this set in a half-space $\{(x, b_j) \leq u\}$ of Gaussian measure $\frac{1}{2}$. Obviously, $t > 0$ if $\Phi(t) > \frac{1}{2}$.

In the statement of the next theorem, notation $G'(u)$ refers to the quantity $\liminf_{h\searrow 0}[G(u+h) - G(u)]/h$ for the points where the *bona fide* derivative does not exist. $\liminf_{\lambda\searrow 1}[G(\lambda u) - G(u)]/[\lambda - 1] = uG'(u)$ in whatever sense this notation be used.

Theorem 1.5.2 *Assume that* $G(u) = \Phi(t)$ *for some* $u > u_0$ *and* $t \in \mathbf{R}$.
If $\Phi(t) > \frac{1}{2}$ *(and consequently* $u, t > 0$*), then*

$$\forall \lambda > 1 \;\; G(\lambda u) > \Phi(\lambda t), \;\; uG'(u) \geq t\Phi'(t) = te^{-\frac{1}{2}t^2}/\sqrt{2\pi}.$$

The inequalities above hold without the restriction $\Phi(t) > \frac{1}{2}$ *if in representation* (1.4.13) $\alpha_j = 0$ *for all* j.

Proof of Theorem 1.5.2. Consider first the case of $t > 0$. Since the origin is an interior point of the level set $\{q(x) \leq u\}$, the following inclusions hold for $\lambda > 1$:

$$\{q(x) \leq u\} \subseteq \lambda\, \{q(x) \leq u\} \subseteq \{q(x) \leq \lambda u\}. \tag{1.5.4}$$

The last of them is a consequence of the assumption that all α_j are positive.

For the half-space $H = H_{e,t}$ (see (1.2.3)), $N(H) = \mathbf{P}\{\eta \in H\} = \Phi(t)$. Hence it follows from Theorem (1.5.4) and (1.5.4) that

$$G(\lambda u) \geq N(\lambda\{q \leq u\}) \geq N(\lambda H) = N(H_{e,\lambda t}) = \Phi(\lambda t), \;\; \lambda > 1.$$

The derivative $(d/(d\lambda)N(\lambda\{q \leq u\})|_{\lambda=1}$ exists by Theorem 1.3.1. The inequality of the theorem is a consequence of inclusions (1.5.4): indeed,

$$uG'(u) = \liminf_{\lambda\searrow 1} \frac{G(\lambda u) - G(u)}{\lambda - 1} \geq \frac{d}{d\lambda}N(\lambda\{q \leq u\})\bigg|_{\lambda=1}$$

$$\geq \left.\frac{d}{d\lambda} N(\lambda H)\right|_{\lambda=1} = t\Phi'(t).$$

If all the numbers α_j are equal to zero, the function q is positively homogeneous of order 1, and the equality $\lambda\{x : q(x) \leq u\} = \{x : q(x) \leq \lambda u\}$ holds for all $\lambda > 0$. For this reason the estimates of the theorem remain valid even if it is not stipulated that the origin lies inside $\{q \leq u\}$. ○

The first inequality of Theorem 1.5.2 can be written in the form

$$1 - G(\lambda u) = \mathbf{P}\{q(\eta) > \lambda u\} \leq 1 - \Phi(\lambda t), \quad \lambda > 1,$$

where t and u are positive. It shows that, independently of the number of dimensions or that of the "faces" of the level set, the probability that a CPLF attains a very large value is majorated by an appropriately scaled "tail" of a one-dimensional standard normal distribution.

It is useful to compare the estimates of Theorem 1.5.2 with known asymptotic formulae for large deviation probabilities of a standard normal rv (1.1.2) and (1.1.3).

1.5.2 Bounds for Inverses of DF's

Denote by $\Phi^{-1}(p)$ the inverse of the standard normal DF and by

$$G^{-1}(p) = \sup\{u : G(u) \leq p\} \tag{1.5.5}$$

an inverse for DF (1.5.1).

Corollary 1.5.1 *The ratio $G^{-1}(p)/\Phi^{-1}(p)$ is non-increasing for $p > \frac{1}{2}$ under the assumptions of Theorem* 1.5.2.

Proof of Corollary 1.5.1. By Theorem 1.5.1, the function $G(u)$ is continuous and strictly increasing when $G(u) > \frac{1}{2}$. Outside a finite exceptional set, it has a continuous positive derivative. Besides, for $p > \frac{1}{2}$ both $G^{-1}(p)$ and $\Phi^{-1}(p)$ are positive. This proves the existence of the derivatives $(d/dp)G^{-1}(p) = 1/G'\left(G^{-1}(p)\right)$ and $(d/dp)\Phi^{-1}(p) = 1/\Phi'\left(\Phi^{-1}(p)\right)$ outside the image of the exceptional set. Moreover, by Theorem 1.5.1

$$\frac{d}{dp}\ln G^{-1}(p) = \frac{1}{G^{-1}(p)G'\left(G^{-1}(p)\right)} \leq \frac{1}{\Phi^{-1}(p)\Phi'\left(G^{-1}(p)\right)} = \frac{d}{dp}\ln\Phi^{-1}(p).$$

This proves the corollary. ○

Remark 1.5.1 Assertions of Theorem 1.5.1 and Theorem 1.5.2 can be extended to more general spherically symmetrical distributions which have densities of the type specified in Theorem 1.3.1.

Lemma 1.5.1 *Assume that in* (1.4.13)

$$\alpha_j = \alpha, \; j = \overline{1,m}, \quad c = \max_{j=\overline{1,m}} |b_j| > 0.$$

Then the difference $G^{-1}(p) - c\Phi^{-1}(p)$ *is nonnegative and nonincreasing for* $p \in (0,\ 1)$.

Proof of Lemma 1.5.1. For $p < p_0 = \mathbf{P}\{\zeta \le u_0\}$, the inverse function defined in (1.5.5) is constant. Thus, monotonicity of the difference considered in the lemma is evident for $p < p_0$.

Since all α_j are equal, $q(x) = \alpha + q^0(x)$, where the CPLF

$$q^0(x) = \max_{j=\overline{1,m}} (x, b_j)$$

is positively homogeneous: $q^0(\lambda x) = \lambda q^0(x)$ if $\lambda > 0$. The rate of its growth as $|x| \to \infty$ is determined by the quantity $c = \sup_{|x|\le 1} q^0(x) = \max_{j=\overline{1,m}} |b_j|$ from the statement of the lemma. By homogeneity,

$$q^0(x+y) = 2q^0\left(\tfrac{1}{2}x + \tfrac{1}{2}y\right) \le q^0(x) + q^0(y)$$

for arbitrary x and y. Hence

$$\left|q^0(x+y) - q^0(x)\right| \le \max\left\{q^0(y), q^0(-y)\right\} \le c\,|y|\,.$$

Consequently, the δ-neighborhoods of level sets $Q_u = \{x : q(x) \le u\}$ satisfy the inclusions $Q_u^\delta \subseteq Q_{u+c\delta}$, $\delta > 0$.

The condition $c > 0$ implies that q is not constant, so Theorem 1.5.1 is applicable. By the above its DF is strictly increasing for $p \ge p_0$, so the DF G has continuous inverse function G^{-1}.

Assume that $p, p' \ge p_0$ and $p' > p$. Then there exist such u and $\delta > 0$ that $G(u) = p$, $G(u+\delta) = p'$. By Corollary 1.2.1

$$\begin{aligned} G^{-1}(p) - c\Phi^{-1}(p) &= u - c\Phi^{-1}(\mathbf{P}\{\eta \in Q_u\}) \\ \ge u - c\left[\Phi^{-1}\left(\mathbf{P}\left\{\eta \in Q_u^{\delta/c}\right\}\right) - \delta/c\right] &\ge u + \delta - c\Phi^{-1}(\mathbf{P}\{\eta \in Q_{u+\delta}\}) \\ &= G^{-1}(p') - c\Phi^{-1}(p')\,. \end{aligned}$$

Consequently, the difference considered is also monotone for $p > p_0$. To verify that it is nonnegative, put $u = G^{-1}(p)$ and choose l so that $|b_l| = \max_j |b_j| = c$:

$$\Phi(u/c) = \mathbf{P}\{(b_l, \eta) \le u\} \ge \mathbf{P}\{q(\eta) \le u\} = p, \quad G^{-1}(p)/c \ge \Phi^{-1}(p). \ \bigcirc$$

1.5.3 Upper Bounds for Density

By Theorem 1.5.1, the DF of CPLF (1.4.13) of Gaussian RV is differentiable for sufficiently large values of its argument. The density of its absolutely continuous component admits the lower bound of Theorem 1.5.2. It is also possible to evaluate this density from above by expressions which do not depend explicitly either on the number of dimensions k or on the choice of representation (1.4.13).

Theorem 1.5.3 a) *Assume that for some* $t \in \mathbf{R}$ *and* $b > u_0$

$$G(b) = \Phi(t) > 0. \tag{1.5.6}$$

Then the derivative G' of Theorem 1.5.1 *admits the upper bound*

$$G'(u) \le \frac{u + u_+ - b}{(u-b)^2}(1+t_-)^2. \tag{1.5.7}$$

b) *If equality* (1.5.6) *holds with $\Phi(t) > \frac{1}{2}$, so both $b > 0$ and $t > 0$, then for $u > b$*

$$G'(u) \le (u + u_+ - b)\, \tau^2 \left[1 + 2\alpha + 2\alpha^2\right] \left[1 - \Phi(\tau u)\right], \tag{1.5.8}$$

where $\tau = t/b$ and $\alpha = (\tau u)^{-2}(1 - b/u)^{-1}$.

It is worth while to compare the estimates of the theorem with asymptotic formula (1.1.2) for the standard normal DF Φ. Comparison shows that the bound of part a) is fairly rough, while that of part b) establishes the very rapid decay of tails characteristic of Gaussian distributions in a very general situation.

1.5.4 Derivation of Polynomial Bound for Density

Proof of Theorem 1.5.3, a). Abbreviate the notation of standard normal density (1.1.4) to $g(x) \stackrel{def}{=} n_{k,0,I}(x)$. This function is smooth and rapidly decays at infinity. Thus, Lemma 1.4.7 can be applied to calculate the right derivative G'.

Put $\beta_j = |b_j|$, so that $b_j = \beta_j e_j$, $|e_j| = 1$, in representation (1.4.13) of the function q.

The formula of Lemma 1.4.7 for the derivative can be transformed using the identity $(f, \nabla g) = -(f, x)\, g$. The result is the following expression:

$$G'(u) = \sum_{j=1}^{m} \int_{F_j^+(u)} \frac{(n_j, y)}{\beta_j}\, g(y)dy, \tag{1.5.9}$$

where $n_j = e_j$ is the outward normal unit vector to the face $F_j(u)$ of the boundary ∂Q_u.

To evaluate the integrand in (1.5.9), it is convenient to represent the points of the integration domain as (cf. (1.4.22)–(1.4.23))

$$y = x + [t + t_j(u, x)]\, n_j = y_0 + t n_j, \quad y \in F_j^+(u), \tag{1.5.10}$$

where $t = \operatorname{dist}(y, Q_u)$ is the distance from y to Q_u, the point x lies in the hyperplane $\{z : (z, n_j) = 0\}$, and the point $y_0 = x + t_j n_j$ belongs to the boundary of Q_u (see Lemma 1.4.3).

By the choice of the function q and the inclusion $Q_b \subset Q_u$,

$$(n_j, y) = t + (n_j, y_0) \le t + u_+/\beta_j, \quad t = \operatorname{dist}(y, Q_u) \le \operatorname{dist}(y, Q_b).$$

At the same time, by Lemma 1.4.3 and Corollary 1.4.1 there is also inequality

$$\frac{1}{\beta_j} = \frac{1}{|b_j|} \le \frac{\operatorname{dist}(Q_b, F_j^+(u))}{u-b} \le \frac{\operatorname{dist}(y, Q_b)}{u-b}.$$

For this reason, the inequality

$$\frac{(n_j, y)}{\beta_j} \leq \frac{1}{\beta_j}\left(t + \frac{u_+}{\beta_j}\right) \leq [\mathrm{dist}\,(y, Q_b)]^2 \left[\frac{1}{u-b} + \frac{u_+}{(u-b)^2}\right] \tag{1.5.11}$$

holds at all points of the set $F_j^+(u)$ independently of the number of the face.

The union of the sets $F_j^+(u)$ constitutes a part of the complement to $Q_u \supset Q_b$ (see Lemma 1.4.1). Hence it follows from (1.5.9) and (1.5.11) that

$$G'(u) \leq \left[\frac{1}{u-b} + \frac{u_+}{(u-b)^2}\right] \int_{Q_u^c} [\mathrm{dist}\,(y, Q_b)]^2\, g(y)dy, \quad Q_u^c = \mathbf{R}^k \setminus Q_u. \tag{1.5.12}$$

After the integration domain is increased to Q_u^c, inequality (1.2.7) yields the estimate

$$\int_{Q_u^c} [\mathrm{dist}\,(y, Q_b)]^2\, g(y)dy \leq \int_{Q_b^c} [\mathrm{dist}\,(y, Q_b)]^2\, g(y)dy \leq \int_t^\infty (u-t)^2 d\Phi(u).$$

To derive inequality a) of Theorem 1.5.3, it remains to evaluate the one-dimensional integral in the right-hand side using Lemma 1.5.2. ○

1.5.5 Derivation of Exponential Bound for Density

Some additional considerations are necessary to derive inequality b) of Theorem 1.5.3. The additional restriction allows one to obtain a nontrivial bound from below for $\mathrm{dist}\,(y, Q_b)$ on the set $\mathbf{R}^k \setminus Q_u$. The bound is then used to make much smaller the domain of integration in (1.5.12).

Lemma 1.5.2 *Assume that conditions of Theorem* 1.5.3 b) *hold and* $q(x) \geq u > b > u_0$. *Then* $t > 0$, $b > 0$, *and*

$$\mathrm{dist}\,(x, Q_b) \geq (u-b)\; t/b.$$

Proof of Lemma 1.5.2. Fix a point x with $q(x) = u > b$. There is a dense set of points where the CPLF q is continuously differentiable, so for each $\varepsilon > 0$ there is a point $y = y_\varepsilon$ such that q is continuously differentiable in its neighborhood and $|x-y| < \varepsilon$, $|q(x) - q(y)| < \varepsilon$. For small ε

$$\mathrm{dist}\,(y, Q_b) \geq (\hat{u} - b)/\left|\nabla q(y)\right|, \quad \hat{u} = q(y), \tag{1.5.13}$$

by Lemma 1.4.1.

The calculations that follow exploit essentially Lemma 1.4.4. By the additional condition of part b), the origin is an interior point for the set Q_b: otherwise it would be possible to cover this set by a half-space of Gaussian measure $\frac{1}{2}$ with boundary containing the origin. Hence $b > G(0) \geq 0$ and $t > 0$.

The supporting hyperplane to the convex set $Q_{\hat{u}}$ at the boundary point y (where $\hat{u} = q(y)$) is $\{z : (\, z - y, n(y)\,) = 0\}$. It follows that $Q_{\hat{u}}$ is a subset of $\{z : (\, z - y, n(y)\,) \leq 0\}$. This inclusion leads to the equivalent inequalities

$$N(Q_{\hat{u}}) = G(\hat{u}) \leq \Phi\left(\, (y, n(y))\,\right) \Leftrightarrow \left(\, y, n(y)\,\right) \geq \Phi^{-1}[G(\hat{u})].$$

Combined with (1.5.13) and Lemma 1.4.4, these inequalities yield the estimate

$$\operatorname{dist}(y, Q_b) \geq (\hat{u} - b)\left(y, n(y) \right)/\hat{u} \geq (\hat{u} - b)\Phi^{-1}\left[G(\hat{u})\right]/\hat{u}.$$

After a passage to the limit in $y_\varepsilon \to x$, with $\hat{u} = q(y_\varepsilon) \to u = q(x)$, one obtains from this inequality the bound

$$\operatorname{dist}(x, Q_b) \geq \frac{(q(x) - b)}{q(x)}\Phi^{-1}\left[G(q(x))\right] = \frac{u - b}{u}\Phi^{-1}\left[G(u)\right] \tag{1.5.14}$$

for an arbitrary point and satisfying the restrictions of the lemma.

It is possible to simplify (1.5.14) with the aid of Corollary 1.5.1. Indeed, $G[q(x)] > G(b)$ and $G(b) > \frac{1}{2}$, so

$$\frac{\Phi^{-1}\left[G(q(x))\right]}{q(x)} = \frac{\Phi^{-1}\left[G(q(x))\right]}{G^{-1}\left[G(q(x))\right]} \geq \frac{\Phi^{-1}\left[G(b)\right]}{G^{-1}\left[G(b)\right]} = \frac{t}{b}.$$

To obtain the assertion of the lemma, it remains to evaluate the right-hand side of (1.5.14) using the above inequality. ○

Proof of Theorem 1.5.3. b) By Lemma 1.5.2, the inequality $\operatorname{dist}(y, Q_b) \geq (u - b)t/b$ holds for $y \in Q_u$. Thus one can replace the domain of integration in (1.5.12) by the set

$$\{y : \operatorname{dist}(y, Q_b) \geq T - t\}, \quad T = t + (u - b)t/b,$$

and then apply Corollary 1.2.1 to pass to an integral with respect to the standard normal distribution on the real line:

$$G'(u) \leq \left[\frac{1}{u - b} + \frac{u_+}{(u - b)^2}\right] \int_T^\infty (v - t)^2 d\Phi(v).$$

This estimate is transformed to that given in the statement of the theorem by application of Lemma 1.1.2 (see below). ○

1.5.6 Bounds on Upper Derivatives of CPLF Density

The density of the absolutely continuous component in distribution (1.5.1) cannot have positive jumps. One can obtain nontrivial estimates for the upper derivative

$$D_+G' \equiv \limsup_{h \to 0} \frac{1}{h}\left[G'(u + h) - G'(u)\right] \tag{1.5.15}$$

provided that its argument is large enough.

Theorem 1.5.4 *Assume that $\alpha_j \geq 0$ for all j in representation* (1.4.13) *of the function q.*

a) *If condition* (1.5.6) *is satisfied, then for $u > b$*

$$D_+G'(u) \leq \left(\frac{1 + t_-}{u - b}\right)^2. \tag{1.5.16}$$

b) *If condition* (1.5.6) *is replaced by the more restrictive one* $G(b) = \Phi(t) > \frac{1}{2}$ *of Theorem* 1.5.3,b), *then the above bound can be changed to*

$$D_+G'(u) \le \tau^2 [1 - \Phi(\tau u)] \left(1 + 2\alpha + \alpha^2\right). \tag{1.5.17}$$

Notation above is that of Theorem 1.5.3. The proof of Theorem 1.5.4 is given in subsection 1.5.8 below after some preparations in subsection 1.5.7.

1.5.7 Estimates for Derivative of Density: Smooth Case

Throughout this subsection the function $q : \mathbf{R}^k \to \mathbf{R}$ satisfies the assumptions of subsections 1.4.1–1.4.3.

Lemma 1.5.3 *Under the assumptions of Lemma* 1.4.2

$$G''(u) \le -\int_{Q_u^c} \frac{B(y)}{|\nabla q(x)|^2} dy, \tag{1.5.18}$$

where y, $x = x(y)$ *are related as in* (1.4.1) *and*

$$\begin{aligned} B(y) \;=\; (n(x), \nabla)^2 g(y) \;&+\; 2\,(n(x), \nabla g(y))\,(\partial/\partial t) \ln J(t,x) \\ &+\; g(y)\,[(\partial/\partial t) \ln J(t,x)]^2 . \end{aligned}$$

Proof of Lemma 1.5.3. To derive an expression for the second derivative of G, one can start with differentiation of (1.4.11) for $v > u$. This yields the equality

$$\begin{aligned} G''(v) \;=\; & -\int_{\partial Q_u} ds(x) \int_0^\infty \frac{B(y) J(t+T,x)}{(n(x), \nabla q(y_0))^2} dt \\ & + \int_{\partial Q_u} ds(x) \frac{(n(x), \nabla)^2 q(y_0)}{(n(x), \nabla q(y_0))^3} \int_0^\infty \frac{\partial}{\partial t} [g(y) J(t+T,x)]\, dt, \end{aligned} \tag{1.5.19}$$

where the notations are those of (1.4.11) and (1.5.18).

In the right-hand side, the second summand is nonpositive for $v = u$ because g is nonnegative and q convex: indeed, the second derivative along the outer normal, which it contains, is nonnegative. The desired result is then obtained by integration in the variable t with subsequent passage to the limit as $v \searrow u$.

After the passage to the limit in $v \searrow u$, the first summand in the right-hand side is transformed into a "volume" integral using formula (1.4.8). This yields the estimate of the lemma. ○

Remark 1.5.2 In the integrand of (1.5.18)

$$2\,(n, \nabla g)\, \frac{\partial}{\partial t} \ln J + \left[\frac{\partial}{\partial t} \ln J\right]^2 g \ge -g\,(n, \nabla \ln g)^2$$

at the points with $g(y) > 0$ (the sum of left- and right-hand sides is square of a real number).

Thus, at the points with $g(y) > 0$ the integrand in (1.5.18) admits the estimate

$$\begin{aligned} B(y) &\geq (n(x), \nabla)^2 g(y) - g(y)\,(n(x), \nabla \ln g(y))^2 \\ &= g(y)\,(n(x), \nabla)^2 \ln g(y). \end{aligned} \tag{1.5.20}$$

Further in this subsection, the random argument of the smooth convex function q has standard Gaussian distribution, and its DF is

$$\hat{G}(u) = \int_{\{q(x)\leq u\}} g(x)dx, \quad g(x) \equiv n_{k,0,I}(x). \tag{1.5.21}$$

Lemma 1.5.4 *If* $u > b > q_0 = \inf q(x)$, *then function* (1.5.21) *is twice differentiable and its second derivative admits the bound*

$$\hat{G}''(u) \leq \int_{\mathbf{R}^k\setminus Q_u} \left(\operatorname{dist}(y, Q_b)/(u-b) \right)^2 g(y)dy, \quad Q_u = \{q \leq u\}, \quad u \in \mathbf{R}.$$

Proof of Lemma 1.5.4. The fact that the first and second derivatives of (1.5.21) do exist is deduced from Lemma 1.5.3 with the aid of an approximation of the standard Gaussian density by smooth functions with compact support (see Section 1.4). An estimate similar to (1.5.18) is derived in the same way.

The normal density is positive in all the space. For this reason the integrand in the estimate mentioned above satisfies, moreover, inequality (1.5.20). In the special case considered here, $-(n(x), \nabla)^2 \ln g(y) = 1$. This yields the inequality

$$\hat{G}''(u) \leq \int_{\{q>u\}} \frac{g(y)dy}{|\nabla q(x)|^2}, \tag{1.5.22}$$

with x the point of the boundary of the integration domain nearest to y.

By Lemma 1.4.4,

$$1/|\nabla q(x)|^2 \leq \left(\operatorname{dist}(x, Q_b)/(u-b) \right)^2, \quad Q_b = \{q \leq b\}.$$

If $y = x + tn(x)$, $t > 0$, is an interior point of the set $\{q > u\} = \mathbf{R}^k \setminus Q_u$, then

$$\operatorname{dist}(y, Q_b) \geq \operatorname{dist}(x, Q_b).$$

Indeed, let z be the point of the set Q_b which is nearest to y. Then

$$|y-z|^2 = |x-z|^2 + t^2 + 2t\,(x-z, n(x)) \geq |x-z|^2 \geq [\operatorname{dist}(x, Q_b)]^2,$$

because the tangential hyperplane which contacts ∂Q_u at x is, at the same time, the boundary of a half-space containing both Q_u and Q_b. Hence the inequality of the lemma follows from (1.5.22). ○

Corollary 1.5.2 *Under the restrictions of Lemma* 1.5.4, *there is inequality*

$$\begin{aligned} \delta^2_{h,h'}\hat{G}(u) &\equiv \hat{G}(u+h+h') - \hat{G}(u+h) - \hat{G}(u+h') + \hat{G}(u) \\ &\leq hh' \int_{\{q>u\}} \left(\operatorname{dist}(y, q_b)/(u-b) \right)^2 g(y)dy. \end{aligned}$$

Proof of Corollary 1.5.2. To establish the assertion of the corollary, represent the second-order difference considered in the form

$$\delta^2_{h,h'}\hat{G}(u) = \int_0^h \int_0^{h'} \hat{G}''(u+t+t')\,dt\,dt'.$$

It remains to note that the right-hand side in the estimate of Lemma 1.5.4 for the second derivative is a non-increasing function. ○

1.5.8 Evaluation of Upper Derivative: CPLF Density

Proof of Theorem 1.5.4. a) Assume first that q is a CPLF with compact level sets satisfying the conditions of the theorem. This function can be approximated, to an arbitrary degree of accuracy, by smooth strictly convex ones (see subsection 1.4.6). An analog of the estimate of Corollary 1.5.2 for the original CPLF is derived from the estimate for the approximating smooth function by a passage to the limit:

$$\begin{aligned}\frac{1}{hh'}\delta^2_{h,h'}G(u) &\equiv \frac{1}{hh'}\left[G(u+h+h') - G(u+h') - G(u+h) + G(u)\right] \\ &\leq \int_{\{q>u\}} \left(\operatorname{dist}(y, Q_b)/(u-b)\right)^2 g(y)dy \qquad (1.5.23)\end{aligned}$$

for $h, h' > 0$. Using Corollary 1.2.1, one can pass from (1.5.23) to an estimate in terms of an integral with respect to the one-dimensional standard normal distribution:

$$\frac{1}{hh'}\delta^2_{h,h'}G(u) \leq \int_t^\infty \left(\frac{v-t}{u-b}\right)^2 d\Phi(v). \qquad (1.5.24)$$

To derive the inequality of Theorem 1.5.4 it remains to apply Lemma 1.1.2 and then to pass to the limit subsequently in $h \searrow 0$ and $h' \searrow 0$.

The sign of h' is, in fact, unimportant: for $h' < 0$ one can have recourse to the identity

$$\frac{1}{hh'}\delta^2_{h,h'}G(u) = \frac{1}{h|h'|}\delta^2_{h,|h'|}G(u+h').$$

The right-hand side depends on u continuously. For this reason, the above described passage to the limit yields the estimate of the theorem also for the left-sided upper derivative.

For a CPLF whose level sets may be non-compact, the estimate of (1.5.23) is derived from the one above by a passage to the limit. The function q can be approximated for the purpose, e.g., by the functions

$$q_n(x) \equiv \max\left\{q(x), \tfrac{1}{n}|x|_+\right\}, \quad |x|_+ = \max_{j=\overline{1,k}}\left|x^{(j)}\right|.$$

After an analog of estimate (1.5.23) has been derived, the proof of the theorem is completed by an application of Corollary 1.2.1 and a passage to the limit in $h \searrow 0$, $h' \to 0$. Existence of the (right-sided) derivative G' follows from Theorem 1.5.1. ○

Proof of Theorem 1.5.4, part b) is based on Lemma 1.5.2. Applying this proposition, it is possible to restrict the domain of integration in (1.5.23) (cf. proof of Theorem 1.5.3, b). This permits one to replace (1.5.24) by the estimate

$$\frac{1}{hh'}\delta^2_{h,h'}G(u) \leq \int_T^\infty \left(\frac{v-t}{u-b}\right)^2 d\Phi(v),$$

where $T = t + (u-b)t/b$.

Further calculations repeat the proof of Theorem 1.5.4, part a), the only difference being that Lemma 1.1.2 is used instead of Lemma 1.1.1. ○

1.6 Some Related Results

1. Let $\xi \in \mathbf{R}^k$ be a Gaussian RV and A a symmetric real matrix (not necessarily a positive definite one). Denote by $Q_A = (A\xi, \xi)$ the value of the corresponding quadratic form with argument ξ, and put

$$\mathbf{p}(x) = \sup_A \mathbf{P}\left\{|Q_A - \mathbf{E}Q_A| > x\sqrt{\mathbf{D}\,(Q_A)}\right\}.$$

The following inequality holds for all $x > 0$ [Po78a]:

$$\frac{1}{\sqrt{\pi}}\frac{\sqrt{e}-1}{\sqrt{1+x\sqrt{2}}} \leq \exp\left\{\frac{x}{\sqrt{2}}\right\}\mathbf{p}(x) \leq 2\sqrt{1+x\sqrt{2}},$$

as well as an analog of the upper bound for the so-called sub-Gaussian distributions. Besides [Po78b],

$$\tfrac{1}{2}\left[1-\Phi\left(x/\sigma\right)\right] \leq \inf_A \mathbf{P}\left\{|Q_A - \mathbf{E}Q_A| \geq x\sqrt{\mathbf{D}Q_A}\right\} \leq 2\left[1-\Phi(x)\right],$$

where Φ is, as usual, the standard normal DF and

$$\sigma = 2^{-5/2}\sigma_1\left(1-4\sigma_1^2\right), \quad \sigma_1 = \tfrac{3}{8}\left[1-\Phi\left(1/4\right)\right] \approx 0.024192.$$

2. Let $\left(\xi^{(1)}, \ldots, \xi^{(n)}\right) \in \mathbf{R}^n$ be a Gaussian RV, and assume that the functions $f, g : \mathbf{R}^n \to \mathbf{R}$ are non-decreasing in each coordinate. If

$$\mathbf{E}\xi^{(j)} = 0, \quad \mathbf{E}\xi^{(j)}\xi^{(l)} \geq 0, \quad j, l = \overline{1,n},$$

then (see [Pitt82], [JoPP83])

$$\mathbf{E}f\left(\xi^{(1)}, \ldots, \xi^{(n)}\right) g\left(\xi^{(1)}, \ldots, \xi^{(n)}\right) \geq \mathbf{E}f\left(\xi^{(1)}, \ldots, \xi^{(n)}\right) \mathbf{E}g\left(\xi^{(1)}, \ldots, \xi^{(n)}\right).$$

3. For a Gaussian process ξ_t, $t \in T$, with $\mathbf{E}\xi_t = 0$, put $\Delta\left(s,t\right) = \sqrt{\mathbf{E}\left(\xi_s - \xi_t\right)^2}$. For each finite set $T_0 \subset T$,

$$\mathbf{E}\max_{s,t\in T}|\xi_s - \xi_t| \geq \sqrt{2\pi\log_2\left(T/2\right)}\,\min_{s,t\in T_0, s\neq t}\Delta\left(s,t\right),$$

where $|T_0|$ is the number of elements in T_0. See [Su69] and [Fer74].

4. Let ξ_{ij} be independent rv's with the standard normal distribution $N_{1,0,1} = \Phi$, $b = (b_{rj})$ some collection of real numbers, and σ_j some positive real numbers satisfying the condition $1 = \sigma_1^2 > \sigma_2^2 > \cdots > 0$. Put

$$S(b) \equiv \sum_{r=1}^{m} \sigma_r^2 \sum_{j=1}^{n_r} (\xi_{rj} - b_{rj})^2, \quad |b_r|^2 = \sum_{j=1}^{n_r} b_{rj}^2.$$

Denote, moreover,

$$G(x) \equiv \mathbf{P}\{S(b) > x\}, \quad G_{r_1,r_2,\ldots,r_k}(x) \equiv \mathbf{P}\left\{\sum_{r=1}^{m} \sigma_r^2 \sum_{j=1}^{\tilde{n}_r} \xi_{rj}^2 > x\right\},$$

where $\tilde{n}_r = \tilde{n}_r(k) = n_r + 2\sum_{j=1}^{k} \delta_{rr'}$ and $\delta_{rr'}$ is the Kronecker symbol.

In the above notation,

$$\mathbf{P}\{S(b) > x\} = \sum_{k=1}^{\infty} \frac{(-1)^k}{k!} \sum_{r_1,\ldots,r_k=1}^{m} \prod_{j=1}^{k} \left[\sigma_{r_j}^2 \left|b_{r_j}\right|^2\right] \frac{d^k}{dx^k} G_{r_1,\ldots,r_k}(x).$$

Moreover,

$$\lim_{x\to\infty} \frac{\mathbf{P}\{S(b) > x\}}{\mathbf{P}\left\{\sum_{j=1}^{n_1} (\xi_{1j} - b_{1j})^2 > x\right\}} = A(b),$$

where

$$A(b) = \prod_{r=2}^{m} \left(1 - \sigma_r^2\right)^{-\frac{1}{2}n_r} \exp\left\{\frac{1}{2}\sum_{r=2}^{m} \frac{\sigma_r^2}{1 - \sigma_r^2} |b_r|^2\right\}.$$

See [Ber75]; a passage to the limit for $m \to \infty$ permits one to extend this result to infinite-dimensional Gaussian distributions (see Chapter 2 below).

5. Let $N = N_{2,0,I}$ be the standard Gaussian distribution in $\mathbf{R}^2$, and assume that the sets $A, B \subset \mathbf{R}^2$ are convex, with $A = -A$, $B = -B$. Then

$$N\left(A \bigcap B\right) \geq N(A)N(B).$$

See [Pitt77]; some multidimensional generalizations can be found in [Bor81].

6. Let η be a RV with the standard Gaussian distribution $N_{k,0,I}$.

If C is an arbitrary convex subset of $\mathbf{R}^k$, l a nonnegative integer, and ε, δ arbitrary positive numbers, then the measure of its boundary strip (cf. (A.1.2)) satisfies the relation

$$\mathbf{E}\left\{|\eta|^l;\ \eta \in C^{\varepsilon} \setminus C^{-\delta}\right\} \leq c(\varepsilon + \delta),$$

where the constant $c = c(k, l)$ is independent of the choice of the set C and admits the estimate

$$c(k, l) \leq (k + 2l - 1)\, 2^{(l-1)/2} \Gamma\left([k + l - 1]/2\right) / \Gamma(k/2).$$

See[*BhRR76] (for $l = 0$ the estimate was established in[RRa60]). For estimates for the constants with explicit dependence on dimension number, see also [Bahr67a, Bahr67b, Saz68].

The constant in the above estimate is not independent of the dimension number unless one stipulates some restrictions on the shape of the set. For the cube $C_r = \left\{x \in \mathbf{R}^k : \left|x^{(j)}\right| \leq r,\ j = \overline{1,k}\right\}$, it was proved [Bl73] that

$$\sup_{\varepsilon \geq 0} (1/\varepsilon)\, \mathbf{P}\left\{\eta \in C_r^\varepsilon \setminus C_r\right\} \geq (1 - 1/k)^{k-1}\, r_k/2$$

if r is large enough, with r_k satisfying the equation $2k[1 - \Phi(r) = 1$ and the asymptotic relation $r_k \sim c'\sqrt{\ln k}$. Thus,

$$\sup_{r,\varepsilon \geq 0} (1/\varepsilon)\, \mathbf{P}\left\{\eta \in C_r^\varepsilon \setminus C_r\right\} \geq c''\sqrt{\ln k}.$$

Chapter 2

Seminorms of Gaussian Vectors in Infinite Dimensions

This chapter considers the distributions of seminorms of Gaussian RV's with values in infinite-dimensional spaces. These distributions conserve the features typical of their finite-dimensional counterparts — rapid decay of large deviation probabilities and a certain degree of smoothness.

The exposition employs three distinct approaches. The rapid decay of "tail" is derived from the characteristic property of Gaussian distribution (Section 2.1). The geometric apparatus developed in Chapter 1 is used in Section 2.2 to establish the presence of an absolutely continuous component in the distribution. For Gaussian RV's with values in a Hilbert space, one can also use the fact that for the distribution of squared norm the CF can be written out explicitly. This approach is adopted in Sections 2.3 and 2.4.

2.1 Exponential Summability of Seminorms

2.1.1 Summability of Quadratic Exponential

In this section $\langle X, \mathcal{X} \rangle$ is a measurable real linear space.

A distribution N in X is called ***centered Gaussian*** when it has the following characteristic property: if the RV's ξ and η are independent and have same distribution $\mathcal{L}(\xi) = \mathcal{L}(\eta) = N$, then the RV's

$$\xi_\alpha = \cos\alpha \cdot \xi + \sin\alpha \cdot \eta, \quad \eta_\alpha = -\sin\alpha \cdot \xi + \cos\alpha \cdot \eta \tag{2.1.1}$$

are also independent and $\mathcal{L}(\xi_\alpha) = \mathcal{L}(\eta_\alpha) = N$ for each choice of the constant "angle" $\alpha \in (0, \frac{1}{2}\pi)$. A ***centered Gaussian RV*** is a RV whose distribution is centered Gaussian.

Theorem 1.1.3 shows that the value of a non-random measurable linear functional on a centered Gaussian RV is a centered normal rv. This justifies the above definition — e.g., in a situation where the measurable sets belong to some cylindric σ-algebra generated by a family of linear functionals, all its finite-dimensional distributions are Gaussian.

A measurable function $|\cdot| : X \to [0, \infty]$ on a measurable linear space $\langle X, \mathcal{X} \rangle$ is a *measurable seminorm* if the set $X_f = \{x : |x| < \infty\}$ is a linear subspace of X and the restriction of $|\cdot|$ to this set has the usual properties of a seminorm.

Suppose that ξ is a centered Gaussian RV in a measurable linear space X and $|\cdot|$ is a measurable seminorm.

Theorem 2.1.1 *If $|\xi| < \infty$ a.s., then there exists a number $\alpha > 0$ such that*

$$\mathbf{E} \exp\left\{\alpha |\xi|^2\right\} < \infty. \tag{2.1.2}$$

This result can be made more precise if the seminorm is defined in a more specific way. The assumption that it is a.s. finite can be relaxed. As before, ξ is a centered Gaussian RV in a measurable linear space, and $|\cdot|$ a measurable seminorm.

Theorem 2.1.2 *Assume that $|\xi| = \sup_{k \in \mathbf{N}} |\langle \xi, t_k \rangle|$ a.s., where t_k are measurable linear functionals over X. Put $\sigma^2 = \sup_{k \in \mathbf{N}} \mathbf{E} \langle \xi, t_k \rangle^2$.*

If $\mathbf{P}\{|\xi| < \infty\} > 0$, then $|\xi| < \infty$ a.s., and the expectation in (2.1.2) is finite for each $\alpha < \frac{1}{2}\sigma^{-2}$.

Theorem 2.1.2 applies, e.g., to centered Gaussian RV's in a separable Banach space.

Remark 2.1.1 The quantity σ^2 in the statement of Theorem 2.1.2 is finite. Indeed, under the converse assumption one can select a sequence $n(k) \nearrow \infty$ so as to make convergent the series

$$\sum_{k=1}^{\infty} 1/\sigma(k), \quad \sigma(k) = \sqrt{\mathbf{E} \langle \xi, t_{n(k)} \rangle^2}.$$

All the rv's $\langle \xi, t_{n(k)} \rangle / \sigma(k)$ have standard normal distribution, so

$$\forall\, L > 0 \;\; \mathbf{P}\left\{|\langle \xi, t_{n(k)} \rangle| \le L\right\} \le \sqrt{2/\pi} L/\sigma(k).$$

It follows that $|\xi| = \infty$ a.s., because by the Borel–Cantelli lemma

$$\forall\, L > 0 \;\; |\xi| \ge \sup_{k \in \mathbf{N}} |\langle \xi, t_{n(k)} \rangle| > L \text{ a.s.}$$

Remark 2.1.2 Suppose that $\xi^{(k)}$ are Gaussian real rv's with nonnegative expectations $\alpha_k \equiv \mathbf{E}\xi^{(k)} \ge 0$ for all k.

If the rv $\zeta = \sup_{k \in \mathbf{N}} \xi^{(k)}$ is finite a.s., then the quantity $\sigma = \sup_{k \in \mathbf{N}} \sigma_k$, where $\sigma_k = \sqrt{\mathbf{D}\xi^{(k)}}$, is finite. Indeed, the inequality

$$\mathbf{P}\{\zeta \le L\} \le \mathbf{P}\left\{\xi^{(k)} \le L\right\} = \Phi([L - \alpha_k]/\sigma_k) \le \Phi(L/\sigma_k)$$

holds for all L and k. (As usual, Φ is the standard normal DF.) If the variances σ_k^2 were unbounded, the above relation would imply that $\mathbf{P}\{\zeta \le L\} \le \frac{1}{2}$ and $\mathbf{P}\{\zeta < \infty\} \le \frac{1}{2}$.

Remark 2.1.3 If $2\sigma^2\alpha > 1$, then the expectation of the exponential in Theorem 2.1.1 is infinite: this folows from formula (1.1.2) combined with the estimate $|\xi| \geq \langle \xi, t_k \rangle$.

2.1.2 Proof of Summability of Quadratic Exponential

Proof of Theorem 2.1.1. Let η be an independent copy of the RV ξ from the statement of the theorem.

It is possible to select the number $a > 0$ so that

$$q = \mathbf{P}\{|\eta| < a\} = \mathbf{P}\{|\xi| < a\} > 0.$$

Because of independence of ξ and η, this yields the equality

$$\mathbf{P}\{|\xi| \geq x\} = \frac{1}{q}\mathbf{P}\{|\xi| \geq x, |\eta| < a\}.$$

Besides, the triangle inequality leads to the inclusion

$$\begin{aligned}\{|\xi| \geq x, |\eta| < a\} &\subset \{|\xi + \eta| \geq x - a, |-\xi + \eta| \geq x - a\} \\ &= \left\{|\xi'| \geq (x-a)/\sqrt{2}\right\} \bigcap \left\{|\eta'| \geq (x-a)/\sqrt{2}\right\},\end{aligned}$$

where by definition (2.1.1) the RV's $\xi' = \frac{1}{\sqrt{2}}(\xi + \eta)$ and $\eta' = \frac{1}{\sqrt{2}}(-\xi + \eta)$ are independent RV's with the distribution $\mathcal{L}(\xi)$.

Since all the RV's ξ, η, ξ', and η' are identically distributed, the inclusion above yields the estimate

$$\mathbf{P}\{|\xi| \geq x\} = \frac{1}{q}\mathbf{P}\{|\xi| \geq x, |\eta| < a\} \leq \frac{1}{q}\left[\mathbf{P}\left\{|\xi| \geq \frac{x-a}{\sqrt{2}}\right\}\right]^2.$$

Define the sequences x_n and u_n by the equations

$$x_0 = a, \quad x_{n+1} = a + x_n\sqrt{2}, \; n = 0, 1, \ldots, \quad u_n = \frac{1}{q}\mathbf{P}\{|\xi| \geq x_n\}.$$

In this notation the above-derived estimate assumes the form $u_{n+1} \leq u_n^2$, $n \in \mathbf{N}$. Consequently, u_n rapidly converges to zero if a is so chosen that $u_0 = (1-q)/q < 1$, or, which is the same thing, $q < \frac{1}{2}$. For such choice of a,

$$u_n \leq \exp\{-2^n \ln 1/u_0\}.$$

It follows easily that

$$x_n = \left(1 + 2^{1/2} + \cdots + 2^{n/2}\right) a = \frac{2^{(n+1)/2} - 1}{\sqrt{2} - 1} a, \quad n \geq 1.$$

It is convenient to transform the bound selecting the number n so as to satisfy the inequalities $x_n \leq x \leq x_{n+1}$ for a given value of $x > a$. For this choice,

$2^{n+2} \geq \left(1+[\sqrt{2}-1]x/a\right)^2$, and the result is the inequality

$$\mathbf{P}\{|\xi| \geq x\} \leq q \exp\left\{-\frac{(\sqrt{2}-1)^2}{4}\frac{x^2}{a^2}\ln 1/u_0\right\}.$$

This formula and the well-known expression for moments of a positive rv through its "tails", $\mathbf{E}f(X) = f(0) + \int_0^\infty f'(x)\mathbf{P}\{X \geq x\}\,dx$, yield the assertion of the theorem. In fact, the above inequality provides a more explicit bound on the exponential moment: there is a positive constant c_0 such that for $\beta \leq c_0$

$$\mathbf{E}\exp\left\{\beta\frac{|\xi|^2}{a^2}U\right\} \leq c_1(\beta), \quad U = \ln\frac{1-\mathbf{P}\{|\xi| \geq a\}}{\mathbf{P}\{|\xi| \geq a\}}, \tag{2.1.3}$$

and the value of $c_1(\beta)$ is completely determined by the choice of β. ○

2.1.3 Highest Order of Finite Exponential Moment

The proof of Theorem 2.1.2 requires some preparations.

Below the sequence $\xi = \left(\xi^{(k)}\right)$, $\xi^{(k)} = \langle\xi, t_k\rangle$, $k \in \mathbf{N}$, is considered as a RV with values in the space $\mathbf{R}^{\mathbf{N}}$ of infinite sequences of real numbers. Its measurable sets are ones from the cylindric σ-algebra generated by the family of coordinate linear functionals

$$t_k : \mathbf{R}^{\mathbf{N}} \ni x = \left(x^{(j)}\right) \mapsto x^{(k)} \in \mathbf{R}.$$

As usual, l_∞ is the space of bounded sequences with the norm defined as $|x|_\infty = \sup_{k\in\mathbf{N}}\left|x^{(k)}\right|$.

The linear hull of the family of Gaussian rv's $\xi^{(k)} \in \mathbf{R}$ is denoted Ξ. Each of its elements is a finite linear combination of rv's $\xi^{(k)}$:

$$\zeta \in \Xi \iff \zeta = \textstyle\sum_{k=1}^m a_k\xi^{(k)}, \ m \in \mathbf{N}, \ a_k \in \mathbf{R}.$$

Proof of Theorem 2.1.2 is divided into a number of steps.

(A) Under the assumptions of the theorem it is possible to construct a sequence of independent standard Gaussian rv's $\eta^{(j)} \in \Xi$ and a collection of bounded sequences $b_j = \left(b_j^{(k)}, k \in \mathbf{N}\right) \in l_\infty$, in such a way that for each n the $\mathbf{R}^\infty$-valued RV

$$\xi_{-n} \equiv \left(\xi_{-n}^{(k)}, k \in \mathbf{N}\right) = \xi - \textstyle\sum_{j=1}^n \eta^{(j)}b_j \tag{2.1.4}$$

does not depend on $\eta^{(j)}$ with $j \leq n$, and each its coordinate $\xi_{-n}^{(k)}$ is a finite linear combination of the "new" rv's $\eta^{(j)}$. It is easy to see that the RV ξ_{-n}, if it exists, is a centered Gaussian RV in $\mathbf{R}^{\mathbf{N}}$, and that

$$\xi_{-(n+1)} = \xi_{-n} - \eta^{(n+1)}b_{n+1}.$$

It is convenient first to describe a recursive algorithm for constructing sequences of (2.1.4) starting from $\xi_0 = \xi$, and then to show that it does indeed provide decompositions with stipulated properties.

(A1) Suppose that $b_j \in l_\infty$ and $\eta^{(j)}$ have already been selected for $j \le n$.

If the RV ξ_{-n} of (2.1.4) has no nonzero coordinates, the process stops. In this case ξ is a finite linear combination of bounded nonrandom sequences with Gaussian coefficients, so $|\xi|_\infty < \infty$ a.s.

If the "residual vector" ξ_{-n} has some nonzero coordinates, denote the smallest number of nonzero coordinate by

$$m = m(n) = \min\left\{k : \sigma_{nk}^2 \neq 0\right\}, \quad \sigma_{nk} \equiv \sqrt{\mathbf{D}\xi_{-n}^{(k)}}.$$

Then $\eta^{(n+1)} = \xi_{-n}^{(m)}/\sigma_{nm} \in \mathbf{R}$ is a standard normal rv. By the construction, it does not depend on $\eta^{(j)}$, $j = \overline{1,n}$. Yet it clearly belongs to Ξ (as well as all the coordinates of ξ_{-n} and the coefficients $\eta^{(j)}$, $j = \overline{1,n}$). The next nonrandom vector in decomposition (2.1.4) can now be chosen as

$$b_{n+1} = \left(b_{n+1}^{(k)}, \ k \in \mathbf{N}\right), \quad b_{n+1}^{(k)} = \mathbf{cov}\left(\eta^{(n+1)}, \xi_{-n}^{(k)}\right).$$

The new RV $\xi_{-(n+1)}$ is obtained by application of (2.1.4) in the form of recurrence relation.

(A2) It remains to verify that the decomposition of step (A1) has the prescribed properties.

By the choice of $\eta^{(n+1)}$ and b_{n+1}, the coordinates of the new residual $\xi_{-(n+1)}$ are uncorrelated with the Gaussian coefficients: $\mathbf{cov}\left(\xi_{-(n+1)}^{(k)}, \eta^{(l)}\right) = 0$ for $l \le n+1$. Hence the Gaussian RV $\xi_{-(n+1)}$ is independent from $\eta^{(l)}$, $l = \overline{1, n+1}$.

For all n and k

$$\sum_{l=1}^{n+1} \left[b_l^{(k)}\right]^2 \le \sum_{l=1}^{n+1} \left[b_l^{(k)}\right]^2 + \mathbf{E}\left[\xi_{-n-1}^{(k)}\right]^2 = \mathbf{E}\left[\xi^{(k)}\right]^2 \le \sigma^2. \tag{2.1.5}$$

The quantity σ^2 is finite (see Remark 2.1.1). Consequently,

$$b_{n+1} = \left(b_{n+1}^{(k)}\right) \in l_\infty, \quad |b_{n+1}|_\infty \le \sigma < \infty.$$

By the construction, $\xi_{-n}^{(k)} = \xi_{-(n+1)}^{(m)} = 0$ for $k \le m$ (recall that $m = m(n)$ is the number of the first nonzero coordinate of ξ_{-n}). Hence the sequence $m(n)$ is strictly increasing, and for each k there is a number $n_0(k)$ such that $\xi_{-n}^{(k)} = 0$ if $n \ge n_0 = n_0(k)$. Thus

$$\forall k \ \xi^{(k)} = \sum_{l=1}^{n(k)} b_l^{(k)} \eta^{(l)},$$

so the Gaussian rv's $\eta^{(k)}$ generate all the Gaussian family Ξ.

(B) Denote the σ-algebras generated by parts of the sequence $\eta^{(k)}$ by

$$\mathcal{F}_n = \sigma\left(\eta^{(k)}, k \le n\right), \quad \overline{\mathcal{F}}_{-n} = \sigma\left(\eta^{(k)}, k > n\right), \quad \mathcal{F} = \sigma\left(\bigcup \mathcal{F}_n\right).$$

The sequence of independent rv's $\eta^{(k)}$ obeys the zero-one law, so the residual σ-algebra $\overline{\mathcal{F}}_{-\infty} = \bigcap_{n=1}^{\infty} \overline{\mathcal{F}}_{-n}$ contains only events of probability one or zero (it is contained in $\mathcal{F}$ and independent from this σ-algebra).

In decomposition into independent components obtained on step (A)

$$\forall\, n \in \mathbf{N} \ \left|\eta^{(1)} b_1 + \cdots + \eta^{(n)} b_n\right|_\infty < \infty \text{ a.s.}$$

Thus the event $F = \{|\xi|_\infty < \infty : \}$ belongs to the residual σ-algebra. By the assumptions of the theorem, the probability of F is not zero, so it equals one by the zero-one law, i.e., the norm of ξ is a.s. finite (even if there is an infinite set of linearly independent coordinates).

(C) It follows from Theorem 2.1.1 and known estimates for one-dimensional normal laws that for each j and a sufficiently small $\alpha > 0$

$$\mathbf{E} \exp\left\{\alpha |\xi|^2\right\} < \infty, \ \mathbf{E} \exp\left\{\alpha |b_j|_\infty^2 \left[\eta^{(j)}\right]^2\right\} < \infty.$$

Consequently, arbitrarily high powers of the rv's $|\xi_{-n}|_\infty$ are summable.

The sequence $\left\langle |\xi_{-n}|_\infty, \overline{\mathcal{F}}_{-n}\right\rangle$ is a submartingale: indeed, $|\xi_{-n}|_\infty$ is $\overline{\mathcal{F}}_{-n}$-measurable and

$$\mathbf{E}\left\{|\xi_{-n}|_\infty \,\middle|\, \overline{\mathcal{F}}_{-(n+1)}\right\} \geq \sup\nolimits_{k \in \mathbf{N}} \mathbf{E}\left\{\left|\xi_{-n}^{(k)}\right| \,\middle|\, \overline{\mathcal{F}}_{-(n+1)}\right\} \geq |\xi_{-(n+1)}|_\infty$$

by the Jensen inequality. Hence for each $p > 1$

$$\mathbf{E} \sup_{n \geq 1} |\xi_{-n}|^p \leq c\mathbf{E}\, |\xi_0|_\infty^p < \infty, \ \xi_0 = \xi.$$

It follows that $\limsup_{n\to\infty} |\xi_{-n}|_\infty < \infty$ a.s. Clearly, this rv is measurable with respect to the residual σ-algebra and, consequently, degenerate. Thus, there exists a nonnegative constant u_0 such that a.s. $\limsup_{n\to\infty} |\xi_{-n}|_\infty = u_0$.

(D) By the choice of u_0,

$$\lim_{n\to\infty} \mathbf{P}\left\{|\xi_{-n}|_\infty \geq 1 + u_0\right\} = 0.$$

Consequently, inequality (2.1.3) shows that for each $\varepsilon > 0$ and $n > n_0(\varepsilon)$

$$\mathbf{E} \exp\left\{\frac{1}{\varepsilon} |\xi_{-n}|_\infty^2\right\} < \infty.$$

The estimate of (2.1.3) should be used with

$$a = 1 + u_0, \ U = \text{const} \cdot \ln \frac{1 - \mathbf{P}\left\{|\xi_{-n}| \geq a\right\}}{\mathbf{P}\left\{|\xi_{-n}| \geq a\right\}}.$$

It follows from (2.1.5) that for each n the kth coordinate of the vector $\eta^{(1)} b_1 + \cdots + \eta^{(n)} b_n$ admits the estimate

$$\left|\sum\nolimits_{l=1}^n \eta^{(l)} b_l^{(k)}\right| \leq |\eta|_n\, |b^{(k)}|_n,$$

where $\eta = \left(\eta^{(j)}, j = \overline{1, n}\right)$, $b^{(k)} = \left(b_j^{(k)}, j = \overline{1, n}\right) \in \mathbf{R}^n$, and $|\cdot|_n$ is the Euclidean norm. Thus by (2.1.5)

$$\left|\sum_{l=1}^{n} \eta^{(l)} b_l\right|_\infty^2 = \sup_k \left|\sum_{l=1}^{n} \eta^{(l)} b_l^{(k)}\right|^2 \leq \sigma^2 \sum_{l=1}^{n} \left[\eta^{(l)}\right]^2.$$

Combined with Theorem 1.1.7, this estimate yields the inequality

$$\forall\, \alpha < \frac{1}{2\sigma^2} \quad \mathbf{E} \exp\left\{\alpha \left| \sum_{l=1}^{n} \eta^l b_l \right|_\infty^2\right\} \leq \left(1 - 2\alpha\sigma^2\right)^{-n/2} < \infty.$$

The function $\exp\left\{\alpha \left|x\right|_\infty^2\right\}$ is convex, so by Jensen's inequality

$$\exp\left\{\alpha \left|\xi\right|_\infty^2\right\} \leq \varepsilon \exp\left\{\frac{\alpha}{\varepsilon^2} \left|\xi_{-n}\right|_\infty^2\right\} + (1-\varepsilon) \exp\left\{\frac{\alpha}{(1-\varepsilon)^2} \left| \sum_{l=1}^{n} \eta^{(l)} b_l \right|_\infty^2\right\}$$

for $0 < \varepsilon < 1$ and each n.

If $2\alpha\sigma^2 < 1$, the number $\varepsilon > 0$ can be chosen so that $2\alpha\sigma^2/(1-\varepsilon)^2 < 1$. By the part of the theorem already proved, one can select n so large that the right-hand side of the last inequality becomes summable. This completes the proof of the theorem. ○

2.2 Suprema of Gaussian Sequences

2.2.1 Absolute Continuity of "Tail"

This section considers the distribution of the rv

$$\zeta = \sup\left\{\xi^{(j)} : j \in \mathbf{N}\right\}, \tag{2.2.1}$$

where the joint distributions of $\xi^{(j)}$, $j = \overline{1,m}$, are Gaussian for each $m \in \mathbf{N}$.

Random variables (2.2.1) appear naturally in problems about Gaussian RV's with infinite-dimensional values. Thus, the norm of a Gaussian RV with values in a separable Banach space can be represented in form (2.2.1), with $\xi^{(j)} = \langle \xi, t_j \rangle$ being the values of appropriately chosen elements from the conjugate space (see the proof of Proposition A.2.6 in Section A.2).

Below, the study of rv (2.2.1) employs a method of finite-dimensional projections — assertions about them are derived from similar theorems for the rv's

$$\zeta_m = \max\left\{\xi^{(j)} : j = \overline{1,m}\right\}. \tag{2.2.2}$$

A finite-dimensional Gaussian RV with arbitrary covariance matrix and mean value can be constructed by an affine transformation of a standard Gaussian vector with same or greater number of coordinates (see Section 1.1). For this reason, rv (2.2.1) has the same distribution as the CPLF

$$q_m(\eta) \equiv \max_{j \leq m} \left[(\eta, b_j) + \alpha_j\right], \quad \eta, b_j \in \mathbf{R}^k, \tag{2.2.3}$$

where $k = k(m)$, η is a standard Gaussian RV with distribution (1.1.4), b_j are constant vectors, and $\alpha_j = \mathbf{E}\xi^{(j)}$.

Representation (2.2.3) allows one to apply the results of Sections 1.2–1.3 and 1.5 to study the distribution of ζ. Below, as in (1.5.1), (1.5.2), Theorem 1.5.2, and (1.5.15),

$$\begin{aligned} G(u) &= \mathbf{P}\{\zeta \le u\}, \quad G'(u) = \lim_{h \searrow 0} \frac{1}{h}[G(u+h) - G(u)], \qquad (2.2.4) \\ D_+G'(u) &= \limsup_{h \to 0} \frac{1}{h}[G'(u+h) - G'(u)], \quad u_0 = \inf\{u : G(u) > 0\}. \end{aligned}$$

For the finite-dimensional approximation (2.2.2)–(2.2.3) to ζ, the DF is

$$G_m(u) = \mathbf{P}\{\zeta_m \le u\}. \qquad (2.2.5)$$

Notations for its derivatives, finite differences, etc. repeat those for G with the subscript m added. It is easy to see that in all points where G is continuous

$$G_m(u) \searrow G(u). \qquad (2.2.6)$$

In the rest of this section it is always assumed that

$$\forall j \ \alpha_j = \mathbf{E}\xi^{(j)} \ge 0. \qquad (2.2.7)$$

Besides, the sequence of Gaussian rv's in (2.2.1) obeys the conditions

$$\mathbf{P}\{\zeta < \infty\} = 1 \qquad (2.2.8)$$

and

$$0 < \sigma \equiv \sup_{j \in \mathbf{N}} \sqrt{\mathbf{D}\xi^{(j)}} < \infty. \qquad (2.2.9)$$

(By Remark 2.1.3 conditions (2.2.7) and (2.2.8) imply (2.2.9).)

Theorem 2.2.1 *Under assumptions* (2.2.7)–(2.2.9), *the function G is absolutely continuous and has a right-continuous right-sided derivative G' on the half-line (u_0, ∞). Moreover, if $u_0 < u' < u''$, then*

$$G(u'') - G(u') = \int_{u'}^{u''} G'(v)\, dv.$$

On each smaller half-line $[U, \infty) \subset (u_0, \infty)$, the right-sided derivative G' has bounded variation. The set of its discontinuities is at most countable, and all the discontinuities of G' are negative jumps: if $G'(u-0) \ne G'(u) = G'(u+0)$, then $G'(u) - G'(u-0) \ge 0$.

The proof of the theorem is in subsection 2.2.4. It employs estimates similar to those of Section 1.5. Both G and its derivative can indeed have discontinuities (see subsection 2.2.3).

2.2.2 Properties of Density: Upper Bounds

Theorem 2.2.2 a) *Assume that*

$$t = \Phi^{-1}(G[b]), \ G[b] > 0. \qquad (2.2.10)$$

Then for $u > b$ the right-sided derivative G' of Theorem 2.2.1 and its upper derivative (2.2.4) admit the bounds

$$G'(u) \le (1+t_-)^2 \frac{u+u_+-b}{(u-b)^2}, \quad D_+G'(u) \le \frac{(1+t_-)^2}{(u-b)^2}.$$

b) *If condition (2.2.10) is replaced by the stronger restriction*

$$t = \Phi^{-1}(G[b]), \; G[b] > \tfrac{1}{2}, \tag{2.2.11}$$

then

$$\begin{aligned} G'(u) &\le [1-\Phi(\tau u)]\,\tau^2\,(u+u_+-b)\left(1+2\alpha+2\alpha^2\right), \\ D_+G'(u) &\le [1-\Phi(\tau u)]\,\tau^2\left(1+2\alpha+2\alpha^2\right), \end{aligned} \tag{2.2.12}$$

where $\tau = t/b$ and $\alpha = (\tau u)^{-2}(1-b/u)^{-1}$.

The proof of this theorem is placed in subsection 2.2.4.

Remark 2.2.1 By (2.2.9), the rv ζ can exceed arbitrarily large numbers with positive probability. Hence always $G(b) < 1$ if $G(b) > 0$, and the quantity t in (2.2.10) continuously depends on $G(b)$.

2.2.3 Examples of Discontinuities

The following examples show that the functions G and G' considered in Theorems 2.2.1–2.2.5 may indeed have discontinuities.

Let $\eta_0, \eta_1, \ldots$ be independent Gaussian rv's with the standard normal distribution. Put

$$\zeta_1 \equiv \sup_{j\ge 1}[\beta_j\eta_j], \quad \beta_j = 1/\sqrt{2\ln j + 4\ln\ln\max(j,3)}, \quad j=1,2,\ldots$$

The DF of ζ_1 is

$$G_1(u) \equiv \mathbf{P}\{\zeta_1 \le u\} = \prod_{j=1}^{\infty}\mathbf{P}\{\eta_j \le u/\beta_j\} = \prod_{j=1}^{\infty}\Phi(u/\beta_j).$$

The asymptotic formula for normal tails (1.1.2) shows that

$$G_1(u) = 0 \text{ if } u<1, \quad G_1(u) > 0 \text{ if } u \ge 1.$$

Hence G_1 makes a positive jump at the point $u = 1$, and $\mathbf{P}\{\zeta_1 = 1\} > 0$. For $u > 1$, the DF G_1 is continuously differentiable and the derivative $G_1'(u)$ converges to the right-sided derivative at $u = 1$ as $u \searrow 1$. In the calculation of derivative, the convergent series $\sum \Phi'(1/\beta_j)/\beta_j$ serves as a majorant.

The rv $\zeta_2 = \max\{2\eta_0, \eta_0 + \zeta_1\}$ has a density which is discontinuous at the point $u = 2$. Indeed, by the independence of η_0 and ζ_1,

$$G_2(u) = \mathbf{P}\{\zeta_2 \le u\} = \mathbf{P}\{\zeta_1 = 1\}\,\Phi\left(\max\left[u-1, \tfrac{1}{2}u\right]\right) + Q(u),$$

where

$$Q(u) = \Phi\left(\tfrac{1}{2}u\right)\mathbf{P}\left\{1<\zeta_1\le\max\left[1,\tfrac{1}{2}u\right]\right\} + \int_{\max[1,u/2]}^{\infty}\Phi(u-y)G_1'(y)dy.$$

The right-sided derivative Q' is continuous. This is verified using the fact that G_1' is continuous from the right for $u \geq 1$. Yet the derivative of the first summand in the definition of G_2 experiences a negative jump at the point $u = 2$.

2.2.4 Absolute Continuity and Upper Bounds: Proofs

Proof of Theorems 2.2.1 and 2.2.2 is done in a number of steps.

(A) Let $[a_1,\, a_2]$ be an arbitrary bounded segment contained in the half-line $(u_0,\, \infty)$, and $b \in (u_0,\, a_1)$ a point where the function G is continuous. Then $\lim_{m\to\infty} G_m(u) = G(b)$ with $G(b) \in (0,\, 1)$ (see Remark 2.2.1). Consequently, if t_m is chosen from the relation $\Phi(t_m) = G_m(b)$, then $\lim_{m\to\infty} t_m = t = \Phi^{-1}[G(b)]$.

By parts a) of Theorems 1.5.3 and 1.5.4, the following two inequalities are valid for $u > b$:

$$0 \leq G_m'(u) \leq \left(1+[t_m]_-\right)^2 \frac{u+u_+-b}{(u-b)^2}, \quad D_+G_m'(u) \leq \frac{\left(1+[t_m]_-\right)^2}{(u-b)^2}. \tag{2.2.13}$$

Consequently, one can select numbers A_1, A_2, and m_0 such that for $m \geq m_0$ and $u > b$ the density and its upper derivative satisfy the inequalities

$$0 \leq G_m'(u) \leq A_1, \quad D_+G_m'(u) \leq A_2. \tag{2.2.14}$$

The first of these inequalities implies for G the Lipschitz condition

$$0 \leq G(u'') - G(u') \leq A_1\,(u''-u'), \quad u', u'' \in [a_1, a_2].$$

If the function G is continuous at both points u' and u'', this inequality follows immediately from (2.2.14) after a passage to the limit because of (2.2.6). To derive the inequality for an arbitrary pair of points, note that G is right continuous everywhere and cannot have more than a countable set of discontinuities, so the set of continuity points is everywhere dense. Hence the inequality follows from the preceding argument: one should approximate the points considered by some sequences of continuity points which converge to them from the right.

In the above, the segment $[a_1,\, a_2]$ and the point $b \in (u_0,\, a_1)$ were arbitrary. Thus it has been proved, in particular, that G is continuous on the half-line $(u_0,\, \infty)$.

(B) The second inequality (2.2.14) for the upper derivative of G_m' shows that the function $A_2 u - G_m'(u)$ does not decrease on the segment $[a_1,\, a_2]$ and, hence, has bounded variation. The variation of G_m' itself admits an upper bound which is uniform in $m \geq m_0$:

$$\mathsf{V}_{a_1}^{a_2} G_m' \leq \mathsf{V}_{a_1}^{a_2} A_2 u + [A_2 u - G_m'(u)]\big|_{a_1}^{a_2} \leq A_1 + A_2\,(a_2 - a_2)\,.$$

Since the functions G_m' have uniformly bounded variation, there is a convergent subsequence such that

$$G_{\tilde{m}}'(u) \to g(u) \tag{2.2.15}$$

at all points where the limit function g is continuous. This latter is itself of necessity right continuous, and its variation is bounded on $[a_1, a_2]$.

The set of discontinuities of g is, at most, countable. Hence (2.2.15) holds almost everywhere on $[a_1, a_2]$. It follows from the above bound on variation that all the functions G'_m are uniformly bounded. Thus it is a corollary of Theorem 1.5.1 that

$$\begin{aligned} G(u'') - G(u') &= \lim_{\tilde{m}\to\infty} \left[G_{\tilde{m}}(u'') - G_{\tilde{m}}(u')\right] \\ &= \lim_{\tilde{m}\to\infty} \int_{u'}^{u''} G'_{\tilde{m}}(u)du = \int_{u'}^{u''} g(u)du \end{aligned}$$

for each pair $u', u'' \in [a_1, a_2]$, $u'' > u'$. It follows from this equality that the right-continuous function g is indeed the right-sided derivative of G.

No function of bounded variation can have discontinuities other than jumps — hence the possible discontinuities of G' are also jumps. The choice of the segment $[a_1, a_2]$ being arbitrary, it follows that G' enjoys the local properties established above on all the half-line (u_0, ∞). All jumps are nonpositive because of the second inequality (2.2.14).

(C) It is a consequence of the first inequality in (2.2.13) that for $h > 0$

$$\begin{aligned} G(u+h) - G(u) &= \lim_{\tilde{m}\to\infty} \left[G_{\tilde{m}}(u+h) - G_{\tilde{m}}(u)\right] \\ &\leq h(1+t_-)^2 \frac{u+u_+-b}{(u-b)^2}. \end{aligned} \tag{2.2.16}$$

The first of the estimates in the statement of Theorem 2.2.2 is now easily obtained by a passage to the limit in $h \searrow 0$.

The derivation of the bound on the upper derivative of G' is similar. For small values of $h > 0$ and $h' \in \mathbf{R}$, there is representation (cf. (1.5.23))

$$\frac{1}{hh'}\delta^2_{h,h'}G(u) = \lim_{\tilde{m}\to\infty} \frac{1}{hh'}\delta^2_{h,h'}G_{\tilde{m}}(u) \leq \frac{(1+t_-)^2}{(\tilde{u}-b)^2}, \tag{2.2.17}$$

where $\tilde{u} = \min[u, u+h]$. A passage to the limit, first in $h \searrow 0$ and then in $h' \to 0$, yields the first estimate of Theorem 2.2.2 for D_+G'.

To derive the more precise estimates under condition (2.2.11), one can proceed by analogy. The sole difference is application of the stronger bounds of parts b) in Theorems 1.5.3 and 1.5.4 instead of (2.2.13).

(D) It remains to ascertain that the variation of G' remains bounded on each half-line $[U, \infty)$ interior to (u_0, ∞). Note that variation on bounded subsegments is bounded: this was established on step (B).

Let A be a fixed number satisfying $A > \max[u_0, 1]$. By the already established bound on the upper derivative of density, $D_+G'(u) \leq C/u^2$ for $u \geq A$ (the constant C is determined by choice of A). For this reason the function $-[G'(u) + C/u]$ does not decrease and, since it is bounded on the half-line $[A, \infty]$, has bounded variation on this half-line. The variation of C/u is also bounded on the same half-line. Hence

$$\bigvee_A^\infty G'(u) \leq \bigvee_A^\infty C/u \leq \bigvee_A^\infty [-C/u - G'(u)] < \infty.$$

This completes the proof of Theorem 2.2.1 and Theorem 2.2.2. ○

2.2.5 Monotonicity of Inverse DF Ratio

Below G^{-1} is the inverse of G defined in (1.5.5).

Theorem 2.2.3 *If b and t satisfy* (2.2.11), *then* $bG'(b) \geq t\Phi'(t)$ *and*

$$\forall\ u > b\ \ 1 - G(u) \leq 1 - \Phi(ut/b).$$

The ratio of inverse functions $G^{-1}(p)/\Phi^{-1}(p)$ does not increase for $p > \frac{1}{2}$.

Proof of Theorem 2.2.3. The estimate for the "tail" of rv (2.2.1) is deduced by passage to the limit using (2.2.6) from the corresponding inequality of Theorem 1.5.2 for the approximating "finite-dimensional" function G_m.

If b is a point of continuity for the function G', the inequality for densities is, in fact, a relation between the derivatives of the inverse functions G^{-1} and Φ^{-1}. It follows from the corresponding inequality for G_m (see Theorem 1.5.2). The passage to the limit which can be used for the purpose is that described in detail on step (C) of the argument in subsection 2.2.4.

If b is not a continuity point for G', one should consider a convergent sequence of continuity points, $b_m \searrow b$, together with the corresponding solutions t_m of equation (2.2.10) with G replaced by G_m. Since both G and G' are right continuous, the inequality follows after a passage to the limit (see Remark 2.2.1).

Verification of the monotonicity property of the ratio of inverse functions follows the proof of Corollary 1.5.1. ○

2.2.6 Sequences with Constant Variance

In the important special case of identically distributed (but possibly dependent) Gaussian coordinates $\xi^{(j)}$, the bounds of Theorem 2.2.2 and Theorem 2.2.3 can be supplemented with a lower bound for D_+G'.

Theorem 2.2.4 *Assume that in* (2.2.1)

$$\mathbf{E}\xi^{(j)} = \alpha, \quad \mathbf{D}\xi^{(j)} = \sigma^2 > 0 \tag{2.2.18}$$

for all $j \in \mathbf{N}$. Then the function G is continuous on the whole real line and continuously differentiable for $u > u_0$. Its derivative G' has finite limit $G'(u_0 + 0) = \lim_{h\searrow 0} G'(u_0+h)$ and satisfies a Lipschitz condition on all smaller half-lines, i.e., for $u \geq U > u_0$. Moreover,

$$\limsup_{h\to 0} \frac{1}{h}\left[G'(u+h) - G'(u)\right] \geq \frac{1}{\sigma^2}(u-\alpha)G'(u).$$

The following lemma will be necessary to prove Theorem 2.2.4.

Lemma 2.2.1 *If the variances of all the rv's in* (2.2.1) *are equal (see* (2.2.1)*), then there exists a nondecreasing function* ψ *such that for all* $u \in \mathbf{R}$

$$G(u) = \int_{-\infty}^{u} \psi(v) d\Phi(v/\sigma).$$

Proof of Lemma 2.2.1. For brevity, put $\sigma = 1$.

It follows from the additivity of probability that the approximating sequence (2.2.5) satisfies the equality

$$G_m(u) = \sum_{j=1}^{m} \int_{-\infty}^{u} \widehat{G}_j(v)\, d\Phi(v - \alpha_j) = \int_{-\infty}^{u} \psi_m(v) d\Phi(v),$$

where $\widehat{G}_j(v) = \mathbf{P}\left\{\xi^{(i)} \le v, i \ne j \,\middle|\, \xi^{(j)} = v\right\}$ and

$$\psi_m(v) = \sum_{j=1}^{m} \widehat{G}_j(v) \exp\left\{\tfrac{1}{2}\left(\alpha_j v - \alpha_j^2\right)\right\}.$$

The conditional probabilities $\widehat{G}$ in this representation do not decrease. This circumstance is verified for $j = 1$ only, verification for other coordinates being quite similar.

The conditional distribution of $\xi^{(2)}, \ldots, \xi^{(m)}$ given $\xi^{(1)} = v$ is Gaussian. Its mean value and covariance matrix are described by (1.1.10) and (1.1.11). It is important that the conditional covariances do not depend on v. Hence

$$\widehat{G}_1(v) = N_{m-1,0,\sigma}(D_v),$$

where the domain D_v is

$$D_v = \left\{x = \left(x^{(2)}, \ldots, x^{(m)}\right) : x^{(i)} \le (1 - \rho_{1i})\, v + \rho_{1i}\alpha_1,\ i = 2, \ldots, m\right\}$$

and ρ_{ij} is the coefficient of correlation between $\xi^{(i)}$ and $\xi^{(j)}$, so $1 - \rho_{1i} \ge 0$ for all i. For the latter reason the domain D_v becomes larger if v is increased, and the function $\widehat{G_1}$ is, consequently, a nondecreasing one.

For a fixed value of u, the sequence $\psi_m(u)$ is bounded. Indeed,

$$\int_{u}^{\infty} \psi_m(v) d\Phi(v) \le \int_{-\infty}^{\infty} \psi_m(v) d\Phi(v) \le 1.$$

At the same time, by the monotonicity property of ψ_m already proved,

$$\psi_m(u) \le \frac{1}{1 - \Phi(u)} \int_{u}^{\infty} \psi_m(v)\, d\Phi(v).$$

Hence the functions ψ_m are uniformly bounded on each half-line $(-\infty, u)$. Consequently, one can select a sequence $\widehat{m}(m) \in \mathbf{N}$ such that, whatever the choice of the segment $[-A, A]$, the functions $\psi_{\widehat{m}}$ converge to a nonnegative nondecreasing function ψ (in the space of functions on $[-A, A]$ square-summable with respect to $d\Phi$). This latter meets the requirements specified in the lemma. ○

Proof of Theorem 2.2.4. Without loss of generality, one can assume that in (2.2.1)

$$\forall j\ \mathbf{E}\xi^{(j)} = 0,\ \mathbf{E}\left[\xi^{(j)}\right]^2 = 1.$$

By Lemma 2.2.2, G is a continuous function. By the same lemma, the discontinuities of its derivative G' might only be upward jumps. Yet on the half-line (u_0, ∞) this possibility is excluded by Theorem 2.2.1. Consequently, G' is continuous for $u > u_0$.

The function

$$\psi(u) = \sqrt{2\pi} G'(u) \exp\{u^2/2\} = G'(u)/\Phi'(u)$$

is nondecreasing by Lemma 2.2.1, so

$$\begin{aligned} &\liminf_{h\to 0} \frac{1}{h} [G'(u+h) - G'(u)] \\ = \; & \Phi'(u) \left[\liminf_{h\to 0} \frac{1}{h} [\psi(u+h) - \psi(u)] - \frac{uG'(u)}{\Phi'(u)} \right] \geq -uG'(u). \end{aligned}$$

Combined with the upper estimates of Theorem 2.2.2 for D_+G', this bound implies the Lipschitz condition for G' on the half-line $[U, \infty)$ if $U > u_0$. This proves the theorem. ○

2.2.7 Large Deviation Asymptotics for Density

Theorem 2.2.5 *Assume that the conditions of Theorem 2.2.4 are fulfilled and that the function $b(u)$ obeys the restrictions*

$$\lim_{u\to\infty} [u - b(u)] = 0, \quad \lim_{u\to\infty} u[u - b(u)] = \infty.$$

Define $t = t(u)$ by equation (2.2.10) with $b = b(u)$. Then the following asymptotic relations hold for the derivative of G as $u \to \infty$:

$$G'(u) = \frac{t}{b} \Phi' \left(\frac{t}{b} u \right) [1 + o(1)] = \frac{1}{\sigma} \Phi'(t) \exp\left\{ -\frac{u(u-b)}{\sigma^2} \right\} [1 + o(1)],$$

where $b = b(u)$ and σ is as in (2.2.9).

To derive the asymptotic relations of Theorem 2.2.5, it will be necessary to make use of the following property of the inverse function (1.5.5).

Lemma 2.2.2 *If rv (2.2.1) satisfies condition (2.2.7), then the difference of inverse DF's*

$$H(p) = G^{-1}(p) - \sigma\Phi^{-1}(p)$$

is nonnegative and nonincreasing for $0 < p < 1$.

Remark 2.2.2 Lemma 2.2.2 permits one to evaluate the density G' from below.

Under the assumptions of the lemma $H(p) \searrow h \geq 0$ as $p \nearrow 1$. It follows from the inequality

$$\frac{d}{dp} [G^{-1}(p) - \sigma\Phi^{-1}(p)] \leq 0$$

with $p = G(u)$ and the definition of H that for large $u > 0$ with $G(u) > 0$

$$\begin{aligned} G'(u) &\geq \frac{1}{\sigma}\Phi'\left(\Phi^{-1}(G(u))\right) = \frac{1}{\sigma}\Phi'\left(\frac{u - H(G(u))}{\sigma}\right) \\ &\geq \frac{1}{\sigma}\Phi'\left(\frac{u-h}{\sigma}\right) = \frac{\exp\left\{-\frac{1}{2}(u-h)^2/\sigma^2\right\}}{\sigma\sqrt{2\pi}}. \end{aligned}$$

Proof of Lemma 2.2.2. For a fixed m, the maximal quadratic deviation of Gaussian rv in the finite-dimensional approximation (2.2.3) to (2.2.1) is

$$c(m) = \max\nolimits_{j\leq m}\sqrt{\mathbf{D}\xi^{(j)}} = \max\nolimits_{j\leq m}|b_j|.$$

By Lemma 1.5.1 the function $H_m \overset{def}{=} G_m^{-1} - c(m)\Phi^{-1}$ is nonnegative and nonincreasing on (0, 1). By Remark 2.1.3 $\sigma < \infty$, and it is easy to see that $c(m) \nearrow \sigma$ as $m \to \infty$.

The function G is continuous. By Theorem 2.2.1 it satisfies (2.2.6) at the points where $G(u) > 0$.

Hence there is convergence of inverse functions: $G_m^{-1} \nearrow G^{-1}$. The monotonicity property of H survives the passage to the limit. ○

Proof of Theorem 2.2.5. a) Suppose first that $\alpha = 0$ and $\sigma^2 = 1$ in condition (2.2.18).

By Lemma 2.2.2, for large values of u the quantities $b = b(u)$ and $t = t(u)$ satisfying (2.2.10) obey the inequality

$$0 \leq b(u) - t(u) \leq \text{const}, \tag{2.2.19}$$

so b and t are asymptotically equivalent. Asymptotic equivalence of u and b is presupposed by the conditions of the theorem, so $t/b = 1 + o(1)$ and $u/b = 1 + o(1)$ as $u \to \infty$.

The function $G^{-1} - \Phi^{-1}$ is nonincreasing by Lemma 2.2.2. Differentiating it, one can conclude that for large values of u

$$\Phi'(t) \leq G'(b). \tag{2.2.20}$$

In addition, Theorem 2.2.2 yields an upper bound for G' which is valid for large u:

$$G'(u) \leq u\left[1 - \Phi\left(tu/b\right)\right]\left[1 + o(1)\right] = \Phi'\left(tu/b\right)\left[1 + o(1)\right] \tag{2.2.21}$$

if $u - b > 0$ and $G(b) > \frac{1}{2}$. The last equality follows from (1.1.2) and asymptotic equivalence of u, t, b.

Under the restrictions of the theorem, condition $u - b > 0$ is fulfilled if u is large. Hence it follows from Lemma 2.2.1 that

$$G'(u)/\Phi'(u) \geq G'(b)/\Phi'(b),$$

so (2.2.20) and the explicit expression for the normal density yield the estimate

$$G'(u) \geq \Phi'(u)\Phi'(t)/\Phi'(b) = \Phi'\left(\sqrt{t^2u^2/b^2 + \kappa}\right),$$

where

$$\kappa = u^2 + t^2 - b^2 - t^2u^2/b^2 = (u^2 - b^2)(b^2 - t^2)/b^2 = O\left[(u-b)(b-t)\right] \to 0.$$

Hence $G'(u) \geq \Phi'(tu/b)[1 + o(1)]$. Combined with (2.2.21), this estimate leads to the relation

$$G'(u) = \Phi'(tu/b)[1 + o(1)].$$

To obtain the asymptotic formula of the theorem, it is necessary to transform the argument of the normal density in the right-hand side of the last expression.

In the situation considered here,

$$t^2u^2/b^2 = t^2\left(1 + [u-b]/b\right)^2 = t^2 + 2u(u-b) + o(1).$$

To verify this, one should use asymptotic relations $t \sim b \sim u$, formula (2.2.19), and the conditions of the theorem which stipulate proximity of u and b. When the right-hand side of this last formula is substituted into the explicit expression of the standard normal density, it yields the asymptotic equality

$$G'(u) = \Phi'(t)\exp\{u(u-b)\}[1 + o(1)].$$

b) To derive the formula of the theorem for arbitrary mean $\alpha = \mathbf{E}\xi^{(j)}$ and variance $\sigma^2 = \mathbf{E}\left[\xi^{(j)}\right]^2$, change from the variable u to $v = (u-\alpha)/\sigma$ and put

$$G_1(v) = G(\sigma v + \alpha), \quad b_1(v) = [b(\sigma v + \alpha) - \alpha]/\sigma, \quad t_1(v) = t(\sigma v + \alpha).$$

The DF $G_1(v)$ is that of a rv satisfying conditions of part a), and t_1, b_1 solve the corresponding equation $\Phi(t_1) = G_1(b_1)$. It follows from the assumptions of the theorem that the remaining conditions of part a) hold, viz., both $v - b_1 \to 0$ and $v(v - b_1) \to \infty$ as $v \to \infty$.

Changing variables in the corresponding relation of part a), one obtains the equality

$$G'(u) = \frac{1}{\sigma}G_1'\left(\frac{u-\alpha}{\sigma}\right) = \frac{1}{\sigma}\Phi'(t(u))\exp\left\{\frac{1}{\sigma^2}(u-\alpha)(u-b)\right\}[1 + o(1)].$$

The second estimate of Theorem 2.2.5 is verified using an analogue of (2.2.19), the equality $b - \sigma t = o(1)$; it also follows from Lemma 2.2.2. ○

2.3 Norm of Gaussian RV in Hilbert Space

Below X is a separable real Hilbert space with inner product $(\cdot,\cdot)$ and norm $|\cdot|$.

2.3.1 Asymptotic Expressions for "Tails" of Norm

A RV $\eta \in X$ is called *Gaussian* if all its finite-dimensional projections have Gaussian distributions. This latter condition is equivalent to assuming that the rv (η, t) is normal for each $t \in X$, i.e., has finite variance and CF

$$\mathbf{E}\exp\{i(\eta, t)\} = \exp\left\{i\mathbf{E}(\eta, t) - \tfrac{1}{2}\mathbf{D}(\eta, t)\right\}. \tag{2.3.1}$$

This definition agrees with that of Section 2.1 if the RV is centered, $\mathbf{E}\eta = 0$. Thus, if η' and η'' are independent copies of η, the symmetrized RV $\tilde{\eta} = \eta' - \eta''$ is centered Gaussian. By the results of Section 2.1 $\mathbf{E}\exp\left\{\alpha\,|\tilde{\eta}|^2\right\} < \infty$ if $\alpha > 0$ is sufficiently small. The inequalities

$$\mathbf{P}\{|\eta| \geq r\} = (p_b)^{-1}\mathbf{P}\{|\eta' - \eta''| \geq r,\ |\eta''| < b\} \leq (p_b)^{-1}\mathbf{P}\{|\eta' - \eta''| \geq r - b\},$$

where $p_b = \mathbf{P}\{|\eta''| < b\}$, show that $\mathbf{E}\exp\left\{\alpha'\,|\eta|^2\right\}$ is finite for some $\alpha' > 0$.

Since a Gaussian RV has summable norm, its expectation can be defined as the Bochner integral

$$\mathbf{E}\eta = \int_X x F_\eta(dx), \quad F_\eta(\cdot) = \mathbf{P}\{\eta \in \cdot\}, \tag{2.3.2}$$

with respect to the distribution of η. Evidently, $(\mathbf{E}\eta, t) = \mathbf{E}(\eta, t)$ for all t.

The ***covariance operator*** of η is defined by the equality

$$(\mathcal{V}f, g) = \mathbf{cov}\left[(\eta, f), (\eta, g)\right], \quad f, g \in X. \tag{2.3.3}$$

It is nonnegative definite: $(\mathcal{V}f, f) \geq 0$ for each $f \in X$.

The eigenvalues of the covariance operator $\mathcal{V}$ of the Gaussian RV η are denoted by

$$\sigma_1^2 = \cdots = \sigma_\nu^2 > \sigma_{\nu+1}^2 \geq \sigma_{\nu+2}^2 \geq \ldots \geq 0, \tag{2.3.4}$$

where ν is the multiplicity of the principal eigenvalue σ_1^2. In what follows, $(e_j, j \in \mathbf{N})$ is an ***eigenbasis*** of $\mathcal{V}$ (i.e., an orthonormal basis of unit eigenvectors) and E the ***principal eigenspace***:

$$\forall j \ \ \mathcal{V}e_j = \sigma_j^2 e_j, \ |e_j| = 1, \quad E = \left\{x \in X : \mathcal{V}x = \sigma_1^2 x\right\}. \tag{2.3.5}$$

(The choice of basis is not unique if, e.g., $\nu > 1$).

The covariance operator has finite trace: for $\eta_0 = \eta - \mathbf{E}\eta$,

$$\mathrm{tr}(\mathcal{V}) = \sum_{j=1}^{\infty}(\mathcal{V}e_j, e_j) = \sum_{j=1}^{\infty}\mathbf{E}(\eta_0, e_j)^2 = \sum_{j=1}^{\infty}\sigma_j^2 = \mathbf{E}\,|\eta_0|^2, \tag{2.3.6}$$

since by the Bessel identity $\sum_{j=1}^{\infty}(\eta_0, e_j)^2 = |\eta_0|^2$. (One can use an arbitrary orthonormal basis in X in this calculation.)

Each vector $a \in X$ and each nonnegative definite symmetrical operator with finite trace $\mathcal{V}$ generate a Gaussian distribution in X having CF equal to $\exp\left\{i(a,t) - \frac{1}{2}(\mathcal{V}t, t)\right\}$ (see subsection A.2.6). To avoid trivial complications, it is assumed below that

$$\mathbf{E}\eta = 0, \quad \sigma_1^2 = \sup_{|t|=1}(\mathcal{V}t, t) = 1. \tag{2.3.7}$$

It is the principal eigenvalue of the covariance operator and its multiplicity which determine the asymptotic behavior of the probability that a centered Gaussian RV lands outside of a large sphere with centre at the origin.

Theorem 2.3.1 *It is possible to choose such positive constants c_i that for $r \geq c_1$*

$$\mathbf{P}\{|\eta| \geq r\} = (1 + \varepsilon(r))\frac{\mathcal{K}}{\Gamma(\nu/2)}\left(r^2/2\right)^{\nu/2-1}\exp\left\{-\tfrac{1}{2}r^2\right\}, \tag{2.3.8}$$

where $|\varepsilon(r)| \le c_2/r^2$ *and*

$$\mathcal{K} = \prod_{j=1}^{\infty} 1/\sqrt{1-\sigma_{\nu+j}^2/\sigma_1^2}. \tag{2.3.9}$$

Theorem 2.3.1 is proved by the method of characteristic functions in subsection 2.3.3 (see subsection 2.3.3 for an alternative proof). The argument employs a version of the saddlepoint method for evaluating integrals with large parameter in combination with the well-known inversion formula for characteristic functions. A similar argument (see subsection 2.3.4) proves Theorem 2.3.2 — an estimate of stability for the asymtotic representation of Theorem 2.3.1.

Assume the normalization of (2.3.7) and put

$$\begin{aligned} \mu_{0,f,r}(\vartheta) &\equiv \exp\left\{\tfrac{1}{2}r^2[\cos\vartheta]^2\right\} \\ &\times \mathbf{E}\exp\{-r(\eta,f)\sin\vartheta\}\mathbf{1}_{\{|\eta+[r\sin\vartheta]|\ge r\}}, \end{aligned} \tag{2.3.10}$$

where $\vartheta \in \left(-\frac{\pi}{2},\frac{\pi}{2}\right]$, $f \in X$ is orthogonal to E, $|f|=1$, and $r>0$.

Theorem 2.3.2 *There exist positive constants* c_i *such that if* $r \ge c_1$ *and* $|\vartheta| \le c_2$, *then* $\mu_0 = \mu_{0,f,r}$ *satisfies the relation*

$$\mu_0(\vartheta) = \mu_0(0)(1+\varepsilon), \quad \mu_0(0) = \exp\left\{\tfrac{1}{2}r^2\right\}\mathbf{P}\{|\eta|\ge r\},$$

where $\varepsilon=\varepsilon(r,f,\vartheta)$ *and* $|\varepsilon|\le c_3\vartheta^2$.

(Recall that the constants are enumerated locally, so the values of similarly denoted constants may differ in different parts of arguments.)

2.3.2 An Auxiliary Measure

Notations of this subsection are same as above, in particular, the unit vector f is one from the definition of μ_0.

Consider the measure

$$G(B;f,r,\vartheta) \equiv e^{\frac{1}{2}r^2[\cos\vartheta]^2}\,\mathbf{E}e^{-r(\eta,f)\sin\vartheta}\mathbf{1}_{\left\{\frac{1}{2}|\eta+r\sin\vartheta\, f|^2\in B\right\}}, \tag{2.3.11}$$

which is defined on the Borel subsets of $\mathbf{R}$. Its Fourier transform is

$$\begin{aligned} g(t) &= \textstyle\int_{\mathbf{R}} e^{itu}G(du) \\ &= \exp\left\{\tfrac{1}{2}r^2[\cos\vartheta]^2\right\}\mathbf{E}\exp\left\{\tfrac{1}{2}|\eta+r\sin\vartheta f|^2 - r(\eta,f)\sin\vartheta\right\}. \end{aligned} \tag{2.3.12}$$

Lemma 2.3.1 *For all* $t\in\mathbf{R}$ *function* (2.3.12) *admits the representation*

$$g(t) = \mathcal{K}\left(r^2/2\right)^{\nu/2}\exp\left\{\tfrac{1}{2}r^2\right\}\frac{\widehat{\pi}\left(z/r^2\right)}{z^{\nu/2}}\exp\left\{-[\sin\vartheta]^2 z\widehat{\psi}\left(z/r^2\right)\right\},$$

where $z=\frac{1}{2}r^2(1-it)$, $\mathcal{K}$ *is the constant* (2.3.9) *of Theorem* 2.3.1,

$$\widehat{\pi}(w) = \prod_{j=1}^{\infty}\frac{1}{\sqrt{1+2\lambda_j w}}, \quad \widehat{\psi}(w) = \sum_{j=1}^{\infty}\frac{f_j^2}{1+2\lambda_j w}, \quad \lambda_j = \frac{\sigma_{\nu+j}^2}{1-\sigma_{\nu+j}^2},$$

and $f_j = (f, e_j)$ *are the coordinates of* f *in the eigenbasis of* $\mathcal{V}$.

In the statement of the lemma and further, $\sqrt{z}$ is the continuous branch of square root that is positive for positive values of its argument.

Proof of Lemma 2.3.1. By the choice of the basis, the rv's $\eta_j = (\eta, e_j)/\sigma_j$ are independent and have standard normal distribution. For this reason

$$g(t) = h_0^{\nu} \prod_{j=1}^{\infty} h_j,$$

where $h_0 = \mathbf{E} \exp\left\{\frac{it}{2}\eta_1^2\right\}$ and for $j \geq 1$

$$h_j = \mathbf{E} \exp\left\{\tfrac{i}{2}t\sigma_{\nu+j}^2\eta_{\nu+j}^2 + [it-1][r\sin\vartheta]\sigma_{\nu+j}f_j\eta_{\nu+j} + \tfrac{i}{2}t\left[rf_j\sin\vartheta\right]^2\right\}.$$

A standard calculation shows that

$$h_0 = 1/\sqrt{1-it} = \left(r^2/2\right)^{1/2}/\sqrt{z}.$$

A similar calculation shows (cf. proof of Theorem 1.1.7) that for $j \geq 1$

$$\begin{aligned} h_j &= \frac{1}{\sqrt{1-it\sigma_{\nu+j}^2}} \exp\left\{\frac{1}{2}\left[rf_j\sin\vartheta\right]^2\left[it - \frac{\sigma_{\nu+j}^2}{it\sigma_{\nu+j}^2-1}(it-1)^2\right]\right\} \\ &= \frac{1}{\sqrt{(1-\sigma_{\nu+j}^2)(1+2\lambda_j z/r^2)}} \exp\left\{\frac{1}{2}\left[rf_j\sin\vartheta\right]^2\left[1 - \frac{z/r^2}{1+2\lambda_j z/r^2}\right]\right\}. \end{aligned}$$

The assertion of the lemma follows from these equalities. It is essential that the unit vector f belongs to the subspace of X spanned by $e_{\nu+j}$, $j \geq 1$, so $\sum_{j=1}^{\infty} f_j^2 = 1$. ○

Remark 2.3.1 In Lemma 2.3.1 the functions $\widehat{\pi}$ and $\widehat{\psi}$ admit the following bounds from above: if $4\lambda_1 \mathbf{Re}\, w \geq -1$, then

$$|\widehat{\pi}(w)| \leq c,\ |\widehat{\pi}'(w)| \leq c\,|\widehat{\pi}(w)|,\ \left|\widehat{\psi}(w)\right| \leq c\ \left|\widehat{\psi}'(w)\right| \leq c. \tag{2.3.13}$$

Remark 2.3.2 If $4\lambda_1 \mathbf{Re}\, z \geq -r^2$, then the inequality

$$\mathbf{Re}\left[z\widehat{\psi}\left(\frac{z}{r^2}\right)\right] \geq -c\,|\mathbf{Re}\, z|,$$

holds with a constant c which is independent of the choice of f. Indeed,

$$\begin{aligned} \mathbf{Re}\left[z\widehat{\psi}\left(\frac{z}{r^2}\right)\right] &= \mathbf{Re}\left(\sum_{j=1}^{\infty} \frac{\left(z + 2\lambda_j |z|^2/r^2\right) f_j^2}{|1+2\lambda_j z/r^2|^2}\right) \\ &\geq \mathbf{Re}\left(\sum_{j=1}^{\infty} \frac{z f_j^2}{|1+2\lambda_j z/r^2|^2}\right). \end{aligned}$$

In the right-hand side zf_j^2 is multiplied by bounded factors:

$$\left|1 + 2\lambda_j z/r^2\right|^{-2} \leq \left|1 - \tfrac{1}{2}\lambda_j/\lambda_1\right|^{-2} \leq 4.$$

2.3.3 Derivation of Asymptotic Formulae

A large part of calculations used to prove Theorems 2.3.1–2.3.2 is common for both arguments. Inversion formula for the CF of measure (2.3.11) yields the equality

$$G\left[\tfrac{1}{2}r^2, \tfrac{1}{2}\Lambda^2 r^2\right] = \frac{1}{2\pi}\int_{-\infty}^{\infty}\left(e^{-itr^2/2} - e^{-it\Lambda^2 r^2/2}\right)\frac{g(t)dt}{it}.$$

It is convenient to change to the variable $z = \frac{1}{2}r^2(1-it)$. After this the equality reduces to

$$G\left[\tfrac{1}{2}r^2, \tfrac{1}{2}\Lambda^2 r^2\right] = \boldsymbol{\mathcal{K}}\left(r^2/2\right)^{\nu/2-1} J(r,\Lambda), \tag{2.3.14}$$

where

$$J(r,\Lambda) = \frac{1}{i2\pi}\int_{C_r}\left[e^z - e^{z\Lambda^2 - r^2(\Lambda^2-1)/2}\right]g^+(z)dz,$$

$$g^+(z) = \frac{1}{z^{\nu/2}}\frac{\widehat{\pi}\left(z/r^2\right)}{1-2z/r^2}\exp\left\{-\left[\sin\vartheta\right]^2 z\widehat{\psi}\left(z/r^2\right)\right\}.$$

The integration path is the vertical straight line $C_r = \left\{z \in \mathbf{C} : \mathbf{Re}\, z = \frac{1}{2}r^2\right\}$ directed from negative to positive values of the imaginary part. The integrand is analytic for $\mathbf{Re}\, z > 0$. In this latter domain, by Remarks 2.3.1–2.3.2

$$\left|g^+(z)\right| \leq c' \left|\mathbf{Im}\, z\right|^{-1-\nu/2} r^2 \exp\left\{c'' \left|\mathbf{Re}\, z\right| \left[\sin\vartheta\right]^2\right\}. \tag{2.3.15}$$

This permits one to change the integration contour to the straight line $C_1 = \{z : \mathbf{Re}\, z = 1\}$. By (2.3.15) the function $g^+(1+it)$ is absolutely integrable on the real line. If $r^2 > 2$ and $\Lambda \to \infty$, there is convergence on C_1:

$$\left|\exp\left\{z\Lambda^2 - \tfrac{1}{2}r^2\left(\Lambda^2 - 1\right)\right\}\right| = \exp\left\{\tfrac{1}{2}r^2 - \Lambda^2\left(\tfrac{1}{2}r^2 - 1\right)\right\} \to 0.$$

Consequently, for $r^2 > 2$

$$\lim_{\Lambda\to\infty} J(r,\Lambda) = J_0(r) \stackrel{def}{=} \frac{1}{i2\pi}\int_{C^1} e^z g^+(z)dz$$

and $G\left[\frac{1}{2}r^2, \infty\right) = \boldsymbol{\mathcal{K}}\left(r^2/2\right)^{\nu/2-1} J_0(r)$.

Unless otherwise specified, it is assumed further on that $\lambda_1 \neq 0$, i.e., some of the smaller eigenvalues are positive.

While calculating J_0, one can change the contour of integration to the broken line C_+ consisting of the segments which connect the point $z = +1$ with $z = (-1 \pm i)a$, $a = \frac{1}{4}r^2/\lambda_1$, and the vertical half-lines $\{z = (-1 \pm is)a, s \geq 1\}$.

Proof of Theorem 2.3.1. Put $\vartheta = 0$. On the slanted parts of the integration path C_+, the estimates of Remarks 2.3.1–2.3.2 imply that

$$\left|z^{-\nu/2}e^z - g^+(z)e^z\right| \le \frac{c'}{r^2}|z|\exp\{c''\mathbf{Re}\, z\},\quad c''>0.$$

On the vertical half-lines of C_+ the integrand admits the estimate

$$\left|z^{-\nu/2}e^z\right| + \left|g^+(z)e^z\right| \le \hat{c}r^2e^{-\check{c}r}/\left|\mathbf{Im}\, z\right|^{1+\nu/2},$$

with the function $1/u^{1+\nu/2}$ summable over the half-line $[1,\infty)$. Hence

$$|J_0(r)-L| \le c'''\frac{1}{r^2}\int_0^\infty (1+u)e^{-c''u}du + \hat{c}r^2e^{-\check{c}r}\int_1^\infty \frac{du}{u^{1+\nu/2}} \le \frac{\bar{c}}{r^2},$$

where

$$L = (i2\pi)^{-1}\int_{C_+} z^{-\nu/2}e^z dz = 1/\Gamma(\nu/2)$$

is the value of the Gamma-density at the origin. The last equality follows after the path of integration is changed to C_1. This completes the proof of Theorem 2.3.1. ○

Proof of Theorem 2.3.2. a) First assume that the covariance operator $\mathcal{V}$ has at least three nonzero eigenvalues, viz., either $\nu \ge 2$ and $\lambda_1 > 0$ or $\nu = 1$ and $\lambda_2 > 0$. Under this additional assumption the derivation of the bound of Theorem 2.3.2 is quite similar to the proof of Theorem 2.3.1 above.

Put

$$g_0(z) = \frac{1}{z^{\nu/2}}\frac{\hat{\pi}\left(z/r^2\right)}{1-2z/r^2}.$$

On the slanted parts of the path C_+

$$\left|e^z g_0(z) - e^z g^+(z)\right| \le c\left[\sin\vartheta\right]^2(1+|\mathbf{Re}\, z|)\exp\{c''\mathbf{Re}\, z\},\quad c''>0.$$

(No attempt is made to keep track of the values of constants, so c, c', ... may differ from the corresponding constants in earlier parts of the proof.) On the vertical half-lines

$$\left|e^z g_0(z) - e^z g^+(z)\right| \le c\left[\sin\vartheta\right]^2 r^2\exp\left\{-c''r^2\right\}/\left|\mathbf{Im}\, z\right|^{3/2}.$$

Under the assumptions of the theorem, the above bounds imply the estimate

$$|\mu_0(\vartheta)-\mu_0(0)| \le c\vartheta^2\left(r^2/2\right)^{\nu/2-1} \le c'\vartheta^2\mu_0(0).$$

b) If $\lambda_1 = 0$, i.e., the very first "nonprincipal" eigenvalue of $\mathcal{V}$ is zero, the assertion of Theorem 2.3.1 is trivial, and its asymptotic expression is the well-known asymptotic formula for the χ^2-distribution with ν degrees of freedom.

If $\nu = 1$ and $\lambda^2 = 0$, then $\left|z\hat{\psi}(z)\right| \le r^2$ on the vertical parts of the integration path. This permits one to evaluate the integrand from above by a summable function:

$$\left|e^z g_0(z) - e^z g^+(z)\right| \le c\left[\sin\vartheta\right]^2 r^4e^{-c''r^2}/\left|\mathbf{Im}\, z\right|^{1+\nu/2}.$$

The proof is completed as in part a). ○

2.3.4 Direct Derivation of Large Deviation Asymptotics

The asymptotic formula of Theorem 2.3.1 can be obtained without having recourse to the method of CF's. The following more elementary calculations provide additional information about the probabilities considered.

Under the assumptions of Theorem 2.3.1, the rv's $(\eta, e_j)/\sigma_j$ are independent and have the standard normal law. By (2.3.7)) $\sigma_1^2 = 1$, so the rv $\chi^2 = \sum_{j=1}^{\nu} (\eta, e_j)^2$ has the Gamma-density

$$q(x) = \begin{cases} (x/2)^{\nu/2-1} e^{-x/2}/\Gamma(\nu/2), & x > 0, \\ 0, & x \le 0. \end{cases} \tag{2.3.16}$$

In the representation

$$|\eta|^2 = \chi^2 + \zeta, \quad \zeta = \sum_{j=1}^{\infty} (\eta, e_{\nu+j})^2, \tag{2.3.17}$$

the summands are independent. Hence the rv $|\eta|^2$ also has the density $p(x) = \mathbf{E}q(x-\zeta)$. It follows from the above that

$$p(x)/q(x) = \mathbf{E}(1-\zeta/x)^{\nu/2-1} \exp\{\tfrac{1}{2}\zeta\} \mathbf{1}_{\{\zeta<x\}}.$$

If $\nu \ge 2$ (i.e., the covariance operator has multiple principal eigenvalue), a.s. convergence $\lim_{x\to\infty}(1-\zeta/x) = 1$ is dominated:

$$0 \le (1-\zeta/x)^{\nu/2-1} \le 1, \quad \zeta \le x.$$

The Lebesgue theorem on dominated convergence implies that for $\nu \ge 2$

$$\lim_{x\to\infty} p(x)/q(x) = \mathbf{E}\exp\{\tfrac{1}{2}\zeta\} = \mathcal{K}, \tag{2.3.18}$$

where the right-hand side is defined by (2.3.9). (The last equality follows from Theorem 1.1.7 and an obvious passage to the limit in number of dimensions.)

If $\nu = 1$ and the distribution of ζ is not degenerate, the calculation is slightly more complicated. Using the above expression for $\mathbf{E}\exp\{\frac{1}{2}\zeta\}$, it is easy to see that

$$\begin{aligned} p(x)/q(x) - \mathcal{K} &= \mathbf{E}\left[1/\sqrt{(1-\zeta/x)_+} - 1\right] \exp\{\tfrac{1}{2}\zeta\} \\ &- \mathbf{E}\mathbf{1}_{\{\zeta \ge x\}} \exp\{\tfrac{1}{2}\zeta\}. \end{aligned} \tag{2.3.19}$$

The second term in the right-hand side goes to zero as $x \to \infty$.

Choose $\delta \in (0, 1)$. For $\zeta < \delta x$

$$\sqrt{1-\zeta/x} - 1 \le 1/\sqrt{1-\delta} - 1, \quad 1/\sqrt{1-\delta} - 1 = \tfrac{1}{2}\delta[1+o(1)], \ \delta \to 0.$$

Consequently, for $x \to \infty$ the value of the right-hand side of (2.3.19) is determined mainly by the quantity

$$r(\delta) = \mathbf{E}\exp\{\tfrac{1}{2}\zeta\}\left[1/\sqrt{1-\zeta/x} - 1\right] \mathbf{1}_{\{\delta \le \zeta/x < 1\}}$$

if δ is chosen small enough.

The measure

$$\mu(A) = \mathbf{E}\exp\{\tfrac{1}{2}(1+\gamma)\zeta\}\,\mathbf{1}_A(\zeta), \quad A \in \mathcal{B}(\mathbf{R}),$$

has a density $\widehat{q}$ if $\gamma < 1/\sigma_2^2 - 1$. This density is bounded on the half-line $x \geq 1$ if the distribution of ζ is nondegenerate. The latter easily follows from (2.3.17) if $\mathcal{V}$ has only one or two nonzero "junior" eigenvalues; if χ^2 includes three or more nondegenerate summands, the fact that this rv has a bounded density is the consequence of integrability of its CF (see next section).

Thus the quantity $r(\delta)$ admits the bound

$$\begin{aligned} r(\delta) &\leq \exp\{-\tfrac{1}{2}\gamma\delta x\}\,\mathbf{E}\exp\{\tfrac{1}{2}(1+\gamma)\zeta\}\,\mathbf{1}_{\{\delta\leq\zeta/x<1\}}\sqrt{1-\zeta/x} \\ &= \exp\{-\tfrac{1}{2}\gamma\delta x\}\int_{\delta x}^{x}\frac{\widehat{q}(y)dy}{\sqrt{1-y/x}} \leq x\exp\{-\tfrac{1}{2}\gamma\delta x\}\max_{y\geq 1}\widehat{q}(y)\int_0^1\frac{dz}{\sqrt{z}} \end{aligned}$$

with some $\gamma > 0$. Its right-hand side goes to zero as $x \to \infty$ for each fixed positive δ.

Hence for each $\delta \in (0,\,1)$

$$\limsup_{x\to\infty}[p(x)/q(x) - \mathcal{K}] \leq \left(1/\sqrt{1-\delta} - 1\right)\mathcal{K}$$

and

$$p(x)/q(x) - \mathcal{K} \geq -\mathbf{E}\,\mathbf{1}_{\{\zeta\geq x\}}\exp\{\tfrac{1}{2}\zeta\}.$$

This implies (2.3.18) also in the case of non-multiple principal eigenvalue of the covariance operator $\mathcal{V}$.

The asymptotic formula of Theorem 2.3.1 for the DF is derived by integration of the above asymptotic representation of the density. It is readily seen from the calculations of this subsection that the inequality

$$p(x) = \frac{d}{dx}\mathbf{P}\left\{|\eta|^2 < x\right\} \leq \mathcal{K}q(x) \tag{2.3.20}$$

holds with $q(x)$ given by (2.3.16) if $\nu \geq 2$.

2.4 Probabilities of Landing in Spheres

2.4.1 Asymptotics for "Small" and "Displaced" Spheres

This section studies the asymptotic behavior at $\varepsilon \to \infty$ of the probabilities

$$\mathbf{P}\{|\eta| \leq \varepsilon\}, \quad \mathbf{P}\{|\eta - a| \leq \varepsilon\}, \quad \mathbf{P}\left\{\left|\eta - \tfrac{1}{\varepsilon}a\right| \leq \delta/\varepsilon\right\},$$

where η is a centered Gaussian RV with values from a separable real Hilbert space X. Notations are mainly those of Section 2.3.

The distribution of η is supposed to be essentialy infinite-dimensional: the eigenvalues of its covariance operator $\mathcal{V}$ satisfy the condition

$$\forall j \in \mathbf{N}\ \sigma_j^2 > 0. \tag{2.4.1}$$

The distribution of the rv $|\eta|^2$ has a bounded density if $\sigma_3^2 > 0$ (see subsection 2.4.3 below). Its asymptotic behavior at $x \to \infty$ depends on the spectrum of the covariance operator $\mathcal{V}$ in a rather more complicated way than the large deviation asymptotics of Section 2.3. To describe the "small deviation" asymptotics, it is necessary to employ the resolvent of the covariance operator

$$\mathcal{R}[z] = (\mathcal{I} + 2z\mathcal{V})^{-1}, \quad z \in \mathbf{C}, \tag{2.4.2}$$

where $\mathcal{I}$ is the identity operator in X.

Theorem 2.4.1 *The density* $p_0(x) = (d/dx)\mathbf{P}\left\{|\eta|^2 < x\right\}$ *satisfies the asymptotic relation*

$$p_0(x) = \frac{[1+o(1)]}{\sqrt{4\pi \operatorname{tr}(\mathcal{V}^2\mathcal{R}^2[\gamma])}} \exp\left\{-\int_0^\gamma \operatorname{tr}(\mathcal{V}\{\mathcal{R}[u] - \mathcal{R}[\gamma]\})\, du\right\}, \quad x \searrow 0,$$

where $\gamma = \gamma(x)$ *is the positive solution of the equation*

$$x = \operatorname{tr}(\mathcal{V}\mathcal{R}[\gamma]) \equiv \textstyle\sum_{j=1}^{\infty} \sigma_j^2 / \left(1 + 2\gamma\sigma_j^2\right). \tag{2.4.3}$$

The asymptotic expression for the DF is $\mathbf{P}\left\{|\eta|^2 < x\right\} = [1/\gamma + o(1)]\, p_0(x)$.

If the "gravity centre" of the distribution of η does not coincide with the centre of the sphere, the asymptotic behavior of the probability of hitting this small sphere is described by

Theorem 2.4.2 *Assume that the vector* $\mathcal{V}^{-1/2}a$ *is well defined. Then the rv* $|\eta - a|^2$ *has absolutely continuous distribuiton. As* $x \to 0$, *the asymptotic expression for its density is*

$$p_a(x) = \frac{d}{dx}\mathbf{P}\left\{|\eta - a|^2 \le x\right\} = [1 + o(1)] \exp\left\{-\tfrac{1}{2}\left|\mathcal{V}^{-1/2}a\right|^2\right\} p_0(x),$$

and one for the DF is

$$\mathbf{P}\left\{|\eta - a|^2 \le x\right\} = [1 + o(1)] \exp\left\{-\tfrac{1}{2}\left|\mathcal{V}^{-1/2}a\right|^2\right\} \mathbf{P}\left\{|\eta|^2 \le x\right\}.$$

The proofs of both above theorems are in subsection 2.4.3. Preparations occupy subsection 2.4.2. The method of proof is closely related to that of Section 2.3. It is based on the possibility to write out explictly the CF of the rv's considered. The following result is closely affiliated to Theorem 2.4.1 and Theorem 2.4.2.

Theorem 2.4.3 *Fix a number* $\delta \le |a|$. *As* $\varepsilon \to \infty$,

$$\mathbf{P}\{|\eta - (1/\varepsilon)a| \le \delta/\varepsilon\} = \frac{[1+o(1)]}{8\pi}\frac{\varepsilon}{\delta}\left|\mathcal{R}^{3/2}[\widehat{\gamma}]\,\mathcal{V}^{1/2}a\right|^{-1}$$
$$\times \exp\left\{-2(\varepsilon/\widehat{\gamma})^2\left|\mathcal{R}[\widehat{\gamma}]\,\mathcal{V}^{1/2}a\right|^2 - \textstyle\int_0^{\widehat{\gamma}} \operatorname{tr}(\mathcal{V}\mathcal{R}[u])\, du\right\},$$

where $\widehat{\gamma} = \widehat{\gamma}(\delta)$ *is the solution of the equation*

$$\delta^2 = \left|\mathcal{R}\left[\widehat{\gamma}\right] a\right|^2 \equiv \sum_{j=1}^{\infty} (a, e_j)^2 \,/\, \left(1 + 2\widehat{\gamma}\sigma_j^2\right)^2 . \qquad (2.4.4)$$

The proof of this theorem is in subsection 2.4.4 below.

2.4.2 Laplace Transform of Squared Norm Density

As also the calculations of Section 2.3, the proofs of Theorems 2.4.1–2.4.2 are based upon the following known explicit expression for the Laplace transform of the distribution of the rv $|\eta - a|^2$.

Lemma 2.4.1 *If* $2\mathrm{Re}\, z\sigma_1^2 > -1$, *then*

$$\mathbf{E} \exp\left\{-z\left|\eta - a\right|^2\right\} = g(z) \exp\left\{-H(z, a)\right\},$$

where

$$\begin{aligned} g(z) &= \exp\left\{-\tfrac{1}{2}\mathrm{tr} \ln \mathcal{R}\left[z\right]\right\} \equiv \prod_{j=1}^{\infty} 1/\sqrt{1 + 2z\sigma_j^2}, \\ H(z, a) &= z\left(\mathcal{R}\left[z\right] a, a\right) \equiv z\sum_{j=1}^{\infty} (a, e_j)^2 \,/\, \left(1 + 2z\sigma_j^2\right). \end{aligned}$$

Proof of Lemma 2.4.1. Let $\mathcal{P}_n$ be the orthogonal projection onto the invariant subspace of $\mathcal{V}$ spanned by the unit vectors e_j, $j = \overline{1, n}$, from the eigenbasis of $\mathcal{V}$. Whatever the choice of $n \in \mathbf{N}$, the Laplace transform of the distribution of $\left|\mathcal{P}_n(\eta - a)\right|^2$ can be calculated applying Theorem 1.1.7, with $\mathcal{V}$ and its resolvent replaced by the finite-dimensional operator $\mathcal{P}_n\mathcal{V}\mathcal{P}_n$ and its resolvent respectively.

As $n \to \infty$, there is monotone convergence $\left|\mathcal{P}_n(\eta - a)\right|^2 \nearrow |\eta - a|^2$. Hence the assertion of the lemma follows for negative values of z by a passage to the limit using the monotone convergence theorem of real analysis. For a positive or complex value of z the passage to the limit can then be substantiated using the Lebesgue dominated convergence theorem. This completes the proof of the lemma. ○

In further calculations it is convenient to separate the real and imaginary parts of the function H in Lemma 2.4.1. The fact that all operators $f(\mathcal{V})$ commute and the well-known resolvent identity, $\mathcal{R}\left[w\right] = \mathcal{R}\left[z\right] + 2(z - w)\mathcal{V}\mathcal{R}\left[w\right]\mathcal{R}\left[z\right]$, yield the equality

$$(\gamma + it)\mathcal{R}\left[\gamma + it\right] = \gamma\mathcal{R}\left[\gamma\right] + 2t^2\mathcal{V}\mathcal{R}\left[\gamma\right]\mathcal{J}\left[\gamma + it\right] + it\mathcal{J}\left[\gamma + it\right], \quad \gamma, t \in \mathbf{R}.$$

In this latter, the operator

$$\mathcal{J}\left[z\right] = \mathcal{R}\left[z\right]\mathcal{R}\left[\bar{z}\right] = \left[(\mathcal{I} + 2\gamma\mathcal{V})^2 + (2t\mathcal{V})^2\right]^{-1}$$

is symmetrical and nonnegative definite, as well as $\mathcal{I} - \mathcal{J}\left[z\right]$. The above formula can be used, e.g., to transform the expression for H to

$$-H(z, a) = -H(\gamma, a) - 2t^2\left(\mathcal{V}\mathcal{R}\left[\gamma\right]\mathcal{J}\left[z\right] a, a\right) + it\left(\mathcal{J}\left[z\right] a, a\right), \qquad (2.4.5)$$

where $a \in X$ and $z = \gamma + it$. The function $|g(z)|$ can also be described in terms of the spectrum of $\mathcal{J}[\cdot]$:

$$|g(z)| = \exp\left\{-\tfrac{1}{4}\mathrm{tr}\left(\ln \mathcal{J}[z]\right)\right\} = \prod_{j=1}^{\infty}\left[\left(1+2\gamma\sigma_j^2\right)^2 + \left(2t\sigma_j^2\right)^2\right]^{-1/4}. \quad (2.4.6)$$

It follows from (2.4.5) and (2.4.6) that the rv $|\eta - a|^2$ has bounded continuous density $p_a(x)$ for each choice of a if the eigenvalue σ_3^2 of the covariance operator $\mathcal{V}$ is strictly positive (recall that in (2.3.4) a multiple eigenvalue is considered as a series of equal ones). Indeed, in this case its CF is absolutely integrable:

$$\left|\mathbf{E}\exp\left\{it\,|\eta - a|^2\right\}\right| = |g(it)| \le \left(1+\sigma_3^2 t^2\right)^{-3/4}.$$

This inequality shows that the density is bounded uniformly in a and x:

$$p_a(x) \le c, \quad c = c\left(\sigma_3^2\right), \quad (2.4.7)$$

where the constant in the bound depends only on the eigenvalue specified.

2.4.3 Derivation of Density Asymptotics

The calculations which prove Theorems 2.4.1–2.4.2 are quite similar. Under the assumptions of these theorems, the distribution of $|\eta - a|^2$ has a continuous density. The Laplace transform inversion formula yields for the DF the expression

$$G_1(x;a) \equiv \mathbf{P}\left\{|\eta - a|^2 \le x\right\} = \frac{1}{i2\pi}\int_{C_\gamma} \kappa_1(z) e^{zx - H(z)} g(z)\,dz, \quad (2.4.8)$$

where $\kappa_1(z) \equiv 1/z$, and a similar equation with κ_1 replaced by $\kappa_2(z) \equiv 1$ for the density $G_2(x;a) = p_a(x) = G_1'(x;a)$. In (2.4.8) the path of integration is the line $C_\gamma = \{z : \mathbf{Re}\, z = \gamma\}$ directed from the negative towards positive values of the imaginary part of the argument; the choice of the parameter $\gamma > 0$ is arbitrary. Asymptotic expressions of both G_1 and G_2 can be obtained by the saddlepoint method.

Proof of Theorem 2.4.1. a) It is natural to begin with the calculation of the density p_0 corresponding to $i = 2$, $a = 0$, and $H(z,a) = 0$. In this case the integrand of (2.4.8) is

$$\exp\{zx\}\, g(z) = \exp\{\chi(z)\}, \quad \chi(z) = \chi(z;x) \equiv zx + \ln g(z). \quad (2.4.9)$$

The function χ has a saddlepoint $\gamma > 0$ which satisfies equation (2.4.3) and grows infinitely as x goes to zero:

$$x = -\left.\frac{d}{dz}\ln g(z)\right|_{z=\gamma} \equiv \mathrm{tr}\left(\mathcal{V}\mathcal{R}[\gamma]\right), \quad \lim_{x\searrow 0}\gamma(x) = +\infty.$$

The vertical line passing through this point of the real axis can be used as the steepest descent path.

Write (2.4.8) in the form

$$p_0(x) = \frac{1}{2\pi} g(\gamma) e^{\gamma x} J_\gamma(x), \quad J_\gamma(x) = \int_{-\infty}^{\infty} g_0\left(it/\gamma; \gamma\right) e^{itx} dt, \qquad (2.4.10)$$

where

$$g_0(w;\gamma) = g(\gamma + \gamma w)/g(\gamma) = \prod_{j=1}^{\infty} 1/\sqrt{1 + w\lambda_j(\gamma)}, \quad \lambda_j(\gamma) = \frac{2\gamma\sigma_j^2}{1 + 2\gamma\sigma_j^2}.$$

Using (2.4.3), the first two factors in the right-hand side of (2.4.10) can be transformed into

$$e^{\gamma x} = \exp\left\{\gamma \operatorname{tr}\left(\mathcal{V}\mathcal{R}\left[\gamma\right]\right)\right\}, \quad g(\gamma) = \exp\left\{-\int_0^{\gamma} \operatorname{tr}\left(\mathcal{V}\mathcal{R}\left[u\right]\right) du\right\}.$$

This leads to the equality

$$e^{\gamma x} g(\gamma) = \exp\left\{-\int_0^{\gamma} \operatorname{tr}\left(\mathcal{V}\left[\mathcal{R}\left[u\right] - \mathcal{R}\left[\gamma\right]\right]\right) du\right\}. \qquad (2.4.11)$$

To verify the first assertion of the theorem, one should find the principal term of the asymptotics of the integral $J_\gamma(x)$ in (2.4.10) as $\gamma \to \infty$.

While calculating this integral, it is convenient to change the integration variable to $u = t/\gamma$ and transform the integral according to the formula

$$(1/\gamma) J_\gamma(x) = \int_{-\infty}^{\infty} e^{h(u;x)} du, \quad h(u;x) = iux\gamma(x) + \ln g_0(iu, \gamma(x)). \qquad (2.4.12)$$

The function h vanishes for $u = 0$, and by (2.4.3) also $h'(0) = 0$. Some of its derivatives of higher orders at the origin are also easily computed: e.g.,

$$-h''(0) = \tfrac{1}{2} \textstyle\sum_{j=1}^{\infty} \lambda_j^2\left(\gamma(x)\right). \qquad (2.4.13)$$

Moreover, $-h''(0) \nearrow \infty$ as $x \searrow 0$ because in this case both $\gamma(x) \nearrow \infty$ and $\lambda_j = \lambda_j(\gamma(x)) \nearrow 1$ for each j. Since all λ_j are uniformly bounded,

$$|h'''(u)| \le \textstyle\sum_{j=1}^{\infty} \lambda_j^3(\gamma(x)) \le 2\,|h''(0)|. \qquad (2.4.14)$$

Relation (2.4.13) makes plausible the hypothesis that it is a small neighborhood of zero, $[-\tau, \tau]$ with appropriately chosen infinitesimal $\tau = \tau(x) > 0$, which determines the asymptotic behavior of the integral in (2.4.12).

The choice of τ above should meet certain requirements.

By the Taylor formula, for $|u| \le \tau$ and some $\vartheta \in [-1, 1]$

$$h(u) = -\tfrac{1}{2}\,|h''(0)|\, u^2 + O\left(\tau^3\, |h'''(\vartheta u)|\right) = -\tfrac{1}{2}\,|h''(0)|\, u^2 + O\left(\tau^3\, |h''(0)|\right).$$

Thus one can employ a calculation quite typical of the Laplace method to establish the asymptotic relation

$$\int_{-\tau}^{\tau} e^{h(u)} du = [1 + o(1)]\, \sqrt{2\pi/\,|h''(0)|} \qquad (2.4.15)$$

if only

$$\tau^2 \left|h''(0)\right| \to \infty, \quad \tau^3 \left|h''(0)\right| \to 0. \tag{2.4.16}$$

Outside the segment $[-\tau, \tau]$ the function g_0 admits the following estimates. As already noted, $\lim_{x\to 0} \lambda_j(\gamma(x)) \to 1$ for each fixed value of j. Consequently, one can select positive constants c' and c'' in such a way that the restriction $0 \le x \le c'$ implies the estimate

$$|g_0(iu;\gamma)| \le \prod_{j=1}^{8} \left(1+u^2\lambda_j^2\right)^{-1/4} \le c''/(1+u^2)^2. \tag{2.4.17}$$

At the same time, the function $|g_0(iu;\gamma)|$ is monotone decreasing as the absolute value of its argument u increases. Hence for $|u| \ge \tau$

$$|g_0(iu;\gamma)| \le |g_0(i\tau;\gamma)| = \prod_{j=1}^{\infty} \left(1+\tau^2\lambda_j^2\right)^{-1/4} \le \exp\left\{ -c\tau^2 \sum_{j=1}^{\infty} \lambda_j^2 \right\} \tag{2.4.18}$$

if the constant c is chosen appropriately. The second inequality is a consequence of the fact that $\lambda_j \le 1$ for all j if τ is small.

From (2.4.17) and (2.4.18) follows the inequality

$$|g_0(iu;\gamma)| \le \frac{\widehat{c}}{1+u^2} \exp\left\{-\check{c}\tau^2 \left|h''(0)\right|\right\}. \tag{2.4.19}$$

The first factor in its right-hand side is integrable on all real line. Hence there is inequality

$$\int_{|u|>\tau} |g_0(iu;\gamma)|\, du \le \tilde{c} \exp\left\{-\check{c}\tau^2 \left|h''(0)\right|\right\} = O\left(1/\sqrt{|h''(0)|}\right)$$

if

$$\tau^2 \left|h''(0)\right| / \ln \left|h''(0)\right| \to 0. \tag{2.4.20}$$

Restrictions (2.4.16) and (2.4.20) are satisfied, e.g., if $\tau = |h''(0)|^{2/5}$. This establishes the asymptotic equality

$$\tfrac{1}{\gamma} J_\gamma(x) = [1+o(1)] \sqrt{2\pi / \left|h''(0)\right|}.$$

To derive the asymptotic equality of Theorem 2.4.1 for the density p_0, it remains to substitute the above-obtained expession into (2.4.10).

b) For the probability $\mathbf{P}\left\{|\eta|^2 < x\right\}$, the derivation of an asymptotic representation is, in its main features, a repetition of part a) of the proof. The only important difference is that for $i = 1$ the integrand in (2.4.8) contains the additional factor $1/(\gamma + it)$ with $t = \mathbf{Im}\, z$. If the integral is calculated using the steepest descent method with the path of (2.4.12), this factor adds the coefficient $1/\gamma$ to the asymptotic expression of density. This completes the proof of Theorem 2.4.1. ○

Proof of Theorem 2.4.2. For a ball centered at $a \ne 0$, function (2.4.9) is replaced by

$$\chi_+(z) = \chi(z) - H(z,a). \tag{2.4.21}$$

To calculate the asymptotics of the integral, the method of steepest descent can be employed once again. It is unnecessary to search the new saddlepoint for (2.4.21) because the additional summand is almost constant on the integration path used in the proof of Theorem 2.4.1.

Let γ be the solution of (2.4.3). It follows from (2.4.5) that the function

$$\delta H(t) \equiv -H(\gamma + it, a) + H(\gamma, a) \tag{2.4.22}$$

obeys the restriction $\mathbf{Re}\,\delta H(t) \le 0$. In notation of (2.4.10),

$$\delta H(t) = \sum_{j=1}^{\infty} \left[-\frac{t^2}{\gamma^2} + \frac{it}{2\gamma\sigma_j^2} \right] \frac{(a, e_j)^2 \lambda_j}{(1 + 2\gamma\sigma_j^2)\left(1 + \lambda_j^2 t^2/\gamma^2\right)}.$$

The series $\sum_{j=1}^{\infty} (a, e_j)^2 / \sigma_j^2$ is convergent. Consequently, as $x \to 0$ (and, hence, $\gamma \to \infty$), there is convergence $\delta H(\gamma u) \to 0$ that is uniform on each bounded set. Thus, along the integration path of the proof of the preceding theorem the last factor in the right-hand side of the formula

$$\exp\{\chi_+\} = \exp\{-H(\gamma, a)\} \exp\{\chi\} \exp\{\delta H(t)\}$$

is uniformly bounded and, moreover, $\delta H(t) \to 0$ as $x \to 0$ uniformly on each segment $[-\gamma u, \ \gamma u]$, so $e^{\delta H} \to 1$ as well.

Repeating the calculations of the proof of Theorem 2.4.1, one arrives at the expression for density

$$p_a(x) = p_0(x) \exp\{-H(\gamma, a\} [1 + o(1)] \tag{2.4.23}$$

and a similar expression for the DF.

It remains to note that $\lambda_j \to 1$ for each j under the assumptions of Theorem 2.4.2, so

$$H(\gamma, a) = \textstyle\sum_{j=1}^{\infty} (a, e_j)^2 \lambda_j / \sigma_j^2 \to \left|V^{-1/2} a\right|^2.$$

This completes the proof. ○

2.4.4 "Distant Ball" Asymptotics

As in subsection 2.4.3, it follows from the inversion formula for the Laplace transform and Lemma 2.4.1 that there is integral representation

$$\mathbf{P}\left\{\left|\eta - \tfrac{1}{\varepsilon} a\right| \le \delta/\varepsilon\right\} = \frac{1}{i2\pi} \int_{C_\gamma} \exp\left\{\chi^0(z)\right\} dz, \tag{2.4.24}$$

where the path of integration is $C_\gamma = \{z : \mathbf{Re}\, z = \gamma\}$ and $\chi^0(z) = \varphi(z) + z\delta^2/\varepsilon^2$, while

$$\varphi(z) = \ln[g(z)/z] - H\left(z, (1/\varepsilon) a\right). \tag{2.4.25}$$

The asymptotic behavior of the integral in (2.4.24) is studied using the saddlepoint method. The saddlepoint is found from the equation

$$\frac{\delta^2}{\varepsilon^2} - \frac{1}{\gamma} - \operatorname{tr}\left(\mathcal{V}\mathcal{R}\left[\gamma\right]\right) - \frac{1}{\varepsilon^2}\left(\mathcal{R}^2\left[\gamma\right]a, a\right) = 0.$$

This equation can be replaced by the more convenient approximate equation (2.4.4) whose solution no longer depends on the small parameter ε.

In what follows, the integration path in (2.4.24) is replaced by the vertical straight line passing through $\widehat{\gamma}$. This transforms the formula to

$$\mathbf{P}\left\{\left|\eta - \tfrac{1}{\varepsilon}a\right| \le \delta/\varepsilon\right\} = AK, \tag{2.4.26}$$

where (recall that g was defined in Lemma 2.4.1)

$$A = \frac{1}{2\pi\widehat{\gamma}}g(\widehat{\gamma})\exp\left\{\frac{1}{\varepsilon^2}\left[\widehat{\gamma}\delta^2 - H\left(\widehat{\gamma}, a\right)\right]\right\}, \quad K = \int_{-\infty}^{\infty}\widehat{k}(t)e^{k(t)}dt,$$

$$\widehat{k}(t) = \frac{1}{1 + it/\widehat{\gamma}}\prod_{j=1}^{\infty}\frac{1}{\sqrt{1 + it\mu_j}}, \quad \mu_j = \frac{2\sigma_j^2}{1 + 2\widehat{\gamma}\sigma_j^2},$$

and the exponent in the expression for K is

$$\begin{aligned} k(t) &= (1/\varepsilon^2)\left[it\delta^2 - H(\widehat{\gamma} + it, a) + H(\widehat{\gamma}, a)\right] \\ &= \frac{it}{\varepsilon^2}\left\{\delta^2 - (\mathcal{J}[\widehat{\gamma} + it]a, a) - 2t^2\left(\mathcal{V}\mathcal{R}\left[\widehat{\gamma}\right]\mathcal{J}(\widehat{\gamma} + it)a, a\right)\right\} \qquad (2.4.27) \\ &= -\frac{2t^2}{\varepsilon^2}\left(\mathcal{V}\mathcal{R}\left[\widehat{\gamma}\right]\mathcal{J}[\widehat{\gamma} + it]a, a\right) + i2t^3\left(\mathcal{V}^2\mathcal{R}^2\left[\widehat{\gamma}\right]\mathcal{J}[\widehat{\gamma} + it]a, a\right). \end{aligned}$$

(Transformation of the above expression uses (2.4.4) and (2.4.5).)

While computing K in (2.4.26), it is convenient to use $u = t/\varepsilon$ as the integration variable. In this notation

$$K = \varepsilon\int_{-\infty}^{\infty}\widehat{k}(\varepsilon u)\exp\left\{k(\varepsilon u)\right\}du. \tag{2.4.28}$$

The remaining calculations are typical of the Laplace method. In a neighborhood of zero the integrand of (2.4.28) has the form

$$(1 + \kappa)\exp\left\{-2u^2\left(\mathcal{V}\mathcal{R}^3\left[\widehat{\gamma}\right]a, a\right)\right\},$$

where the bound $|\kappa| \le c\varepsilon\left(\tau + \tau^3\right)$ holds uniformly in $|u| \le \tau$. Hence for $\tau = \tau(\varepsilon)$ the restrictions

$$\tau \to \infty, \quad \varepsilon\tau^3 \to 0 \tag{2.4.29}$$

imply the equality

$$\varepsilon\int_{-\tau}^{\tau}\widehat{k}(\varepsilon u)e^{k(\varepsilon u)}du = \varepsilon\sqrt{\frac{\pi}{2\left(\mathcal{V}\mathcal{R}^3\left[\widehat{\gamma}\right]a, a\right)}}\left[1 + o(1)\right]. \tag{2.4.30}$$

Outside the segment $[-\tau,\ \tau]$ the following estimates are true. The function

$$\mathbf{Re}\, k(\varepsilon u) = 2u^2 \left(\mathcal{VR}\left[\widehat{\gamma}\right] \mathcal{J}\left[\widehat{\gamma} + i\varepsilon u\right] a, a \right) = \sum_{j=1}^{\infty} \frac{\mu_j u^2 \left(a, e_j\right)^2}{\left(1 + \varepsilon^2 u^2 \mu_j^2\right)\left(1 + 2\widehat{\gamma}\sigma_j^2\right)^2}$$

grows with the growth of u. The quantities μ_j are bounded for $|u| \geq \tau$ and $\varepsilon\tau \leq c'$, so this function admits the estimate

$$-\,\mathbf{Re}\, k(\varepsilon u) \geq 2c'' \sum_{j=1}^{\infty} \frac{\mu_j \tau^2 \left(a, e_j\right)^2}{\left(1 + 2\widehat{\gamma}\sigma_j^2\right)^2} = 2c''\tau^2 \left(\mathcal{VR}^3\left[\widehat{\gamma}\right] a, a\right). \qquad (2.4.31)$$

At the same time, the inequality

$$\left|\widehat{k}(\varepsilon u)\right| \leq c_m / \left(1 + \varepsilon^2 u^2\right)^{m/4} \qquad (2.4.32)$$

holds for each m (cf. (2.4.17)).

The following inequality is a consequence of (2.4.31) and (2.4.32):

$$\int_{|u|\geq\tau} \left|\widehat{k}(\varepsilon u) e^{k(\varepsilon u)}\right| du \leq \frac{c}{\varepsilon} \exp\left\{-c''\tau^2 \left(\mathcal{VR}^3\left[\widehat{\gamma}\right] a, a\right)\right\} \int_{-\infty}^{\infty} \frac{dt}{1+t^2}.$$

The right-hand side of this inequality is an infinitesimal of order $o(\varepsilon)$ provided that

$$\lim_{\varepsilon\to 0} \tau^2 / \ln(1/\varepsilon) = \infty. \qquad (2.4.33)$$

Both restrictions (2.4.29) and (2.4.33) are satisfied, e.g., if $\tau = \varepsilon^{-\alpha}$ with $\alpha \in (0,\ \frac{1}{3})$.

Thus, an asymptotic expression for the factor K in the right-hand side of (2.4.26) is given by the right-hand side of (2.4.30). To establish the assertion of Theorem 2.4.3 it remains to transform the factor A in (2.4.26) along the same lines as in the derivation of (2.4.11). ○

2.5 Related Results

2.5.1 "Geometric" Theorems

Gaussian distributions constitute the class of infinite-dimensional measures which has been studied in most detail. Not being a systematic survey, this section includes the statements of some theorems which complement those of the preceding sections (see also Appendix B). As always, Φ is the standard normal DF.

1. Let ξ be a Gaussian RV with values in a separable real Banach space with norm $|\cdot|$. If $\lambda > 0$, then (see [PiS85])

$$\mathbf{E} e^{\lambda|\xi|} \leq \exp\left\{\lambda \mathbf{E}\,|\xi| + \tfrac{1}{2}\lambda^2 \mathbf{E}\,|\xi - \mathbf{E}\xi|^2\right\}.$$

2. Assertions similar to Theorems 2.4.1–2.4.3 can be made more precise using more refined properties of the covariance spectrum. The theorems below were established in [Ib79] by methods similar to those of Sections 2.3–2.4.

Let η be a centered Gaussian RV with values in the separable real Hilbert space X, whose covariance operator $\mathcal{V}$ has eigenvalues satisfying the condition $\sigma_j^2 = O\left(1/j^{1+\alpha}\right)$ with some $\alpha > 0$.

Consider a nonrandom vector $a = \mathcal{V}^{\bar{\delta}} b$, $b \in X$, $\bar{\delta} \in \left[\frac{1+2\alpha}{4+4\alpha}, \frac{1}{2}\right]$.

If γ is the solution of (2.4.3) and $x \to 0$, then

$$\frac{\mathbf{P}\left\{|\eta - a|^2 \le x\right\}}{\mathbf{P}\left\{|\eta|^2 \le x\right\}} = \exp\left\{-2\gamma\,|\mathcal{R}[\gamma]\,a|^2 - 2\gamma^2\left|\mathcal{V}^{\frac{1}{2}}\mathcal{R}[\gamma]\,a\right|^2\right\}[1 + o(1)].$$

The asymptotic behavior of the densities and probabilities considered in Theorems 2.4.1–2.4.2 can be described in terms of asymptotics at $y \to \infty$ of the quantity $n(y) = \operatorname{card}\left\{j : \sigma_j^2 > 1/y\right\}$.

If $\left|\mathcal{V}^{-1/2} a\right| < \infty$ and $n(y) = \alpha y^{\kappa}[1 + o(1)]$ as $y \to \infty$ for some $\alpha > 0$ and $\kappa \in (0, 1)$, then

$$\ln p_a(x) = \left(\frac{\alpha\pi}{\cos\pi\kappa}\right)^{1/(1-\kappa)} \frac{1-\kappa}{2(\kappa x)^{\kappa/(1-\kappa)}}[1 + o(1)].$$

If $n(y)/y^{\varepsilon} \to \infty$ for some $\varepsilon > 0$, then

$$\limsup_{x\to 0} \frac{\ln\ln p_a(x)}{\ln x} = \lim_{y\to\infty} \frac{\ln n(y)}{\ln y} = \tau \overset{def}{=} \inf\left\{t : \textstyle\sum_{j=1}^{\infty} \sigma_j^{2t} < \infty\right\}.$$

3. In the space of bounded sequences l_∞ there exists a measure, defined on the cylindric σ-algebra generated by the coordinate functionals, which assigns zero measure to each unit ball. See [FrT80].

4. Let ξ and η be Gaussian RV's with values in a separable real Banach space X. Assume that $\mathbf{E}\xi = 0$.

If for each vector a from the unit ball of X there is a number $\delta = \delta(a) > 0$ such that

$$\forall r < \delta \;\; \mathbf{P}\{|\xi - a| < r\} = \mathbf{P}\{|\eta - a| < r\},$$

then the distributions of ξ and η coincide. See [Bor77].

5. A Gaussian density in a Euclidean space has partial derivatives of all orders and can be decomposed into a convergent Taylor series. This result can be extended to some infinite-dimensional Gaussian distributions — one can define derivatives of a measure and construct expansions similar to power series [Be82].

6. The results listed in this paragraph are taken from [Pau82].

Consider the space of infinitesimal sequences c_0 with the norm defined as $|x| = \sup_{j\in\mathbf{N}} |x_j|$.

There is a centered Gaussian RV in c_0 for which the density $(d/du)\mathbf{P}\{|\xi| < u\}$ is unbounded.

Put $\sigma_j = 1/\sqrt{2\ln j}\, b_j$, where the strictly decreasing sequence $b_j = b(j)$ is generated by a function $b: \mathbf{R} \to \mathbf{R}$ which has an inverse b^{-1}. Define $\eta = (\sigma_j\eta_j, j \in \mathbf{N}) \in c_0$, where η_j are independent standard normal rv's. The norm of this vector has bounded density if for some $\varepsilon \in (1, \frac{1}{2})$ the function $x \mapsto g(x) = \left[b^{-1}(x)\right]^{\varepsilon}$ satisfies the relations

$$\lim_{x\to 0} g(x) = \infty, \quad \forall \alpha > 1 \ \lim_{x\to 0} \frac{\ln g(x)}{\ln g(\alpha x)} = 0.$$

If the assumption of centering is left out, it is usually impossible to obtain a uniform upper bound on the density of the norm of a Gaussian RV. Thus, assume that the centered Gaussian RV $\eta = (\eta_j, j \in \mathbf{N})$ in c_0 has independent coordinates and

$$\forall \varepsilon > 0 \ \textstyle\sum_{j=1}^{\infty} \exp\left\{-\varepsilon/\mathbf{E}\eta_j^2\right\} < \infty.$$

Then

$$\forall\, \delta > 0 \ \sup_{|a|\le\delta} \sup_{x} \frac{d}{dx}\mathbf{P}\{|\eta + a| < x\} = \infty.$$

7. The estimates listed in this paragraph are taken from [HofSD79]. They can be useful in constructing examples of infinite-dimensional Gaussian distributions with specific properties in spaces of real sequences.

Let $\xi = (\xi_j, j \in \mathbf{N})$ be a sequence of independent standard normal rv's.

For a positive sequence $\tau = (\tau_j, j \in \mathbf{N})$ put

$$h_\tau(\xi) \stackrel{def}{=} \sqrt{\textstyle\sum_{j=1}^{\infty} \tau_j^2\xi_j^2}, \quad F(t) \stackrel{def}{=} \mathbf{P}\{h_\tau(\xi) \le t\}.$$

The behavior of F near zero is as follows.

Assume that $\tau_j = \tau(j)$, where the function $\tau: \mathbf{R} \to \mathbf{R}_+$ is decreasing, $0 < \tau(x) \le 1/\sqrt{x}$, and $\int_0^\infty \tau^2(x)dx < \infty$. Then the following relation holds for all $x \ge 1$ and $t \in [0, 1]$:

$$F(t) \le A_1 \exp\left\{\ln\phi(x) + (x-1)\ln t + \int_1^x \ln\phi(y)dy\right\},$$

where $\phi(x) = 1/[\tau(x)\sqrt{x}]$ and A_1 is a constant. Moreover,

$$F(t) \ge B_1\left(1 - \frac{\overline{\psi}(x)}{t^2 - s^2}\right)\exp\left\{-\frac{1}{2}\ln x + (x+1)\ln s - \frac{s^2}{2\tau(x)} + \int_1^x \ln\phi(y)dy\right\},$$

where $\overline{\psi}(x) = \int_x^\infty \tau^2(y)dy$, the numbers $s, t \in [0, 1]$, $s < t$, and $x \ge 1$ are arbitrary, and B_1 is a constant.

For a positive sequence $a = (a_j, j \in \mathbf{N})$,

$$q_a(\xi) \stackrel{def}{=} \sup_{j\in\mathbf{N}} |\xi_j|/a_j < \infty \text{ a.s.} \iff \exists\, t \in (0,\infty) \sum_{j=1}^{\infty} \exp\left\{-\tfrac{1}{2}t^2a_j^2\right\} < \infty.$$

For small values of its argument, the DF of $q_a(\xi)$ admits the following upper and, if $a_j \geq a > 0$, lower bounds:

$$\exp\left\{-\tfrac{4}{3}\psi(t)Q(t)\right\} \leq G(t) \stackrel{def}{=} \mathbf{P}\left\{q_a(\xi) \leq t\right\} \leq \exp\left\{-\sqrt{2/\pi}\psi(at)Q(t)\right\},$$

where $Q(t) \stackrel{def}{=} \sum_{j=1}^{\infty} \exp\left\{-\frac{1}{2}t^2a_j^2\right\}/(1+ta_j)$ and

$$\psi(at) = \sqrt{2/\pi}\frac{\ln(1/[\Phi(at)-\frac{1}{2}]) + \ln\frac{1}{2}}{1-2[\Phi(at)-\frac{1}{2}]} \sim \sqrt{2/\pi}\ln(1/t), \quad t \searrow 0.$$

Consider a strictly decreasing sequence $b_j \geq 0$. It is possible to chose the sequence $a_j > 0$ so that there exist numbers $m_j \in (b_{j+1}, b_j)$ such that

$$\forall j \in \mathbf{N} \;\; G(b_j) - G(m_j) \geq 2\left[b_j - m_j\right], \quad G(m_j) \leq m_j - b_{j+1}.$$

8. Let ξ be a Gaussian RV with values in a locally convex linear topological space X. If ξ is centered and the Borel set A is convex and symmetric with respect to zero, then for each $a \in X$

$$\mathbf{P}\{\xi \in A\} \geq \mathbf{P}\{\xi \in a + A\}.$$

This inequality is a corollary to a much more general result. Define the inner measure by the equality

$$\mu_*(A) = \inf\left\{\mathbf{P}\{\xi \in K\} : K \subset A, K \in \mathcal{C}_X\right\}$$

where $\mathcal{C}_X$ is the class of compact subsets of X.

If A and B are measurable sets and $0 < \lambda < 1$, then (see [Bor74, HofSD79])

$$\mu_*\left(\lambda A + (1-\lambda)B\right) \geq \left[\mu_*(A)\right]^{\lambda}\left[\mu_*(B)\right]^{1-\lambda}.$$

2.5.2 Theorems on Gaussian Processes

The next few results refer to more special problems of the theory of Gaussian stochastic processes. They are included to provide a brief reminder of the refinements which can be attained through consideration of a more concrete object with rich structure.

1. Let ξ_t, $t \in \mathbf{R}$, be a stationary Gaussian process with zero mean value and covariance function $r(t-s) = \mathbf{E}\xi_s\xi_t$.

For each $c > 0$

$$\mathbf{P}\left\{\sup_{0\leq t\leq c} \xi_t \geq u\sqrt{r(0)}\right\} = \frac{c+o(1)}{2\pi}e^{-\frac{1}{2}u^2}\sqrt{|r''(0)|/r(0)}.$$

See [Berm71], [Di81].

2. Let w_t, $t \in [0,\ T]$, be a Wiener process. If the function $a(t)$ defines an admissible translation, then (see [Sy79]) for $\varepsilon \to \infty$

$$\begin{aligned} & \mathbf{P}\left\{\sup_{0\le t\le T} |w_t - a(t)| \le \varepsilon\right\} \\ = \ & [1+o(1)] \exp\left\{\tfrac{1}{2}\int_0^T [a'(t)]^2\, dt\right\} \mathbf{P}\left\{\sup_{t\le T} |w_t| \le \varepsilon\right\}. \end{aligned}$$

3. Let T be a compact metric space, Φ the standard normal DF, and ξ_t, $t \in T$, a centered Gaussian process with continuous covariance function. Assume moreover that the paths of this process are bounded functions almost surely. Then [Ta88a]

$$\lim_{u\to\infty} \frac{\mathbf{P}\left\{\sup_{t\in T} \xi_t \ge u\right\}}{1-\Phi(u/\sigma)} = 1$$

if and only if

a) the maximum $\sigma^2 \equiv \max_{\tau\in T} \mathbf{E}\xi_t^2$ is attained at a unique point $\tau \in T$;

b) $\lim_{h\to 0} \frac{1}{h} \mathbf{E} \sup_{t\in A(h)} (\xi_t - \xi_\tau) = 0$, where $A(h) = \left\{t : \mathbf{E}\xi_t\xi_\tau \ge \sigma^2 - h^2\right\}$.

4. Let w_t, $t \in [0,\ 1]$, be a Wiener process. Assume that the function $g(x,t)$ is measurable over $\mathbf{R} \times [0,\ 1]$, and that it is possible to select a value of $\alpha > 0$ so that g is monotone in x in the same sense on the interval $[-\alpha,\ 0)$ simultaneously for all t, aa well as on $(0,\ \alpha]$.

If for some $A > 0$ and $B < \pi^2/8$ the inequality

$$|(\partial/\partial x) g(x,t)| \ \ge\ A \exp\left\{-B/x^2\right\}$$

holds almost everywhere on $[-\alpha,\ \alpha]$ for all $t \in [0,\ 1]$ and the rv $J = \int_0^1 g\left(w_t, t\right) dt$ is finite a.s., then the distribution of J has bounded density.

Assume in addition that g is continuously differentiable in x, and that there exist $a < 0$, $b > 0$, $h > 0$ such that $|(\partial/\partial x) g\,(x,t)| \ge d >\ 0$ for all pairs (x,t) satisfying the conditions $x \in [a-h,\ a+h] \bigcup [b-h,\ b+h]$, $t \in [0,\ 1]$, and

$$M \stackrel{def}{=} \sup_{(x,t)\in F} g(x,t) < \infty, \quad F \ = \ [a-h,\ b+h] \times [0,\ 1].$$

Under the above restrictions, the distribution of J is absolutely continuous with respect to the Lebesgue measure on the set $\{v :\ |v| > M\}$, and its density does not exceed $c(h) \exp\left\{-\min[a^2, b^2]\right\}/(2d)$ on this set. See [Li82b].

5. Let X be a real Hilbert space and K a subset of this space. Define $N(K,\varepsilon)$ as the minimal number of balls of radius ε covering K. The quantity

$$r(K) \ = \ \liminf_{\varepsilon\to 0} \frac{\ln\ln N(K,\varepsilon)}{\ln 1/\varepsilon}$$

is called the entropy exponent of K.

Consider a centered Gaussian process ξ_t, $t \in K$, such that $\mathbf{E}\xi_s\xi_t = (s,t)$. If

$$\int_0^a \sqrt{\ln N(K,\varepsilon)}\, d\varepsilon \ < \ \infty,$$

then the process ξ has a modification with trajectories continuous on the set K (this condition is satisfied if $r(K) < 2$). See [Du67].

The fact that condition $r(K) \leq 2$ is necessary for the existence of continuous modifications was established in [Su71].

Similar results can be obtained in terms of a different geometric characteristic of the set K,

$$V_n(K) = \sup\{\lambda^n(\mathcal{P}K) : \dim(\mathcal{P}X) = n\},$$

where the operators $\mathcal{P} : X \to X$ are orthogonal projections onto subspaces of finite number of dimensions $\dim(\mathcal{P}X)$, and λ^n is the Lebesgue measure in an n-dimensional subspace of X with the metric induced from X.

Let the set K be compact, convex, and symmetric. If there exist such numbers $\delta > 0$ and $C > 0$ that

$$\forall n \ V_n(K) \leq C/n^{1+\delta},$$

then the stochastic process ξ_t, $t \in K$, has a modification with continuous sample functions. See [MiP87].

Chapter 3

Inequalities for Seminorms: Sums of Independent Random Vectors

This chapter exposes some inequalities for the distributions of sums $S_{1,n}[\xi] = \xi_1 + \cdots + \xi_n$, where ξ_j are independent RV's with values in some linear measurable (usually, Banach) space. The inequalities below apply to probabilities $\mathbf{P}\{|S_{1,n}| \geq u\}$, with $|\cdot|$ a measurable seminorm on the range of ξ_j.

For real-valued sums, bounds on the "tails" can often be derived using very modest technical means, while limit theorems need a more complex apparatus. In infinite dimensions, the situation presents similar features, and some proofs carry over directly from the real line.

For instance, let ξ_j be RV's with values in a Hilbert space and $|\cdot|$ the norm in this space. One can easily get a version of the *Chebyshev inequality*: if $\mathbf{E}\xi_j = 0$ for all j, then

$$\mathbf{P}\{|S_{1,n}| \geq x\} \leq \frac{1}{x^2}\mathbf{E}|S_{1,n}|^2, \tag{3.0.1}$$

and

$$\mathbf{E}|S_{1,n}|^2 = \sum_{j,k=1}^{n} \mathbf{E}(\xi_j, \xi_k) = \sum_{j=1}^{n} \mathbf{E}|\xi_j|^2 .$$

Thus, in a sense $|S_{1,n}[\xi]| = \mathcal{O}(\sqrt{n})$ for i.i.d. summands.

However, there are features which in the infinite-dimensional case are drastically different from those typical of distributions on the real line or in a Euclidean space. In the finite-dimensional case, restrictions are usually formulated in terms of individual summands and do not presuppose knowledge of distribution of the sum as a whole. In infinite dimensions, "individual" conditions may not suffice: it is sometimes necessary to include restrictions on the distribution of the sum in order to obtain stronger bounds for its "tails". The following example serves to show the relative independence of restrictions on concentration of the sum and those concerning the distributions of individual summands.

Below, $\boldsymbol{c}_0$ is the Banach space of real sequences convergent to zero, with the norm $|x|_\infty = \sup_{i\in\mathbf{N}} |x^{(i)}|$. It is general knowledge that this space is separable and contains a basis.

Example 3.0.1 Consider an array of independent real rv's $\xi_j^{(i)}$ such that $\xi_j^{(i)} = \pm 1$ with probability $\frac{1}{2}$. The sequences $\xi_j = \left(\xi_j^{(i)}/\ln\ln 10i, i \in \mathbf{N}\right)$ can be considered as RV's in $\boldsymbol{c}_0$.

The i.i.d. RV's $\xi_j \in \boldsymbol{c}_0$ are symmetrical and a.s. bounded: $|\xi_j|_\infty = 1/\ln\ln 10$.

By the construction, all individual coordinates of the sum, the rv's $\eta_n^{(i)} = \xi_1^{(i)} + \cdots + \xi_n^{(i)}$, are independent. It is easy to see that $\mathbf{P}\left\{\eta_n^{(i)} = n\right\} = \left(\frac{1}{2}\right)^n$. For this reason,

$$\mathbf{P}\left\{\max_{i\le 2^n} \eta_n^{(i)} < n\right\} = \prod_{i=1}^{2^n}\left(1 - \mathbf{P}\left\{\eta_n^{(j)} = n\right\}\right) \le \exp\left\{-2^n\left(\tfrac{1}{2}\right)^n\right\} = e^{-1}.$$

Hence, for all n

$$\mathbf{P}\left\{|S_{1,n}|_\infty \ge \frac{n}{\ln\ln\left[10\cdot 2^n\right]}\right\} \ge \mathbf{P}\left\{\max_{i\le 2^n}\eta_n^{(i)} \ge n\right\} \ge 1 - e^{-1}.$$

Thus, the infinite-dimensional sum has norm of order $n/\ln\ln n$ with probability separated from zero, even though its coordinates have variances of order $\sqrt{n}$.

To avoid effects similar to one demonstrated in the example, one may either make additional assumptions about the properties of the norm (see, e.g., subsection A.2.5), or introduce collective conditions mentioned above.

3.1 Bounds on Tails of Maximal Sums

3.1.1 Relations Between Maximal Sums

In this section, $\langle X, \mathcal{X}\rangle$ is a real linear measurable space. The function $|x|$ with values in the set $[0, +\infty]$ is a measurable seminorm on X (cf. subsection 2.1.1): it is measurable and satisfies the conditions

$$|x+y| \le |x| + |y|, \quad |\alpha x| = |\alpha|\,|x|, \quad \alpha \in \mathbf{R},$$

on the set $X_f = \{x : |x| < \infty\}$ which is, moreover, a linear subspace of X.

The X-valued RV's ξ_j, $j \in J_+ \subset \mathbf{Z}$, are defined on some probability space $\langle \Omega, \mathcal{A}, \mathbf{P}\rangle$. The set J_+ is $\mathbf{N}$, $\mathbf{Z}$, or some finite segment of the integer lattice. This sequence of RV's generates the following σ-algebras of events:

$$\begin{aligned} \mathcal{F}_J &= \sigma\left(\xi_j, j \in J\right), \quad J \subset \mathbf{Z}, \\ \mathcal{F}_{\le k} &= \sigma\left(\xi_j, j \le k\right), \quad k \in \mathbf{Z}, \\ \mathcal{F}_{>k} &= \sigma\left(\xi_j, j > k\right), \quad k \in \mathbf{Z}. \end{aligned} \tag{3.1.1}$$

In the sequel the RV's ξ_j are independent (A.2.2), so the σ-algebras $\mathcal{F}_J$ and $\mathcal{F}_K$

are independent if $J \bigcap K = \emptyset$. In particular, the "past" $\mathcal{F}_{\leq k}$ and the "future" $\mathcal{F}_{>k}$ of the sequence are independent.

This section considers relations between the distributions of the rv's

$$S_k^l = |S_{k,l}|, \quad S_{k,l} = S_{k,l}[\xi] = \begin{cases} \sum_{j=k}^l \xi_j, & k \leq l, \\ 0, & k > l, \end{cases} \tag{3.1.2}$$

and those of the maximal sums with "loose" ends

$$S_k^{l*} = \max_{k \leq k' \leq l} S_k^{k'}, \quad S_{k*}^l = \max_{k \leq k' \leq l} S_{k'}^l, \quad S_{k*}^{l*} = \max_{k \leq k' \leq k'' \leq l} S_{k'}^{k''}. \tag{3.1.3}$$

To avoid trivial complications it is assumed below that $|\xi_j| < \infty$ a.s. for all j.

By the triangle inequality, the seminorms of sums satisfy relations

$$S_k^l = |S_{i,l} - S_{i,k-1}| \leq S_i^l + S_i^{k-1}. \tag{3.1.4}$$

Hence the rv's in (3.1.3) admit the evident inequalities

$$S_k^{l*} \leq S_{k*}^{l*}, \quad S_{k*}^{l*} \leq 2S_k^{l*}, \quad S_{k*}^l \leq S_{k*}^{l*}, \quad S_{k*}^{l*} \leq 2S_{k*}^l. \tag{3.1.5}$$

Under additional restrictions it is possible to add to these bounds some very general and much less trivial ones.

3.1.2 Maximal Sums and Maximal Summands

The upper bounds on probabilities $\mathbf{p}_n(x) = \mathbf{P}\{S_1^{n*} \geq x\}$ collected below are in terms of characteristics of individual summands and quantiles of $\mathbf{p}_n$. The simplest result of this kind is

Theorem 3.1.1 *Assume that $|\xi_j| \leq L$ a.s. for all j. If $\mathbf{P}\{S_1^{n*} \geq A\} \leq 1/e$, then for all $x \geq 0$*

$$\mathbf{P}\{S_1^{n*} \geq x\} \leq e \exp\{-x/(2a+L)\}.$$

Proof of Theorem 3.1.1. Suppose that $x > 2A + L$. Define the time when the sequence S_1^k first exceeds the level $x - 2A - L$:

$$\tau = \begin{cases} \min\{k \geq 1 : S_1^k \geq x - 2A - L\}, & \text{if } S_1^{n*} \geq x - 2A - L, \\ n+1, & \text{if } S_1^{n*} < x - 2A - L. \end{cases}$$

The rv S_1^τ cannot exceed this level by more than L because all the values of $|\xi_j|$ are bounded from above by this quantity. Consequently, the following inclusion is valid:

$$\{S_1^{n*} \geq x\} \subset \bigcup_{k=1}^n \left(\{\tau = k\} \bigcap \{S_{k+1}^{n*} \geq 2A\}\right).$$

The events $\{\tau = k\}$ and $\{S_{k+1}^{n*} \geq 2A\}$ are independent: the former belongs to the "past" $\mathcal{F}_{\leq k}$ before time k of the sequence (ξ_j), and the latter lies in the "future" $\mathcal{F}_{>k}$.

Combined with (3.1.5), the inclusion above yields the estimate

$$\begin{aligned} \mathbf{P}\{S_1^{n*} \geq x\} &\leq \textstyle\sum_{k=1}^n \mathbf{P}\{\tau = k\} \mathbf{P}\{S_{1*}^{n*} \geq 2A\} \\ &\leq \mathbf{P}\{S_1^{n*} \geq x - 2A - L\} \mathbf{P}\{S_1^{n*} \geq A\} \leq \tfrac{1}{e} \mathbf{P}\{S_1^{n*} \geq x - 2A - L\}. \end{aligned}$$

Applying the above bound several times, one arrives at the inequality

$$\mathbf{P}\{S_1^{n*} \geq z\} \leq \exp\{-[x/(2A+L)]\}, \quad x > 0,$$

where $[\cdot]$ is the integer part of a real number. The assertion of the theorem follows easily from the last estimate. ○

The probability that the maximal sums S_k^{l*} of (3.1.3) attain large values can be evaluated in terms of their lower quantiles and similar probabilities for the "maximal summand"

$$\bigvee_k^l \xi = \max_{k \leq j \leq l} |\xi_j|, \quad k < l. \tag{3.1.6}$$

The following result of this kind has numerous applications.

Lemma 3.1.1 *If the RV's ξ_j are independent, then the distribution of the maximal sum* (3.1.3) *and that of the maximal summand* (3.1.6) *obey the following relation: for all $x, y, z > 0$*

$$\mathbf{P}\{S_1^{n*}[\xi] \geq x + y + z\} \leq \mathbf{P}\{\bigvee_1^n \xi \geq z\} + \mathbf{P}\{S_1^{n*} \geq x\}\mathbf{P}\{S_1^{n*} \geq \tfrac{1}{2}y\}.$$

Proof of Lemma 3.1.1. The sequence S_1^k first attains the level x at the time

$$\tau = \min\{k \geq 1 : S_1^k \geq x\}.$$

By the triangle inequality, $0 \leq S_1^\tau - x \leq \bigvee_1^n \xi$ — the "overshoot" does not exceed the maximal summand. This leads to the inclusion

$$\{\tau = k,\ S_1^n \geq x + y + z,\ \bigvee_1^n \xi < z\} \subseteq \{\tau = k,\ S_1^k \leq x + z\} \bigcap \{S_{k+1}^{n*} \geq y\}.$$

The events in the right-hand side of the inclusion are independent, since they belong, respectively, to $\mathcal{F}_{\leq k}$ and $\mathcal{F}_{>k}$. It follows that

$$\mathbf{P}\{\tau = k,\ S_1^{n*} \geq x + y + z\} \leq \mathbf{P}\{\tau = k\}\mathbf{P}\{S_{1*}^{n*} \geq y\} + \mathbf{P}\{\tau = k,\ \bigvee_1^n \xi \geq z\}.$$

The inequality is trivially true also for $k = n$.

The assertion of the lemma is obtained by summing in k, since by (3.1.4) $S_{1*}^{n*} \leq 2S_1^{n*}$ and $\{\tau \leq n\} = \{S_1^{n*} \geq x\}$. ○

Remark 3.1.1 If the RV's ξ_j are i.i.d., then

$$\mathbf{P}\{S_{k+1}^{n*} \geq y\} \leq \mathbf{P}\{S_1^{n*} \geq y\}.$$

For this reason it is possible to improve the estimate of Lemma 3.1.1 in this case:

$$\mathbf{P}\{S_1^{n*} \geq x + y + z\} \leq \mathbf{P}\{\bigvee_1^n \xi \geq z\} + \mathbf{P}\{S_1^{n*} \geq x\}\mathbf{P}\{S_1^{n*} \geq y\}.$$

The proof repeats that of Lemma 3.1.1, but $\mathbf{P}\{S_{1*}^{n*} \geq y\}$ is replaced by the probability $\mathbf{P}\{S_1^{n*} \geq y\}$ in its main inequality.

3.1.3 Case of Polynomial Tail of Maximal Summand

Theorem 3.1.2 *Assume that for some $t > 0$ and all $z > 0$*

$$\mathbf{P}\{\bigvee_1^n \xi \geq z\} \leq B/z^t. \tag{3.1.7}$$

Fix an arbitrary number $\delta > 0$, and choose A and p_0 satisfying the conditions

$$\mathbf{P}\{S_1^{n*} \geq A\} \leq p_0 < \tfrac{1}{2}, \quad \mathbf{K} \equiv B/(A\delta)^t \leq \tfrac{1}{4}(3+\delta)^{-t}.$$

There exist positive constants $c_i = c_i(\delta)$ such that the following inequality holds for all $x > (3+\delta)$:

$$\mathbf{P}\{S_1^{n*} \geq xA\} \leq \max\left[c_1\mathbf{K}/x^t, \tfrac{1}{2}\exp\{-c_2x^\alpha\}\right], \quad \alpha = (\ln 2)/(\ln(3+\delta)). \quad (3.1.8)$$

Note that $\alpha \approx \ln 2/\ln 3 \approx 0.631$ for small $\delta > 0$. The constants in Theorem 3.1.2 admit the bounds

$$c_1 \leq 2(3+\delta)^{2t}, \quad c_2 \geq (3+\delta)^{-\alpha}\ln 1/(2p_0).$$

Remark 3.1.2 If the summands ξ_j are independent and also identically distributed, the exponent α in Theorem 3.1.2 can be replaced by the larger number $\widehat{\alpha} = \ln 2/\ln(2+\delta)$. In the resulting inequality the constants c_i are also changed, as well as the restrictions on p_0 and r.

Proof of Theorem 3.1.2. Consider the sequence $x_k = (3+\delta)^k$, $k = 0, 1, \ldots$ generated by the recurrence equation $x_{k+1} = (3+\delta)x_k$. Put $p_k = \mathbf{P}\{S_1^{n*} \geq x_kA\}$.

By Lemma 3.1.1 with $x = \frac{1}{2}y = Ax_k$ and $z = \delta Ax_k$, these probabilities obey the inequalities

$$p_{k+1} \leq (p_k)^2 + \mathbf{P}\{\mathrm{V}_1^n\xi \geq x_kA\delta\} \leq (p_k)^2 + \mathbf{K}\gamma^k, \quad k \geq 0,$$

where $\gamma = (3+\delta)^{-t} < 1$.

To analyze this recurrence relation, consider the sequence $U_k = (1/p_k)\sqrt{\mathbf{K}\gamma^k}$ which satisfies, instead of the original inequality, the equivalent one

$$U_{k+1} \geq \gamma^{-k/2}f(U_k) \geq f(U_k), \quad f(u) = \sqrt{\frac{\gamma}{\mathbf{K}}}\frac{u^2}{1+u^2} \geq \frac{2u^2}{1+u^2},$$

as $\sqrt{\gamma/\mathbf{K}} \geq 2$ by the choice of $\mathbf{K}$.

The function f increases on the positive half-line, and $f(u) \geq 1$ for $u \geq 1$. Hence the set $K = \{k \in \mathbf{N} : U_k < 1\}$ is either empty or a segment $\{0, \ldots, k_0\}$.

If $k \in K$, then $p_{k'+1} \leq 2p_{k'}^2$ for all $k' \leq k$, and repeated application of this inequality leads to the relation

$$p_{k+1} \leq \tfrac{1}{2}(2p_0)^{2^{k+1}}. \quad (3.1.9)$$

If $k \notin K$, then $p_k \leq \sqrt{\mathbf{K}\gamma^k}$, so

$$p_{k+1} \leq 2\mathbf{K}\gamma^k. \quad (3.1.10)$$

One of the above bounds, (3.1.9) or (3.1.10), holds for an arbitrary value of k. It remains to note that for $x \in [x_{k+1}, x_{k+2}]$

$$k+1 \leq \frac{\ln x}{\ln(3+\delta)} \leq k+2,$$

so also

$$x_{k+1} \geq \frac{x}{3+\delta}, \quad 2^{k+1} \geq \frac{x^\alpha}{(3+\delta)^\alpha}, \quad \gamma^k \leq \frac{(3+\delta)^{2t}}{x^t}.$$

Combined with (3.1.9)–(3.1.10), these estimates prove the theorem. ○

To verify the assertion of Remark 3.1.2, one can use a perfectly similar reasoning. It starts from the more precise estimate of Remark 3.1.1, which can be written as

$$\widehat{p}_{k+1} \leq (\widehat{p}_k)^2 + \widehat{\mathbf{K}}\widehat{\gamma}^k, \quad \widehat{p}_k = \mathbf{P}\{S_1^{n*} \geq \widehat{x}_k A\},$$

where $\widehat{x}_k = (2+\delta)^k$, $\widehat{\gamma} = 1/(2+\delta)^t$.

3.1.4 Bounds on Moments of Maximal Sum

Theorem 3.1.3 *Let $\varphi : \mathbf{R}_+ \to \mathbf{R}_+$ be a nonnegative nondecreasing function which satisfies the condition*

$$\forall v > 0 \ \ \varphi(4v) \leq F\varphi(v) \tag{3.1.11}$$

with some constant $F \geq 1$. If $\mathbf{P}\{S_1^{n*} \geq A/4\} \leq 1/(2F)$, then

$$\mathbf{E}\,\varphi(S_1^{n*}) \leq 2F\left[\varphi(A) + \mathbf{E}\varphi\left(\mathsf{V}_1^n\xi\right)\right]. \tag{3.1.12}$$

Remark 3.1.3 Condition (3.1.11) is satisfied, in particular, by all monomials $\varphi(v) = v^t$, $t > 0$. Hence condition (3.1.12) with an appropriate value of F implies the bound

$$\mathbf{E}\,[S_1^{n*}]^t \leq c\left(A^t + \mathbf{E}\,[\mathsf{V}_1^n\xi]^t\right) \tag{3.1.13}$$

with a constant c dependent on the choice of t only.

This estimate can be used to derive a number of inequalities often used in the study of sums of independent RV's. For instance, suppose that all the moments $\mathbf{E}\,|\xi_j|^t$, $j = \overline{1,n}$, are finite for some $t \geq 1$. Evidently,

$$\mathbf{E}\mathsf{V}_{j=1}^n |\xi_j|^t \leq \sum_{j=1}^n \mathbf{E}\,|\xi_j|^t, \quad \mathbf{P}\{S_1^n \geq C\mathbf{E}S_1^n\} \leq 1/C,$$

so the bound of the theorem implies the inequality

$$\mathbf{E}\,|S_{1,n}[\xi]|^t \leq c\left[\sum_{j=1}^n \mathbf{E}\,|\xi_j|^t + (\mathbf{E}S_1^n)^t\right] \tag{3.1.14}$$

with an appropriately chosen constant $c = c(t)$. This bound can be written in terms of moments of individual summands for RV's in a Hilbert space satisfying the condition $\mathbf{E}\,\xi_j = 0$. In this special case $\mathbf{E}\,(S_1^n)^2 = \sum_{j=1}^n \mathbf{E}\,|\xi_j|^2$ (cf (3.0.1)), so it follows from Hölder's inequality that $\mathbf{E}S_1^n \leq \sqrt{\mathbf{E}\,(S_1^n)^2}$ and

$$\mathbf{E}\,(S_1^n)^t \leq c_1(t)\sum_{j=1}^n \mathbf{E}\,|\xi_j|^t + c_2(t)\left[\sum_{j=1}^n \mathbf{E}\,|\xi_j|^2\right]^{t/2}, \quad t \geq 2. \tag{3.1.15}$$

Proof of Theorem 3.1.3. The bounded function $\widetilde{\varphi}_N(u) = \min\{\varphi(u), N\}$ also satisfies condition (3.1.12). The following equality for a nonnegative rv ζ is verified applying integration by parts:

$$\mathbf{E}\,\widetilde{\varphi}_N(\zeta) = \widetilde{\varphi}_N(0) + \int_0^\infty \mathbf{P}\{\zeta \geq u\}\, d\widetilde{\varphi}_N(u).$$

By Lemma 3.1.1, there is inequality

$$\mathbf{P}\{S_1^{n*} \geq u\} \leq \mathbf{P}\{\mathsf{V}_1^n\xi \geq \tfrac{1}{4}u\} + \mathbf{P}\{S_1^{n*} \geq \tfrac{1}{4}u\}^2.$$

Hence

$$\mathbf{E}\widetilde{\varphi}_N(S_1^{n*}) \leq I_1 + I_2,\ I_1 = \int_0^\infty \mathbf{P}\{4S_1^{n*} \geq u\}^2\, d\widetilde{\varphi}_N(u),\ I_2 = \mathbf{E}\,\widetilde{\varphi}_N(4\mathsf{V}_1^n\xi).$$

The integrand in I_1 is trivially evaluated by 1 for $u \leq 4A$, and for $u \geq 4A$

$$\mathbf{P}\{4S_1^{n*} \geq u\}^2 \leq \mathbf{P}\{S_1^{n*} \geq A\}\,\mathbf{P}\{4S_1^{n*} \geq u\} \leq (2F)^{-1}\mathbf{P}\{4S_1^{n*} \geq u\}.$$

Consequently, the first integral admits the bound

$$I_1 \leq \widetilde{\varphi}_N(4A) - \widetilde{\varphi}_N(0) + \frac{1}{2F}\mathbf{E}\widetilde{\varphi}_N(4S_1^{n*}) \leq F\widetilde{\varphi}_N(A) + \frac{1}{2}\mathbf{E}\,\widetilde{\varphi}_N(S_1^{n*}).$$

It follows that

$$\frac{1}{2}\mathbf{E}\widetilde{\varphi}_N(S_1^{n*}) \leq F\widetilde{\varphi}_N(A) + \mathbf{E}\widetilde{\varphi}_N(4\mathsf{V}_1^n\xi) \leq F[\widetilde{\varphi}_N(A) + \mathbf{E}\widetilde{\varphi}_N(\mathsf{V}_1^n\xi)].$$

The assertion of the theorem is now established using a passage to the limit as $N \to \infty$, which is justified by the monotone convergence theorem. ○

3.1.5 Exponential Moments of Maximal Sums

Theorem 3.1.4 *Assume that $S_1^{n*} < \infty$ a.s. and for some $h > 0$*

$$A \equiv \mathbf{E}\exp\{h\mathsf{V}_1^n\xi\} < \infty.$$

For each $B \geq 0$ and $\beta < \frac{1}{2}$, there exists a constant $c > 0$ such that

$$\mathbf{E}\exp\left\{B\,[S_1^{n*}]^\beta\right\} \leq c < \infty.$$

The value of c may depend on A, h, B, and β, but not on n.

Proof of Theorem 3.1.4. Assume that $A > 1$, because otherwise the result is trivial.

By the Chebyshev inequality,

$$\mathbf{P}\{\mathsf{V}_1^n\xi \geq u\} \leq A\exp\{-hu\},\quad u \geq 0.$$

It follows from Lemma 3.1.1 and the above estimate that (cf. proof of Theorem 3.1.2)

$$\mathbf{P}\{S_1^{n*} \geq 4u\} \leq Ae^{-hu} + [\mathbf{P}\{S_1^{n*} \geq u\}]^2.$$

For large values of u this bound leads to the estimate

$$\psi(4u) \leq \psi(u)^2, \ \psi(u) = \mathbf{P}\{S_1^{n*} \geq u\} + Ae^{-hu/2}$$

where the function $\psi(u)$ is a decreasing one. (Note that $A^2e^{-hu} \geq Ae^{-hu} + Ae^{-2hu}$ for $u \geq u_0(A, h)$ because $A > 1$.)

Repeated application of this latter bound yields the inequality

$$\psi\left(4^k u_0\right) \leq \exp\left\{2^k \ln \psi(u_0)\right\}$$

for large values of $4^k u_0$. Choosing $k = [\log_4(u/u_0)]$ (with $[\cdot]$, as usual, the integer part of a number), one arrives at the inequality

$$\psi(u) \leq \exp\left\{ -\tfrac{1}{2}\gamma\sqrt{u/u_0} \right\}, \ \gamma = \ln 1/\psi(u_0).$$

The quantity $\ln 1/\psi(u_0) = \ln 1/\left(\mathbf{P}\{S_1^{n*} \geq u_0\} + Ae^{-hu_0/2}\right)$ is positive if u_0 is large enough.

For $\beta < \frac{1}{2}$, the assertion of theorem follows from the last bound and the equality

$$\mathbf{E}\varphi\left(S_1^{n*}\right) = \varphi(0) + \int_0^\infty \mathbf{P}\{S_1^{n*} \geq u\}\, d\varphi(u), \ \ \varphi(u) = \exp\left\{Bu^\beta\right\}. \ \bigcirc$$

3.2 Bounds Based on Symmetry

3.2.1 The Lévy Inequality

A RV is *symmetrical* (or has symmetrical distribution) if the distributions of ξ and $-\xi$ coincide.

Let $\xi_j, j = \overline{1,n}$, be independent symmetrical RV's. The joint distribution of $\varepsilon_j\xi_j$, $j = \overline{1,n}$, is that of the original sequence if the coefficients ε_j are either non-random with values ± 1, or if they are random and independent from each other and all ξ_j with $\mathbf{P}\{\varepsilon_j = \pm 1\} = \frac{1}{2}$.

The following proposition is called the Lévy inequality. Notation below is that of subsection 3.1.1 (see, in particular, (3.1.2)–(3.1.3) and (3.1.6)).

Lemma 3.2.1 *If the RV's ξ_j, $j = \overline{1,n}$, are independent and symmetrical, then for all $x \geq 0$*

$$\mathbf{P}\{S_1^{n*} \geq x\} \leq 2\mathbf{P}\{S_1^n \geq x\},$$

$$\mathbf{P}\left\{\mathsf{V}_1^{n*}\xi \geq x\right\} \leq 2\mathbf{P}\{S_1^n \geq x\}.$$

Proof of Lemma 3.2.1 is almost a verbatim repetition of the corresponding argument for real-valued rv's.

Define the time when the sequence S_1^k first attains the level x by the relation

$$\tau = \begin{cases} \min\left\{k : S_1^k \geq x\right\}, & S_1^{n*} \geq x, \\ n+1, & S_1^{n*} < x, \end{cases}$$

so that $\{\tau \leq n\} = \{S_1^{n*} \geq x\}$.

By the triangle inequality, for $k < n$ there are inclusions

$$\{\tau = k\} = \{\tau = k,\ S_1^k \geq x\} \subset A_- \bigcup A_+,\quad A_\pm = \{\tau = k,\ |S_{1,k} \pm S_{k+1,n}| \geq x\}.$$

The events $A_\pm$ have equal probabilities: it was mentioned earlier that because of symmetry the joint distribution of summands does not change if some of these change sign.

Consequently, for $k < n$

$$\mathbf{P}\{\tau = k\} \leq \mathbf{P}\{A_-\} + \mathbf{P}\{A_+\} = 2\mathbf{P}\{A_+\} \leq 2\mathbf{P}\{\tau = k,\ S_1^n \geq x\}.$$

The inequality remains valid also if $k = n$. The proof of the first inequality of the lemma is completed by summing the above inequalities, since evidently

$$\{\tau \leq n\} \bigcap \{S_1^n \geq x\} = \{S_1^n \geq x\}.$$

To prove the second inequality, put $\widetilde{\tau} = \min\{k : |X_k| \geq x\}$ and note that one can use the above argument because by the symmetry of the RV's considered

$$\mathbf{P}\left\{\widetilde{\tau} = k,\ \left|\sum_{i \neq k} \xi_i \pm \xi_k\right| \geq x\right\} = \mathbf{P}\{\widetilde{\tau} = k,\ S_1^n \geq x\}. \bigcirc$$

3.2.2 Symmetrization

It is often convenient to consider, in addition to a family of random variables, a number of its independent copies, viz., independent families of random variables having same joint distribution (see subsection A.2.1).

A *symmetrization* of the sequence of RV's $\left(\xi_j, j = \overline{1,n}\right)$ is the difference of two independent copies of this sequence:

$$\widetilde{\xi}_j = \xi_j^{(1)} - \xi_j^{(2)}, \tag{3.2.1}$$

where

$$\mathcal{L}\left(\xi_j^{(1)}, j = \overline{1,n}\right) = \mathcal{L}\left(\xi_j^{(2)}, j = \overline{1,n}\right) = \mathcal{L}\left(\xi_j, j = \overline{1,n}\right).$$

If the original sequences consists of independent RV's, all the RV's $\xi_j^{(1)}$ and $\xi_k^{(2)}$ used to construct symmetrization are independent. RV's (3.2.1) are also independent and symmetrical. Indeed, the pairs of RV's $\left\langle \xi_j^{(1)}, \xi_j^{(2)} \right\rangle$ and $\left\langle \xi_j^{(2)}, \xi_j^{(1)} \right\rangle$ are identically distributed in X^2, so their images under the mapping $\langle x, y\rangle \mapsto x - y$, the RV's $\widetilde{\xi}_j$ and $-\widetilde{\xi}_j$, are also identically distributed.

Lemma 3.2.2 *Consider a RV* Ξ *such that* $|\Xi| < \infty$ *a.s. Its symmetrization* $\widetilde{\Xi}$ *satisfies the following inequality: for each* $x, y \geq 0$

$$\mathbf{P}\{|\Xi| \geq x + y\} \leq \frac{\mathbf{P}\left\{|\widetilde{\Xi}| \geq x\right\}}{\mathbf{P}\{|\Xi| \leq y\}},\quad \mathbf{P}\left\{|\widetilde{\Xi}| \geq x\right\} \leq 2\mathbf{P}\left\{|\Xi| \geq \tfrac{1}{2}x\right\}.$$

Proof of Lemma 3.2.2. Let $\Xi^{(1)}$ and $\Xi^{(2)}$ be the independent copies of Ξ used for symmetrization. Then

$$\mathbf{P}\{|\Xi| \geq x + y\}\,\mathbf{P}\{|\Xi| \leq y\} = \mathbf{P}\left\{|\Xi^{(1)}| \geq x + y, |\Xi^{(2)}| \leq y\right\} \leq \mathbf{P}\left\{|\tilde{\Xi}| \geq x\right\}.$$

The second inequality of the lemma follows from the obvious inclusion

$$\left\{|\Xi^{(1)} - \Xi^{(2)}| \geq x\right\} \subseteq \left\{|\Xi^{(1)}| \geq \tfrac{1}{2}x\right\} \bigcap \left\{|\Xi^{(2)}| \geq \tfrac{1}{2}x\right\}. \ \bigcirc$$

Remark 3.2.1 Lemma 3.2.2 shows that the probabilities $\mathbf{P}\{S[\xi]_1^n \geq x\}$ and $\mathbf{P}\left\{S[\tilde{\xi}]_1^n \geq x\right\}$ for large values of x behave in the same manner if the distribution of $S_{1,n}[\xi]$ remains concentrated near the origin for large n. When the probabilities $\mathbf{P}\{S_1^n \leq A\}$ are bounded from below, the bound of Lemma 3.2.2 is nontrivial:

$$\mathbf{P}\{S_1^n[\xi] \geq x\} \leq \mathbf{P}\left\{S_1^n[\tilde{\xi}] \geq x - A\right\} / \mathbf{P}\{S_1^n[\xi] \leq A\}. \tag{3.2.2}$$

3.2.3 A Chebyshev-Type Bound

The sums $S_1^n = S_1^n[\xi]$ considered here are always composed of independent RV's. Using the Lévy inequality and symmetrization, one can derive some quite general results for sums of RV's with finite moments from theorems concerning tails of the maximal sums. In a situation where the distribution of $S_1^n = |S_{1,n}|$ is concentrated in a bounded domain which does not grow with the number of summands, a non-trivial bound on the tail of this rv is given by

Theorem 3.2.1 *Assume that ξ_j are independent RV's and for some $t > 0$*

$$B = \max\left[2^t, 2\right] \sum_{j=1}^{n} \mathbf{E}\,|\xi_j|^t < \infty.$$

If $\delta > 0$ is an arbitrary number, and A, p_0 are chosen so that they satisfy the restrictions of Theorem 3.1.2 and the inequality $\mathbf{P}\left\{S_1^n \geq \frac{1}{2}A\right\} \leq \frac{1}{4}p_0$, then the probability $\mathbf{P}\left\{S_1^{n*}[\tilde{\xi}] \geq xA\right\}$ for any symmetrized sequence $\tilde{\xi}_j$ admits the bound (3.1.8) of Theorem 3.1.2.

Moreover, the following inequality holds for $u > \frac{1}{2}$ and the original set of RV's:

$$\mathbf{P}\{S_1^n[\xi] \geq uA\} \leq \left(1 - \tfrac{1}{4}p_0\right)^{-1} \mathbf{P}\left\{S_1^{n*}[\tilde{\xi}] \geq (u - \tfrac{1}{2})A\right\}.$$

Proof of Theorem 3.2.1. Let $\left(\xi_j^{(1)}\right)$ and $\left(\xi_j^{(2)}\right)$ be the independent copies of the original sequence of RV's used to construct symmetrizations. Then

$$\mathbf{E}|\tilde{\xi}_j|^t \leq \mathbf{E}\left[|\xi_j^{(1)}| + |\xi_j^{(2)}|\right]^t \leq \max\left[2^t, 2\right] \mathbf{E}\,|\xi_j|^t,$$

so by the Chebyshev inequality

$$\mathbf{P}\left\{\mathsf{V}_1^n\tilde{\xi} \geq z\right\} \leq \sum_{j=1}^{n} \mathbf{P}\left\{|\tilde{\xi}_j| \geq z\right\} \leq B/z^t, \quad z > 0.$$

An application of Lemma 3.2.1 and Lemma 3.2.2 yields the estimate

$$\mathbf{P}\left\{S_1^{n*}[\xi] \geq A\right\} \leq 2\mathbf{P}\left\{S_1^n[\widetilde{\xi}] \geq A\right\} \leq 4\mathbf{P}\left\{S_1^n[\xi] \geq A/2\right\} \leq p_0.$$

The first assertion of the theorem now follows from Theorem 3.1.2. The second one is a consequence of (3.2.2). ○

A slightly more general result is stated using a function φ satisfying the conditions of Theorem 3.1.3.

Theorem 3.2.2 *If* $\mathbf{P}\left\{S_1^n[\xi] \geq \frac{1}{8}A\right\} \leq 1/(8F)$, *then*

$$\mathbf{E}\,\varphi\left(S_1^n[\xi]\right) \leq \left(1 + \frac{2F^3}{1 - 1/(8F)}\right)\varphi\left(\tfrac{1}{4}A\right) + \frac{4F^3}{1 - 1/(8F)}\sum_{j=1}^n \mathbf{E}\varphi\left(|\xi_j|\right).$$

Proof of Theorem 3.2.2. It follows from Lemma 3.2.1 and Lemma 3.2.2 that

$$\mathbf{P}\left\{S_1^{n*}[\widetilde{\xi}] \geq \tfrac{1}{4}A\right\} \leq 4\mathbf{P}\left\{S_1^n \geq \tfrac{1}{8}A\right\} \leq 1/(2F),$$

where $\widetilde{\xi}_j$ are the symmetrized RV's. Besides, the maximal symmetrized summand can be evaluated in terms of the two sequences used for symmetrization:

$$\forall j\ |\widetilde{\xi}_j| \leq 2\max\left[|\xi_j^{(1)}|, |\xi_j^{(2)}|\right] \implies \mathsf{V}_1^n\widetilde{\xi} \leq 2\max\left[\mathsf{V}_1^n\xi^{(1)}, \mathsf{V}_1^n\xi^{(2)}\right].$$

Hence

$$\begin{aligned}\mathbf{E}\varphi\left(\mathsf{V}_1^n\widetilde{\xi}\right) &\leq \mathbf{E}\left\{\varphi\left(\mathsf{V}_1^n\xi^{(1)}\right) + \varphi\left(\mathsf{V}_1^n\xi^{(2)}\right)\right\} \\ &\leq 2\mathbf{E}\varphi\left(\mathsf{V}_1^n\xi\right) \leq 2F\sum_{j=1}^n \mathbf{E}\varphi\left(|\xi_j|\right).\end{aligned}$$

Combined with Theorem 3.1.3, this inequality yields the estimate

$$\mathbf{E}\varphi\left(S_1^{n*}[\widetilde{\xi}]\right) \leq 2F\left(\varphi(A) + 2F\textstyle\sum_{j=1}^n \mathbf{E}\varphi\left(|\xi_j|\right)\right).$$

At the same time, by (3.2.2)

$$\forall u > 0\ \mathbf{P}\left\{S_1^n - \tfrac{1}{8}A \geq u\right\} \leq \frac{\mathbf{P}\left\{S_1^{n*}[\widetilde{\xi}] \geq u\right\}}{1 - 1/(8F)}.$$

Consequently,

$$\mathbf{E}\varphi\left(\max\left\{S_1^n[\xi] - \tfrac{1}{8}A, 0\right\}\right) \leq \frac{\mathbf{E}\,\varphi\left(S_1^{n*}[\widetilde{\xi}]\right)}{1 - 1/(8F)}.$$

To derive from this last inequality the estimate of the theorem, note that

$$\varphi\left(S_1^n\right) \leq \varphi\left(2\max\left[0, \tfrac{1}{8}A, S_1^n - \tfrac{1}{8}A\right]\right) \leq \varphi\left(\tfrac{1}{4}A\right) + F\varphi\left(\max\left[0, S_1^n - \tfrac{1}{8}A\right]\right). \text{ ○}$$

3.3 Exponential Bounds for Sums

3.3.1 S. N. Bernstein's Bound

Consider the sum $S_{1,n}[X]$ constructed according to (3.1.2) from the independent real-valued rv's $X_1, \ldots, X_n$. By the classical central limit theorem, the distribution of this sum is similar to the normal law with same expectation and variance. The following bound on probabilities of large deviations was established by S. N. Bernstein about 1912.

Proposition 3.3.1 *If the rv's X_j satisfy the conditions*

$$\left|\mathbf{E}X_j^l\right| \leq \tfrac{1}{2} l!\, b_j^2 H^{l-2}, \quad l = 2, 3, \ldots, \quad \mathbf{E}X_j = 0,$$

then for each $x \geq 0$ and $B_n = \sqrt{b_1^2 + \cdots + b_n^2}$

$$\mathbf{P}\left\{S_{1,n}[X] \geq xB_n\right\} \leq \exp\left\{-\tfrac{1}{2}x^2\left[1 + xH/B_n\right]^{-1}\right\}.$$

Proposition 3.3.1 employs a form of the Cramér condition which later became quite usual in theorems on large deviations (see later sections). Its bound is fairly exact: if the sum $S_{1,n}$ were normal with variance B_n^2, the principal term of the asymptotic expression (1.1.2) for the probability considered would be roughly $\mathcal{O}\left(\exp\left\{-\frac{1}{2}x^2\right\}\right)$. Thus, in the range $x = o(n^{1/6})$ the upper bound of Proposition 3.3.1 differs from the known asymptotic expression merely by a factor $\mathcal{O}(x)$. Despite its precision, this bound can be derived by very elementary means.

Proof of Proposition 3.3.1. If $0 \leq h < 1/H$, it follows from the moment restrictions above and the Taylor expansion of the exponential function that

$$\mathbf{E}e^{hX_j} = 1 + \sum_{l=2}^{\infty} \frac{1}{l!} h^l \mathbf{E}X_j^l \leq 1 + \frac{1}{2} b_j^2 h^2 \left[1 - hH\right]^{-1} \leq \exp\left\{\tfrac{1}{2}h^2 b_j^2 \left[1 - hH\right]^{-1}\right\}.$$

By the independence of summands, this estimate leads to an inequality for the exponential moment of the sum:

$$\begin{aligned} \mathbf{E}\exp\left\{hS_{1,n}\right\} &= \mathbf{E}e^{hS_{1,n-1}}\mathbf{E}e^{hX_n} \\ &\leq \exp\left\{\tfrac{1}{2}h^2 b_j^2\left[1 - hH\right]^{-1}\right\}\mathbf{E}\exp\left\{hS_{1,n-1}\right\} \\ &\leq \ldots \leq \exp\left\{\tfrac{1}{2}h^2 B_n^2\left[1 - hH\right]^{-1}\right\}. \end{aligned}$$

The bound on the exponential moment can be used in combination with a form of Chebyshev's inequality, $\mathbf{P}\left\{S_{1,n} \geq xB_n\right\} \leq e^{-xh}\mathbf{E}e^{hS_{1,n}}$, $x, h > 0$. Put $C = hB_n$, then

$$\mathbf{P}\left\{S_{1,n} \geq xB_n\right\} \leq \inf_{0 \leq C < B_n/H} \exp\left\{-xC + \tfrac{1}{2}C^2\left[1 - CH/B_n\right]^{-1}\right\}.$$

In the case of small values of H/B_n, a fairly exact approximation of the minimum corresponds to $C = x(1+xH/B_n)^{-1}$ which solves the equation

$$\tfrac{1}{2}xC = \tfrac{1}{2}C^2[1-CH/B_n]^{-1}.$$

In some ways this choice of the parameter is more convenient than the exact extremal value. It yields the estimate of the proposition. ○

The inequality can be made much stronger with practically no changes in the argument. The sequence $\{S_{1,n}\}$ is a martingale, so $\{\exp\{hS_{1,N}\}\}$ is a nonnegative submartingale, and by Kolmogorov's inequality one can replace the sum in the left-hand side of the inequality in Proposition 3.3.1 by the maximal one:

$$\begin{aligned} \mathbf{P}\{\max_{k\le n} S_{1,n} \ge xB_n\} &\le \exp\{-hxB_n\}\,\mathbf{E}\exp\{hS_{1,n}\} \\ &\le \exp\{\ -\tfrac{1}{2}x^2[1+xH/B_n]^{-1}\ \}. \end{aligned}$$

The bound remains valid if the sequence $S_{1,k}[X]$ is a martingale (even though the summands be dependent), and the conditional exponential moments of its martingale-differences given the "past" admit nonrandom estimates of the kind employed above.

Analogs of S.N.Bernstein's bound on large deviation probabilities can be derived for sums of independent infinite-dimensional RV's. Such estimates are exposed below.

3.3.2 Exponential Bounds for Arbitrary Seminorms

In this subsection, ξ_j are RV's with values in a real linear measurable space X and $|\cdot|$ is a measurable seminorm on their range. Notation is that of the preceding sections, so S_k^l are seminorms (3.1.2) of sums, etc. The RV's are always supposed to have finite moment $\mathbf{E}|\xi_j| < \infty$, hence $|\xi_j| < \infty$ a.s.

Theorem 3.3.1 *Assume that ξ_j are independent X-valued RV's and a_j some nonrandom elements of X such that $|a_j| < \infty$.*

a) *The variance of S_1^n admits the bound*

$$\mathbf{D}S_1^n = \mathbf{E}(S_1^n - \mathbf{E}S_1^n)^2 \le \sum_{j=1}^{n}\mathbf{E}|\xi_j - a_j|^2.$$

b) *For all $h \ge 0$,*

$$\mathbf{E}\exp\{hS_1^n\} \le \exp\{h\mathbf{E}S_1^n\}\exp\left\{\sum_{j=1}^{n}\mathbf{E}\left(e^{h|\xi_j-a_j|} - 1 - h|\xi_j - a_j|\right)\right\}.$$

Note that the parameters a_j in the theorem can be chosen at one's convenience. The following proposition is a corollary of Theorem 3.3.1.

Theorem 3.3.2 *Assume that the X-valued RV's ξ_j are independent and for some nonrandom vectors a_j with finite seminorm $|a_j|$ the following restrictions on the moments hold for all natural numbers $l \ge 2$ with some constants $B > 0$*

and $H > 0$:

$$\sum_{j=1}^{n} \mathbf{E} |\xi_j - a_j|^l \le \tfrac{1}{2} l! B^2 H^{l-2}.$$

Then for all $x > 0$

$$\mathbf{P}\{S_1^n - \mathbf{E}S_1^n \ge xB\} \le \exp\left\{-\tfrac{1}{2}x^2 [1 + xH/B]^{-1}\right\}.$$

Remark 3.3.1 The "collective" term $\mathbf{E}S_1^n$ cannot be excluded from the bound without special assumptions about the geometry of the norm — see the example in the introduction to this chapter. This remark applies also to later theorems of this section.

In the calculation to follow, the conditional expectations given the "past" of the sequence (ξ_j) are abbreviated to $\mathbf{E}_k \cdot \equiv \mathbf{E}\left\{\cdot \,|\mathcal{F}_{[0,k]}\right\}$ (see (3.1.1)); the "unconditional" mean can also be written as $\mathbf{E}_0 \cdot \equiv \mathbf{E}\cdot$.

Proof of Theorem 3.3.1. Consider the real-valued rv's

$$X_j = \mathbf{E}_j S_1^n - \mathbf{E}_{j-1} S_1^n, \ \ \mathbf{E}_{j-1} X_j = 0, \ \ j = 1, \ldots$$

which are martingale-differences with respect to the family of σ-algebras $\left(\mathcal{F}_{[0,k]}\right)$. Obviously,

$$S_1^n = \mathbf{E}S_1^n + \sum_{j=1}^{n} X_j, \ \ \exp\{hS_1^n\} = \exp\{h\mathbf{E}S_1^n\} \exp\left\{h \textstyle\sum_{j=1}^{n} X_j\right\}. \quad (3.3.1)$$

The summands in the right-hand side of (3.3.1) have moments which can be evaluated using the restrictions of the theorem on the moments of $|\xi_j - a_j|$. Indeed, the RV

$$\zeta_k = S_{1,n} - (\xi_k - a_k) = a_k + \textstyle\sum_{j \neq k} \xi_j$$

is independent of ξ_k, so $\mathbf{E}_k |\zeta_k| = \mathbf{E}_{k-1} |\zeta_k|$. Thus, if $Y_k = \mathbf{E}_k (S_1^n - |\zeta_k|)$, then

$$X_k = Y_k - \mathbf{E}_{k-1} Y_k, \quad (3.3.2)$$

and by the triangle inequality

$$|Y_k| \le |\xi_k - a_k|. \quad (3.3.3)$$

Inequality (3.3.3) shows that conditional moments of X_k with respect to $\mathcal{F}_{[0,k-1]}$ can be evaluated from above in terms of the "unconditional" moments of $|\xi_k - a_k|$. To derive the bound of the theorem, calculations have to be done carefully to avoid loss of precision.

a) The bound on variance of S_1^n follows from the obvious equality

$$\mathbf{E}(S_{1,n}[X])^2 = \textstyle\sum_{j=1} \mathbf{E}X_j^2.$$

and the relations (the last of which is a consequence of (3.3.3))

$$\mathbf{E}X_k^2 = \mathbf{E}Y_k^2 - \mathbf{E}(\mathbf{E}_{k-1}Y_k)^2 \le \mathbf{E}|\xi_k - a_k|^2.$$

b) To derive the bound on exponential moments, set $m_k = \mathbf{E}_{k-1}X_k$. Inequality (3.3.3) and the elementary inequality $x \le e^{x-1}$ imply the estimate

$$\begin{aligned}\mathbf{E}_{k-1}e^{hX_k} &\le e^{-hm_k}\mathbf{E}_{k-1}e^{hY_k} = e^{-m_k}\mathbf{E}_{k-1}\left[e^{hY_k} - h(Y_k - m_k)\right] \\ &\le \exp\left\{-hm_k + \mathbf{E}_{k-1}\left[e^{hY_k} - 1 - h(Y_k - m_k)\right]\right\} \\ &= \exp\left\{\mathbf{E}_{k-1}\left[e^{hY_k} - 1 - hY_k\right]\right\}.\end{aligned}$$

The function $e^x - 1 - x$ is a nondecreasing one. Consequently, the above bound combined with (3.3.3) leads to the relation

$$\mathbf{E}_{k-1}e^{hX_k} \le \exp\left\{\mathbf{E}\left[e^{h|\xi_k - a_k|} - 1 - h|\xi_k - a_k|\right]\right\}, \qquad (3.3.4)$$

where the conditional mean in the right-hand side has been replaced by the "unconditional" one because $|\xi_k - a_k|$ is independent of the "past" $\mathcal{F}_{[0,k-1]}$.

To derive inequality b) of the theorem, it remains to use (3.3.1) and (3.3.4) applying repeatedly the formula

$$\mathbf{E}\exp\{h[X_1 + \cdots + X_k]\} = \mathbf{E}\exp\{h[X_1 + \cdots + X_{k-1}]\}\mathbf{E}_{k-1}\exp\{hX_k\}. \ \bigcirc$$

Proof of Theorem 3.3.2. The conditions of the theorem and the Taylor expansion of the exponential function lead to the inequality

$$\sum_{j=1}^{n}\mathbf{E}\left[e^{h|\xi_j - a_j|} - 1 - h|\xi_j - a_j|\right] \le \sum_{l=2}^{\infty}\frac{h^l}{l!}\sum_{j=1}^{n}\mathbf{E}|\xi_j - a_j|^l \le \frac{h^2B^2}{2}[1 - hH]^{-1},$$

which holds for $h \in [0,\ 1/H)$. Hence a form of the Chebyshev inequality and Theorem 3.3.1 yield the estimate

$$\begin{aligned}\mathbf{P}\{S_1^n - \mathbf{E}S_1^n \ge x\} &\le \exp\{-h\mathbf{E}S_1^n - hxB\}\mathbf{E}\exp\{hS_1^n\} \qquad (3.3.5) \\ &\le \exp\left\{-xC + \tfrac{1}{2}C^2[1 - CH/B]^{-1}\right\},\end{aligned}$$

where the parameter $C = hB$ can be chosen from the interval $(0,\ B/H)$. If the ratio H/B is small, the choice $C = x/(1 + xH/B)$ from the condition $\frac{1}{2}xC = \frac{1}{2}C^2[1 - CH/B]^{-1}$ is close to the exact minimum point. $\bigcirc$

3.3.3 Exponential Inequalities for Banach Space

The estimates of Theorem 3.3.1 and Theorem 3.3.2 hold for the maximal sums S_1^{n*} of (3.1.3) if the RV's ξ_j take values from a real separable Banach space and are centered:

$$\mathbf{E}\xi_j = 0, \quad j = \overline{1,n}. \qquad (3.3.6)$$

In this case the sequence S_1^k is a nonnegative submartingale with respect to the σ-algebras $\mathcal{F}_{[0,k]}$ (with $\mathcal{F}_{[0,0]}$ being the trivial σ-algebra). This is verified by an application of Jensen's inequality:

$$\mathbf{E}_kS_1^{k+1} = \mathbf{E}|S_{1,k} + \xi_{k+1}| \ge |S_{1,k} + \mathbf{E}\xi_{k+1}|.$$

The sequence $\exp\left\{hS_1^k\right\}$ is also a nonnegative submartingale. For this reason one can apply Kolmogorov's inequality instead of Chebyshev's:

$$\mathbf{P}\left\{S_1^{n*} \geq \mathbf{E}S_1^n + xB\right\} \leq \exp\left\{-hxB - h\mathbf{E}S_1^n\right\}\mathbf{E}\exp\left\{hS_1^n\right\}. \tag{3.3.7}$$

Starting from this bound one can, e.g., replace S_1^n by the maximal sum S_1^{n*} in the left-hand side of the estimate of Theorem 3.3.2.

The estimate of Theorem 3.3.1 can also be somewhat improved.

Theorem 3.3.3 *Let ξ_j be independent RV' with values from a real separable Banach space X with norm $|\cdot|$. If ξ_j satisfy the centering condition* (3.3.6), *then*

$$\forall\, h \geq 0 \;\; \mathbf{E}\exp\left\{hS_1^n\right\} \leq \exp\left\{h\mathbf{E}S_1^n\right\}\prod_{j=1}^{n}\mathbf{E}\left[e^{h|\xi_j|} - h\,|\xi_j|\right].$$

Proof of Theorem 3.3.3. Let X_k be the rv's of (3.3.1) and define one more set of rv's $Y_k = \mathbf{E}_k\left[S_1^n - |S_{1,n} - \xi_k|\right]$. Denote by

$$\mathbf{M}_k X = \mathbf{E}\left\{X \mid \sigma\left(\mathcal{F}_{[0,k-1]}\bigcup\mathcal{F}_{[k+1,n]}\right)\right\}$$

the conditional distribution given both the "past" and "future" of the sequence.

It follows from Jensen's inequality and independence of the summands that

$$m_k = \mathbf{E}_{k-1}Y_k = \mathbf{E}_{k-1}\mathbf{M}_kY_k \geq \mathbf{E}_{k-1}\left[|\mathbf{M}_kS_{1,n}| - |S_{1,n} - \xi_k|\right] = 0.$$

Indeed, by (3.3.6) there is equality $\mathbf{M}_kS_{1,n} = S_{1,n} - \xi_k$, and $Y_k \leq |\xi_k|$ by the triangle inequality.

Consequently, for $h \geq 0$

$$\mathbf{E}_{k-1}e^{hX_k} = \mathbf{E}_{k-1}\left[e^{h(Y_k - m_k)} - h\left(Y_k - m_k\right)\right],$$

since $e^x - x$ is an increasing function. To complete the proof of the theorem, it remains to repeat the final argument in the proof of part b) of Theorem 3.3.1. ○

3.3.4 Exponential Inequalities for Hilbert Space

The bounds of Theorems 3.3.1–3.3.3 are non-trivial only if it is possible to provide a nontrivial bound for the quantity $\mathbf{E}S_1^n$ (or at least for some level A which S_1^n rarely exceeds — see Sections 3.2–3.1). Estimates of this kind follow automatically from moment restrictions on individual summands when the norm of the space X exhibits specific properties. In particular, the problem does not arise for sums of centered RV's with values in a Hilbert space, since the relation $\mathbf{E}S_1^n = \mathcal{O}(\sqrt{\mathbf{E}(S_1^n)^2})$ follows as in (3.0.1). The results of the preceding subsections can be improved in this special case.

In the rest of this subsection, X is a separable real Hilbert space with inner product and norm $(\cdot,\cdot)$, $|\cdot| = \sqrt{(\cdot,\cdot)}$.

Theorem 3.3.4 *Let ξ_j be independent RV's with values in a Hilbert space X which satisfy condition* (3.3.6).

a) *For $h \geq 0$* $\mathbf{E}\,\mathrm{ch}\,(hS_1^n) \leq \prod_{j=1}^n \mathbf{E}\left[\exp\{h\,|\xi_j|\} - h\,|\xi_j|\right]$.

b) *If in addition to* (3.3.6) *the RV's ξ_j satisfy the moment condition of Theorem* 3.3.2 *with $a_j = 0$, then for $x \geq 0$*

$$\mathbf{P}\left\{S_1^{n*} \geq xB\right\} \leq 2\exp\left\{-\tfrac{1}{2}x^2\left[1 + xH/B\right]^{-1}\right\}.$$

Proof of Theorem 3.3.4. Taylor's formula for the hyperbolic cosine yields the expansion

$$\mathbf{E}\,\mathrm{ch}\,(hS_1^n) = 1 + \sum_{m=1}^{\infty} \frac{h^{2m}}{(2m)!}\mathbf{E}\,(S_1^n)^{2m}.$$

In Hilbert space squared norm is the inner square of a vector. Consequently,

$$(S_1^n)^2 = |S_{1,n}|^2 = \sum_{i,j=1}^{n} (\xi_j, \xi_k),$$

and $(S_1^n)^{2m}$ can be represented by a double m-dimensional sum:

$$(S_1^n)^{2m} = \sum_{\mathbf{i}\in[1,n]^m}\sum_{\mathbf{j}\in[1,n]^m} \mu(\mathbf{i},\mathbf{j}), \quad \mu(\mathbf{i},\mathbf{j}) = \prod_{\iota=1}^{m}\left(\xi_{i(\iota)}, \xi_{j(\iota)}\right), \tag{3.3.8}$$

where summation is over the integer points of the cube $[1,n]^m$, and the indices $\mathbf{i} = \left(i(\iota), \iota = \overline{1,m}\right)$ and $\mathbf{j} = \left(j(\iota), \iota = \overline{1,m}\right) \in \mathbf{Z}_+^m$ are m-dimensional vectors with integer coordinates.

Consider the monomial $\mu(\mathbf{i},\mathbf{j})$ associated with a fixed pair of multiindices, and define by κ_j its degree of homogeneity in the variable ξ_j:

$$\kappa_j = \kappa_j(\mathbf{i},\mathbf{j}) \stackrel{def}{=} \mathrm{card}\left(\{\iota : i(\iota) = j\}\right) + \mathrm{card}\left(\{\iota : j(\iota) = j\}\right).$$

Since the RV's ξ_j are independent and centered (3.3.6), the expectation of μ vanishes if this monomial is homogeneous of order one in at least one of the variables ξ_j:

$$\exists\, l \in \{1, \dots n\} \;\; \kappa_l(\mathbf{i},\mathbf{j}) = 1 \implies \mathbf{E}\mu(\mathbf{i},\mathbf{j}) = 0. \tag{3.3.9}$$

By the Cauchy inequality, all the monomials admit the following bound:

$$|\mu(\mathbf{i},\mathbf{j})| \leq \prod_{j=1}^{n} |\xi_j|^{\kappa_j(\mathbf{i},\mathbf{j})}. \tag{3.3.10}$$

Define the set of vector-valued "powers" corresponding to nonzero summands in the decomposition for $\mathbf{E}\,(S_1^n)^{2m} = \mathbf{E}\,|S_{1,n}|^{2m}$ associated to (3.3.8):

$$K_m = \left\{\kappa = (\kappa_j) \in \mathbf{Z}_+^n : \textstyle\sum_{j=1}^n \kappa_j = 2m \;\&\; \forall j\; \kappa_j \neq 1\right\}.$$

Relations (3.3.8)–(3.3.10) imply the estimate

$$\mathbf{E}\,(S_1^n)^{2m} \;\le\; \sum_{\kappa\in K_m}\;\sum_{\kappa(\mathbf{i},\mathbf{j})=\kappa} \mathbf{E}\,|\xi_1|^{\kappa_1(\mathbf{i},\mathbf{j})}\cdots\mathbf{E}\,|\xi_n|^{\kappa_n(\mathbf{i},\mathbf{j})}$$

$$= \sum_{\kappa\in K_m}(2m)!\prod_{j=1}^{n}\frac{\mathbf{E}\,|\xi_j|^{\kappa_j}}{\kappa_j!}. \tag{3.3.11}$$

To obtain the second inequality in (3.3.11) it suffices to collect the terms with same powers of $\xi_\cdot$ in the intermediate sum. In the cube $[1,\, n]^{2m}$, the number of pairs $\langle\mathbf{i},\mathbf{j}\rangle$ with given homogeneity exponents $\kappa_j\,(\mathbf{i},\mathbf{j}) = \kappa_j$ is $(2m)!/\left[\kappa_1!\cdots\kappa_n!\right]$.

To derive estimate a) of the theorem, note that

$$\{0\}\bigcup\bigcup_{j\ge 1}K_j \subset \left\{\kappa\in\mathbf{Z}_+^n : \forall j\ \kappa_j\neq 1\right\} = \left(\mathbf{Z}_+\setminus\{1\}\right)^n.$$

Indeed, all the summands of (3.3.11) are positive, so its right-hand side becomes greater if additional ones are included:

$$\sum_{m=0}^{\infty}\frac{1}{(2m)!}h^{2m}\mathbf{E}\,(S_1^n)^{2m} \;\le\; 1+\sum_{\kappa\in K_m}(2m)!\prod_{j=1}^{n}\frac{h^{\kappa_j}\mathbf{E}\,|\xi_j|^{\kappa_j}}{\kappa_j!}$$

$$\le \sum_{\kappa_1\neq 1}\cdots\sum_{\kappa_n\neq 1}\frac{h^{\kappa_j}\mathbf{E}\,|\xi_j|^{\kappa_j}}{\kappa_j!} \;=\; \prod_{j=1}^{n}\mathbf{E}\left(e^{h|\xi_j|}-h\,|\xi_j|\right).$$

The inequality b) of the theorem is derived from Kolmogorov's inequality (3.3.7) in the same way as the analogous estimate of Theorem 3.3.2. ○

It is sometimes convenient to use exponential bounds above in the form of an inequality for exponential moment.

Corollary 3.3.1 *Under the assumptions of Theorem* 3.3.4, *part* b), *there exists a positive constant* $c = c(r,\alpha)$ *such that for* $r > 0$ *and* $\alpha < \frac{1}{2}(1+r)^{-1}$

$$\mathbf{E}\exp\left\{\alpha\,(S_1^n/B)^2\right\}\mathbf{1}_{\{S_1^n\le rB^2/H\}} \le c.$$

Proof of Corollary 3.3.1. Integration by parts yields for an arbitrary rv ζ the equality

$$\mathbf{E}e^{\alpha\zeta^2}\mathbf{1}_{\{|\zeta|\le A\}} = -\,e^{\alpha u^2}\mathbf{P}\left\{|\zeta|\ge u\right\}\Big|_0^A + \int_0^A 2\alpha u e^{\alpha u^2}\mathbf{P}\left\{|\zeta|\ge u\right\}du.$$

The assertion of the corollary is derived from that of Theorem 3.3.4 and the above formula with $\zeta = (1/B)S_1^n$ and $A = rB/H$. The constant is no greater than $c = 1/(1-\alpha\,[2r+2])$. ○

3.4 Applications of Exponential Bounds

The bounds on probabilities derived in Section 3.3 use very rough information about the distributions of summands. Yet they can serve as a means towards more exact results about large deviations — e.g., to calculate the principal term of asymptotics of $\log \mathbf{P}\{S_1^n[\xi] \geq r_n\}$. Such calculation is done below for a Banach space containing a basis.

Other results of this section are more detailed versions of Corollary 3.3.1 — bounds on exponential moments for sums of RV's in a Hilbert space.

3.4.1 An Asymptotic Formula for Large Deviations

Suppose that the Banach space X contains a basis (e_j) with the biorthogonal coordinate sequence (f_j). The corresponding projection operators

$$\mathcal{P}_m : x \mapsto \sum_{j=1}^m \langle x, f_j\rangle e_j, \quad \mathcal{R}_m : x \mapsto x - \mathcal{P}_m x, \tag{3.4.1}$$

are bounded uniformly in m (see subsection A.2.5). The basis and its biorthogonal sequence generate the "coordinate" mappings

$$\begin{aligned} \hat{\pi} : \quad & X \ni x \mapsto \left(\langle x, f_j\rangle, j = \overline{1,m}\right) \in \mathbf{R}^m, \\ \check{\pi} : \quad & X^* \ni t \mapsto \left(\langle e_j, t\rangle, j = \overline{1,m}\right) \in \mathbf{R}^m. \end{aligned}$$

The following identities are evident:

$$\begin{aligned} \langle \mathcal{P}_m x, t\rangle &= (\hat{\pi}x, \check{\pi}t)_m = \sum_{j=1}^m \langle x, f_j\rangle \langle e_j, t\rangle, \quad x \in X,\ t \in X^*, \\ |\mathcal{P}_m x| &= \sup_{t\in S} \langle \mathcal{P}_m x, t\rangle = \sup_{t\in S} (\hat{\pi}x, \check{\pi}t)_m, \end{aligned} \tag{3.4.2}$$

where $S = \{t \in X^* : |t| = 1\}$ is the unit sphere. Its image $\check{\pi}S \subset \mathbf{R}^m$ is compact.

Consider the sums $S_{1,n} = \sum_{j=1}^n \xi_{jn}$ for a triangular array of row-wise independent RV's $\left(\xi_{jn}, j = \overline{1,n}\right)$ with values in X.

Theorem 3.4.1 *Suppose that the array $\left(\xi_{jn}, j = \overline{1,n}\right)$ satisfies the conditions*

$$\mathbf{E}\xi_{jn} = 0, \ \sum_{j=1}^n \mathbf{E}|\xi_{jn}|^l \leq \tfrac{1}{2} l! B_n^2 H_n^{l-2}, \ l = 2, 3, \ldots, \quad j = \overline{1,n}, \tag{3.4.3}$$

with some positive constants B_n, H_n, and, moreover,

$$\begin{aligned} \beta_n &= \mathbf{E}|S_{1,n}|/B_n \leq c_1, \\ \sigma_n^2 &= \sup_{|t|=1} \mathbf{E}\langle S_{1,n}, t\rangle^2 \geq c_2 B_n^2, \\ b_n^2(m) &= B_n^{-2}\sum_{j=1}^n \mathbf{E}|\mathcal{R}_m \xi_j|^2 \leq \varphi(m), \end{aligned} \tag{3.4.4}$$

where c_i are positive constants. If

$$\lim_{n\to\infty} H_n/B_n = 0, \ \lim_{m\to\infty} \varphi(m) = 0, \ \lim_{n\to\infty} x_n = \infty, \ \lim_{n\to\infty} x_n H_n/B_n = 0,$$

then

$$\lim_{n\to\infty} x_n^{-2} \ln \mathbf{P}\{|S_{1,n}| \geq \sigma_n x_n\} = -\tfrac{1}{2}.$$

This result anticipates more general theorems of a later chapter.

Remark 3.4.1 The conditions of the theorem are satisfied, e.g., for sums of equidistributed RV's in spaces of type 2 containing a basis under the "individual" restrictions on summands described in subsection 3.3.1. In the general case a "collective" condition is necessary, see Example 3.0.1.

Proof of Theorem 3.4.1 is rather long, so it is sectioned in a number of steps.
(I) *Derivation of the lower bound.* Evidently, for each $t \in S$

$$\mathbf{P}\{|S_{1,n}| \geq x_n\sigma_n\} \geq \mathbf{P}\{|\langle S_{1,n}, t\rangle| \geq x_n\sigma_n\}.$$

Fix $\varepsilon > 0$, and select a vector $t_n = t_n(\varepsilon) \in X^*$, $|t_n| = 1$, so that $\mathbf{E}\langle S_{1,n}, t_n\rangle^2 \geq (1-\varepsilon)\sigma_n^2$. The one-dimensional projection $\langle S_{1,n}, t_n\rangle$ admits the lower exponential bound (due to A.N.Kolmogorov)

$$\mathbf{P}\{\langle S_{1,n}, t_n\rangle \geq x\sigma_n\} \geq \gamma(\varepsilon)\exp\left\{-\frac{x^2}{2}\frac{1+\varepsilon}{1-\varepsilon}\right\} \text{ if } \alpha(\varepsilon) < x < \beta(\varepsilon)\frac{\sigma_n}{H_n}.$$

It follows from the lower exponential bound that

$$\liminf_{n\to\infty} x_n^{-2}\ln \mathbf{P}\{|S_{1,n}| \geq x_n\sigma_n\} \geq \lim_{\varepsilon\to 0}\left(-\frac{1+\varepsilon}{2(1-\varepsilon)}\right) = -\tfrac{1}{2}. \qquad (3.4.5)$$

(II) *Derivation of the upper bound.* a) It is obvious that for each $\varepsilon \in (0, 1)$ and natural m

$$\mathbf{P}\{|S_{1,n}| \geq x_n\sigma_n\} \leq \mathbf{p}_{m,n} + \mathbf{r}_{m,n},$$

where

$$\begin{aligned} \mathbf{p}_{m,n} &= \mathbf{P}\{|\mathcal{P}_m S_{1,n}| \geq (1-\varepsilon)x_n\sigma_n, |\mathcal{R}_m S_{1,n}| \leq \varepsilon\sigma_n x_n\}, \\ \mathbf{r}_{m,n} &= \mathbf{P}\{|\mathcal{R}_m S_{1,n}| \geq \varepsilon\sigma_n x_n\} \\ &= \mathbf{P}\left\{|\mathcal{R}_m S_{1,n}| - \mathbf{E}|\mathcal{R}_m S_{1,n}| \geq \varepsilon\sigma_n x_n\left[1 - \frac{\mathbf{E}|\mathcal{R}_m S_{1,n}|}{\varepsilon\sigma_n x_n}\right]\right\}. \end{aligned}$$

Consequently,

$$\limsup_{n\to\infty} x_n^{-2}\ln \mathbf{P}\{|S_{1,n}| \geq x_n\sigma_n\} \leq \max[A, B] \qquad (3.4.6)$$

where

$$A = \limsup_{n\to\infty} x_n^{-2}\ln \mathbf{p}_{m,n}, \quad B = \limsup_{n\to\infty} x_n^{-2}\ln \mathbf{r}_{m,n}.$$

In the sequel m will be chosen so as to make the infinite-dimensional component $\mathbf{r}_{m,n}$ negligible with respect to the finite-dimensional one $\mathbf{p}_{m,n}$, which is easier to treat.

b) Under the restrictions of the theorem, there exist positive constants such that $c_1 \leq \sigma_n^2/B_n^2 \leq c_2$. It will be convenient to use σ_n as the scale parameter below, and write the bounds of (3.4.4) on the infinite-dimensional "residue" as

$$\begin{aligned} \textstyle\sum_{j=1}^n \mathbf{E}\left|\mathcal{R}_m\xi_j\right|^l &\leq c'\varphi(m)\sigma_n^2\left[c''H_n/\sqrt{\varphi(m)}\right]^{l-2}, \; l\geq 2, \\ \mathbf{E}\left|\mathcal{R}_m S_{1,n}\right| &\leq c'''\sigma_n. \end{aligned}$$

(Recall that projections (3.4.1) are uniformly bounded.) By Theorem 3.3.3

$$\limsup_{n\to\infty} x_n^{-2}\ln\mathbf{P}\left\{\left|\mathcal{R}_m S_{1,n}\right|\geq\varepsilon x_n\sigma_n\right\}\leq-\hat{c}\varepsilon^2/\varphi(m)<-\tfrac{1}{2} \tag{3.4.7}$$

provided that the number of dimensions $m=m(\varepsilon)$ is sufficiently large.

c) Fix the number of dimensions m and some $\delta>0$. Under the assumptions of the theorem, there is a finite collection of points $t_\kappa\in X^*$, $|t_\kappa|=1$, $\kappa=\overline{1,N}$, such that the inequality

$$\forall\, x\in X \quad |\mathcal{P}_m x|\leq\frac{\max_{\kappa=\overline{1,N}}\left\langle\mathcal{P}_m x,t_\kappa\right\rangle}{1-c\delta} \tag{3.4.8}$$

holds with a constant $c=c(m)$ which depends on the number of dimensions only.

Indeed, the coordinate image $\check{\pi}S$ of the unit ball is compact. Hence for each $\delta>0$ there exists a finite set of points $t_j\in S$, $j=\overline{1,N}$, such that $\left\{\check{\pi}t_j, j=\overline{1,N}\right\}$ is a δ-net in $\check{\pi}S$. Denote by $|\cdot|_m=\sqrt{(\cdot,\cdot)_m}$ the Euclidean norm in $\hat{\pi}X$ (cf. (3.4.2)). By (3.4.2)

$$|\mathcal{P}_m x|\leq\delta\left|\hat{\pi}x\right|_m+\max_j\left(\hat{\pi}x,\check{\pi}t_j\right)_m\leq c(m)\delta\left|\mathcal{P}_m x\right|+\max_j\left\langle\mathcal{P}_m x,t_j\right\rangle.$$

This proves (3.4.8) if $\delta\leq\delta_0=\delta_0(m)$.

In addition to m and δ, fix $\varepsilon>0$. Then by (3.4.8) the "finite-dimensional" component of the tail in (3.4.6) satisfies the inequality

$$\begin{aligned} \mathrm{p}_{m,n} &\leq \sum_{j=1}^N\mathbf{P}\left\{\left\langle\mathcal{P}_m S_{1,n},t_j\right\rangle\geq(1-\delta)(1-\varepsilon)\,x_n\sigma_n,\ \left|\mathcal{R}_m S_{1,n}\right|\leq\varepsilon x_n\sigma_n\right\} \\ &\leq N\max_j\mathbf{P}\left\{\left\langle S_{1,n},t_j\right\rangle\geq(1-\gamma)\,x_n\sigma_n\right\},\ \gamma=1-\left[(1-\varepsilon)(1-\delta)-\varepsilon\right]. \end{aligned}$$

In this bound, the number N does not depend on the number of summands n, and all the vectors t_j are unit ones. Hence

$$\begin{aligned} J &= \limsup_{n\to\infty} x_n^{-2}\mathrm{p}_{m,n} \\ &\leq \sup_{t\in S}\limsup_{n\to\infty} x_n^{-2}\ln\mathbf{P}\left\{\left\langle S_{1,n},t\right\rangle\geq(1-\gamma)x_n\sigma_n\right\}. \end{aligned} \tag{3.4.9}$$

The tails of distributions of one-dimensional projections $\langle S_{1,n},t_j\rangle$ can be evaluated using known modifications of the exponential bound of S. N. Bernstein similar to Theorem 3.3.4. The choice of σ_n taken into consideration, this yields the inequality

$$J\leq-\tfrac{1}{2}\left[(1-\varepsilon)(1-\delta)-\varepsilon\right]^2. \tag{3.4.10}$$

d) For the "original" probability (3.4.6) estimates (3.4.7), (3.4.9), and (3.4.10) imply the relation

$$\limsup_{n\to\infty} x_n^{-2} \ln \mathbf{P}\left\{|S_{1,n}| \geq x_n\sigma_n\right\} \leq -\tfrac{1}{2}\left[(1-\varepsilon)(1-\delta)\right].$$

The left-hand side of the last inequality does not depend on ε or δ. Hence a passage to the limit yields for it the bound

$$\limsup_{n\to\infty} x_n^{-2} \ln \mathbf{P}\left\{|S_{1,n}| \geq x_n\sigma_n\right\} \leq -\tfrac{1}{2}. \tag{3.4.11}$$

Inequalities (3.4.11) and (3.4.5) prove the theorem. ○

3.4.2 Bounds on Exponential Moments

A convenient form of exponential bounds on large deviation probabilities are estimates for exponential moments of the norm. The propositions below are versions of Corollary 3.3.1 intended for calculations similar to those of subsection 3.4.1

In what follows, the independent Hilbert-space-valued RV's ξ_j satisfy the conditions

$$\sum_{j=1}^{n} \mathbf{E}\left|\xi_j\right|^l \leq \tfrac{1}{2} l! B_n^2 H^{l-2}, \quad l = 2, 3, \ldots, \quad \mathbf{E}\xi_j = 0, \; j = \overline{1, n}. \tag{3.4.12}$$

The operator $\mathcal{R}' : X \to X$ is an orthogonal projection, and α is positive number such that

$$\mathbf{E}\left|\mathcal{R}' S_{1,n}\right|^2 \leq \alpha^2 B_n^2, \quad S_{1,n} = S_{1,n}[\xi] = \textstyle\sum_{j=1}^{n} \xi_j. \tag{3.4.13}$$

Lemma 3.4.1 *Fix a number $s > 0$ such that*

$$\kappa = 1 - s^2\alpha^2 > 0.$$

There exist constants $c_i = c_i(\kappa) > 0$ such that for $s^2 rH/B_n \leq c_1$ and $s^4 rH/B_n \leq c_1\kappa$

$$I_n(s) = \mathbf{E}\exp\left\{\frac{s^2}{2}\frac{\left|\mathcal{R}' S_{1,n}\right|^2}{B_n^2}\right\} \mathbf{1}_{\{|S_{1,n}| \leq rB_n\}} \leq c_2.$$

Proof of Lemma 3.4.1. Let $Y \in \mathbf{R}$ be a standard normal rv (1.1.1) which is independent of all ξ_j. Denote the conditional distribution given the values of all ξ_j by

$$\mathbf{M} f(Y, \xi_1, \ldots) = \int_{\mathbf{R}} f(y, \xi_1, \ldots) e^{-\frac{1}{2}y^2} dy/\sqrt{2\pi}.$$

It is easily seen that for each $U > 0$

$$\exp\left\{\tfrac{1}{2}U^2\right\} = \mathbf{M}e^{UY} \leq c + \mathbf{M}e^{UY}\mathbf{1}_{\{|Y| \leq 3U\}}$$

with some absolute constant c, so

$$\begin{aligned} I_n(s) &\leq c + \mathbf{EM}\exp\left\{s|Y|\left|\mathcal{R}' S_{1,n}\right|/B_n\right\}\mathbf{1}_{\{|S_{1,n}| \leq rB_n, |Y| \leq 3sr\}} \\ &\leq c + \mathbf{M}\left(\mathbf{E}\exp\left\{s|Y|\left|\mathcal{R}' S_{1,n}\right|/B_n\right\}\right)\mathbf{1}_{\{|Y| \leq 3sr\}}. \end{aligned}$$

The expectation with respect to the distributions of ξ_j can be evaluated using Theorem 3.3.4 and condition (3.4.12). Put $u = s\,|Y|\,/B_n$. Inequality $|\mathcal{R}'x| \le |x|$ taken into consideration, the expression provided by Theorem 3.3.4 leads to the bound

$$\mathbf{E}\exp\{u\,|\mathcal{R}'S_{1,n}|\} \le 2\exp\left\{\tfrac{1}{2}\alpha^2u^2B_n^2 + u^3HB_n^2\right\}$$

if $uH = s\,|Y|\,H/B_n \le \frac{1}{2}$. For $|Y| \le 3sr$, the argument results in the inequality

$$\begin{aligned} I_n(s) &\le c + 2\mathbf{M}\exp\left\{\ \tfrac{1}{2}Y^2\left[s^2\alpha^2 + 6s^4Hr/B_n\right]\ \right\} \\ &\le c + 2\mathbf{M}\exp\left\{\ \tfrac{1}{2}(1-\tfrac{1}{2}\kappa)Y^2\ \right\} \le c + 2^{3/2}/\sqrt{\kappa} \end{aligned}$$

which is valid if $3s^2rH/B_n \le \frac{1}{2}$ and $3s^4Hr/B_n \le \kappa/4$. ○

Consider now a family of finite-dimensional projections

$$\begin{aligned} \mathcal{P}_j &: X \to X,\ \mathcal{P}_j\mathcal{P}_k = \mathcal{P}_k\mathcal{P}_j = 0,\ j \ne k, \\ D = &\ \textstyle\sum_{j=1}^m \dim(\mathcal{P}_jX) < \infty. \end{aligned} \tag{3.4.14}$$

Suppose that there exist positive numbers α_j such that

$$\forall\, s_j \ge 0,\ t \in X\ \ \mathbf{E}\left(\textstyle\sum_{j=1}^m s_j\mathcal{P}_jS_{1,n}, t\right)^2 \le B_n^2\textstyle\sum_{j=1}^m \alpha_j^2 s_j^2\,|\mathcal{P}_jt|^2. \tag{3.4.15}$$

(Note that the RV in the inner product above is in a finite-dimensional subspace of X.)

Lemma 3.4.2 *Let s_j, $j = \overline{1,m}$, be positive numbers such that*

$$\kappa = 1 - \alpha_*^2 > 0,\ \ \alpha_*^2 = \max_{j=\overline{1,m}} \alpha_j^2 s_j^2.$$

There exist constants $c_i = c_i(D,\kappa)$ which depend only on κ and the summary number of dimensions $D = \dim([\mathcal{P}_1 + \cdots \mathcal{P}_m]X)$ such that

$$J_n(s_1,\ldots,s_m) = \mathbf{E}\exp\left\{\frac{1}{2}\sum_{j=1}^m \frac{s_j^2}{B_n^2}\,|\mathcal{P}_jS_{1,n}|^2\right\}\mathbf{1}_{\{|S_{1,n}| \le rB_n\}} \le c_1$$

if $\max_j s_j^2\,rH/B_n \le c_2$ *and* $\max_j s_j^4\,rH/B_n \le c_2\kappa$.

Proof of Lemma 3.4.2 follows the same pattern as that of Lemma 3.4.1. Below, the following two spaces are identified: $\left(\sum_{j=1}^m \mathcal{P}_j\right)X = \mathbf{R}^D$.

Consider a standard Gaussian RV $Y \in \mathbf{R}^D$ independent of all ξ_j. The notation $\mathbf{M}$ refers to integration with respect to the distribution of Y only. In this notation

$$\exp\left\{\frac{1}{2}\sum_{j=1}^m s_j^2\frac{|\mathcal{P}_jS_{1,n}|^2}{B_n^2}\right\} = \mathbf{M}\exp\left\{\left(Y, \widehat{S}\right)\right\},\ \ \widehat{S} = \frac{1}{B_n}\sum_{j=1}^m s_j\mathcal{P}_jS_{1,n} \in \mathbf{R}^D.$$

Projections $\mathcal{P}_j$ being orthogonal, the new vector satisfies the inequality $\left|\widehat{S}\right| \le s_*\,|S_{1,n}|\,/B_n$ with $s_* = \max_j s_j$. The RV $|Y|$ has density (1.1.4), so as in the proof of Lemma 3.4.1

$$\begin{aligned} &\mathbf{E}\exp\left\{\tfrac{1}{2}\textstyle\sum_{j=1}^{m} s_j^2 \left|\mathcal{P}_j S_{1,n}\right|^2 / B_n^2\right\} \mathbf{1}_{\{|S_{1,nb}|\le rB_n\}} \\ \le\ & c(D) + \mathbf{M}\left(\mathbf{E}\exp\left\{\left(Y,\widehat{S}\right)\right\}\right) \mathbf{1}_{\{|Y|\le 3s_* r\}}. \end{aligned}$$

Since the projections correspond to orthogonal subspaces, $\left|\sum_{j=1}^{m} \mathcal{P}_j x\right| \le |x|$, and the condition of the lemma implies that

$$\mathbf{E}\left(Y,\widehat{S}\right)^2 \le \alpha_*^2 \left|Y\right|^2.$$

Consequently, for $|Y| \le 3s_* r$, $3s_*^2 rH/B_n \le \frac{1}{2}$, and $3s_*^4 rH/B_n \le \kappa/4$,

$$\begin{aligned} \mathbf{E}\exp\left\{\left(Y,\widehat{S}\right)\right\} &\le \exp\left\{\tfrac{1}{2}\alpha_*^2 \left|Y\right|^2 + 3\left|Y\right|^2 s_*^4 rH/B_n\right\} \\ &\le \exp\left\{\tfrac{1}{2}(1-\tfrac{1}{2}\kappa)\left|Y\right|^2\right\}. \end{aligned}$$

The expectation of the right-hand side is finite and depends only on κ and the number of dimensions. ○

3.5 Bounds Using Finite-Order Moments

3.5.1 Nagaev–Fuk Inequality

The exponential bounds for sums derived in Sections 3.1–3.3 are fairly insensitive to individual peculiarities of summands' distributions — they reproduce the decay rates of Gaussian "tails", although in a restricted range of deviations. One can derive similar bounds in terms of moments of finite order restricting the range even more.

Notation of Sections 3.1–3.3 is used below without commentary.

Theorem 3.5.1 *Let ξ_j be independent RV's with values in a measurable real linear space X, and $|\cdot|$ a mesurable seminorm in X.*

Suppose that there exist nonrandom vectors $a_j \in X$, $|a_j| < \infty$, and some natural number $m \ge 3$ such that

$$\sum_{j=1}^{n} \mathbf{E}\left|\xi_j - a_j\right|^2 \le B^2, \quad \sum_{j=1}^{n} \mathbf{E}\left|\xi_j - a_j\right|^m \le A.$$

Then the following inequality holds for each $x > 0$:

$$\mathbf{P}\left\{S_1^n \ge \mathbf{E}S_1^n + xB\right\} \le c_1\left[\frac{A/B^m}{x^m} + \exp\left\{-c_2 x^2\right\}\right].$$

The values of the constants depend only on m.

Note that special properties of the seminorm do not influence the values of the constants. Its bound is much more special than a similar result of Section 3.1, but at the same time much more precise. Its possibilities are illustrated by derivation of an estimate similar to ones used to prove the so-called law of the iterated logarithm.

Proposition 3.5.1 *Let $\xi, \xi_1, \ldots$ be symmetrical i.i.d. RV's with values in a separable Banach space. Assume that $\mathbf{E}S_1^n \le c'\sqrt{n}$ for all large n, and that*

$$\mathbf{P}\left\{S_1^n \ge \mathbf{E}S_1^n + x\sqrt{n}\right\} \le c''\left[\frac{1}{n^\gamma x^m} + \exp\left\{-Qx^U\right\}\right]$$

for all $x > 0$ and some $\gamma > 0$, $U \in (0,\ 2]$.

If $\beta U \ge 1$, then $\limsup_{n\to\infty} S_1^n / \left(\sqrt{n}\,[\ln\ln n]^\beta\right) < \infty$ a.s.

Note that the a.s. limit behavior of the sequence S_1^n is determined qualitatively by the order of the exponent U. The estimate of Theorem 3.5.1 provides for β the value $\frac{1}{2}$, while Theorem 3.1.2 would yield a larger exponent because its estimate decays less rapidly.

Both Theorem 3.5.1 and Proposition 3.5.1 are proved in subsection 3.5.2.

3.5.2 Demonstration of Nagaev–Fuk Inequality

The following lemma yields a refinement of the estimate of exponential moment of the sum in a situation where the summands are bounded. It is basic for the proof of Theorem 3.5.1.

Lemma 3.5.1 *In addition to the conditions of Theorem 3.5.1, assume that $|\xi_j - a_j| \le L$ a.s. for some constant $L > 0$. Choose a natural number $m > 2$. There exist positive constants $\hat{c}_i$, depending on m only, such that*

$$\forall\, h \ge 0\ \ \mathbf{E}\exp\left\{h\left[S_1^n - \mathbf{E}S_1^n\right]\right\} \le \exp\left\{\ \hat{c}_0 + \tfrac{1}{2}h^2B^2 + \hat{c}_1 A h^m e^{hL}\ \right\}.$$

If $m = 3$, it is easy to write down explicitly a set of constants which meet the requirements of Lemma 3.5.1: one can, e.g., set $\hat{c}_0 = 0$ and $\hat{c}_1 = 1/3!$.

Proof of Lemma 3.5.1. If the RV's $|\xi_j - a_j|$ are bounded, then the higher moments of the summands admit the estimates

$$h^l\mathbf{E}\left|\xi_j - a_j\right|^l \le \begin{cases} [hL]^{l-m}\, h^m\mathbf{E}\left|\xi_j - a_j\right|^m, & \text{if } l \ge m, \\ (1 - l/m) + (l/m)h^m\mathbf{E}\left|\xi_j - a_j\right|^m, & \text{if } 2 \le l \le m-1. \end{cases}$$

(The bound for moments of order $l < m$ is obtained using the Minkowski inequality, $ab \le a^p/p + b^q/q$, $1/p + 1/q = 1$).

Consider the function

$$g_m(u) = \frac{1}{u^m}\left[\exp\{u\} - \sum_{k=0}^{m-1}\frac{u^k}{k!}\right] = \sum_{k=0}^{\infty}\frac{u^k}{(m+k)!}.$$

It is increasing on the interval $(0,\ \infty)$, does not exceed e^u, and is bounded from below by a positive constant.

Proceeding as in similar arguments (see, e.g., the proof of Theorem 3.3.2), one arrives at the inequality

$$\begin{aligned}
&\sum_{j=1}^{n} \mathbf{E}\left[e^{h|\xi_j - a_j|} - 1 - h\,|\xi_j - a_j|\right] \\
\leq\ & \frac{1}{2}h^2B^2 + \sum_{j=1}^{n}\sum_{l=2}^{m-1}\frac{1}{l!}\mathbf{E}\left(h\,|\xi_j - a_j|^l\right) \\
&+ \sum_{j=1}^{n}\sum_{l=m}^{\infty}\mathbf{E}\left[h^m\,|\xi_j - a_j|^m\right] g_m\left(h\,|\xi_j - a_j|\right) \\
\leq\ & \hat{c}_0 + \frac{1}{2}h^2B^2 + \hat{c}_1 A h^m e^{hL}.
\end{aligned}$$

To complete the proof it suffices to use the estimate of Theorem 3.3.1. ○

Proof of Theorem 3.5.1. a) It is convenient to prove the inequality of the theorem first for the bounded truncated RV's

$$\hat{\xi}_j = \mathbf{1}_{\{|\xi_j - a_j| \leq L\}}\xi_j + \mathbf{1}_{\{|\xi_j - a_j| > L\}}a_j, \tag{3.5.1}$$

where the truncation level $L = L(x)$ will be chosen later in the proof.

By the Chebyshev inequality

$$\mathbf{P}\left\{S_{1,n}[\hat{\xi}] \neq S_{1,n}[\xi]\right\} \leq \sum_{j=1}^{n}\mathbf{P}\left\{\hat{\xi}_j \neq \xi_j\right\} \leq \frac{A}{L^m}.$$

The quantity in the right-hand side is small if L is large. In this case the difference between the averages of norms of the sums $S_{1,n}[\hat{\xi}]$ and $S_{1,n}[\xi]$ is also small:

$$\begin{aligned}
\left|\mathbf{E}S_1^n[\hat{\xi}] - \mathbf{E}S_1^n[\xi]\right| &\leq \textstyle\sum_{j=1}^{n}\mathbf{E}|\hat{\xi}_j - \xi_j| \\
&= \textstyle\sum_{j=1}^{n}\mathbf{E}\,|\xi_j - a_j|\,\mathbf{1}_{\{|\xi_j - a_j| > L\}} \leq A/L^{m-1}.
\end{aligned}$$

It follows from the above that

$$\begin{aligned}
\mathbf{P} &\stackrel{def}{=} \mathbf{P}\left\{S_1^n[\xi] \geq \mathbf{E}S_1^n[\xi] + xB\right\} \\
&\leq \mathbf{P}\left\{S_1^n[\hat{\xi}] \geq \mathbf{E}S_1^n[\xi] + xB\right\} + \mathbf{P}\left\{S_{1,n}[\xi] \neq S_{1,n}[\hat{\xi}]\right\} \qquad (3.5.2) \\
&\leq \mathbf{P}\left\{S_1^n[\hat{\xi}] \geq \mathbf{E}S_1^n[\hat{\xi}] + xB - \frac{A}{L^{m-1}}\right\} + \frac{A}{L^m}.
\end{aligned}$$

One can apply the estimate of Lemma 3.5.1 to the truncated sum in this inequality.

b) By (3.5.1) the RV's $\hat{\xi}_j - a_j$ are bounded. It is easily seen that the moments of their norms do not exceed those of the original RV's, since $|\hat{\xi}_j - a_j| \leq \min[L, |\xi_j - a_j|]$. Hence the Chebyshev inequality, (3.5.2), and Lemma 3.5.1 can be combined to yield the estimate

$$\mathbf{P} \le \frac{A}{L^m} + \exp\{\hat{c}_0 + \mathbf{r}\}, \quad \mathbf{r} = -hxB + \frac{hA}{L^{m-1}} + \tfrac{1}{2}h^2B^2 + \hat{c}_1 A h^m e^{hL}. \tag{3.5.3}$$

In the right-hand side of this inequality, the parameters $h > 0$ and $L > 0$ can be chosen arbitrarily.

It is not too easy to write out explicitly the exact values of h and L which minimize the right-hand side of the above bound. The estimate of the theorem is derived by the following elementary argument.

Denote by $\alpha = A/(xB)^m$ the first summand in the right-hand side of the inequality of the theorem, and define the new parameter $\Lambda = L/(xB)$ that controls truncation. In this notation, inequality (3.5.3) assumes the form

$$\begin{aligned} \mathbf{P} &\le \alpha/\Lambda^m + \exp\{\hat{c}_0 + \hat{\mathbf{r}}\}, \\ \hat{\mathbf{r}} &= -hxB + hxB\Lambda(\alpha/\Lambda^m) + \tfrac{1}{2}h^2B^2 + \hat{c}_1 (hxB)^m \alpha e^{\Lambda x h B}. \end{aligned} \tag{3.5.4}$$

This bound is trivial if $\alpha \ge \Lambda^m$ (as well as the bound of the theorem for $\alpha > c$ for any positive constant c since the choice of c_1 remains free).

Depending on the magnitude of $\alpha/(\ln x)^2$, one of the two summands dominates in the bound of the theorem. For this reason small and large values of this ratio should be treated separately.

If $\alpha = 0$, then the exponent $\hat{\mathbf{r}}$ is minimal for $h = x/B$. It is natural to suppose that this choice of the minimization parameter should prove acceptable also for the small positive values of α.

Assume that

$$x^2 \le 3\ln(1/\alpha), \quad \Lambda < \tfrac{1}{6}, \quad \alpha/\Lambda^m \le 1,$$

and choose $h = x/B$ as the value of the minimization parameter. The function $\alpha^{1-3\Lambda}(\ln 1/\alpha)^m$ is bounded on bounded subsets of the positive half-line. For this reason, the corresponding estimate for the exponent of (3.5.4) is

$$\begin{aligned} \hat{\mathbf{r}} &\le -\left(\tfrac{1}{2} - \Lambda\right) x^2 + \hat{c}_1 x^{2m} \alpha e^{3\Lambda \ln(1/\alpha)} \\ &\le -\left(\tfrac{1}{2} - \Lambda\right) x^2 + \hat{c}_1 (\ln 1/\alpha)^m \alpha^{1-3\Lambda} \le -\left(\tfrac{1}{2} - \Lambda\right) x^2 + \tilde{c}, \end{aligned} \tag{3.5.5}$$

where $\tilde{c}$ is a new positive constant.

If $x^2 > 3\ln 1/\alpha$, one can reduce the problem to one where both summands in the right-hand side of (3.5.4) have the same order of magnitude as α. To do so, transform the minimization parameter to $Y = xhB/(\ln 1/\alpha) > 0$. In this new notation the estimate for $\hat{\mathbf{r}}$ can be written in the form

$$\begin{aligned} \hat{\mathbf{r}} &= -Y[1-\Lambda]\ln 1/\alpha + 2^{-1}(Y/x)^2(\ln 1/\alpha)^2 \\ &\quad + \hat{c}_1 (Y\ln 1/\alpha)^m \alpha\alpha^{-\Lambda Y}/\Lambda^{m-1} \\ &\le -\left(Y[1-\Lambda] - 2^{-1}Y^2x^{-2}(\ln 1/\alpha)\right)(\ln 1/\alpha) \\ &\quad + \hat{c}_1 Y^m (\ln 1/\alpha)^m \alpha^{1-\Lambda Y}(\alpha/\Lambda^m). \end{aligned} \tag{3.5.6}$$

Choose Y to satisfy the restrictions

$$1 - \Lambda Y > 0, \; Y(1-\Lambda) - \tfrac{1}{6}Y^2 \geq 1.$$

(They are compatible with those introduced earlier — one can put $\Lambda = 0.1$ and $Y = 2$). It follows that

$$\hat{\mathbf{r}} \leq -Y(1-\Lambda)\ln 1/\alpha + \tfrac{1}{6}Y^2 \ln 1/\alpha + \hat{c}_1(Y \ln 1/\alpha)^m \alpha^{1-\Lambda Y} \leq c'Y^m + \ln\alpha,$$

where c' is one more positive constant.

One of the above bounds, (3.5.5) or (3.5.6), is valid for each positive x. This proves the inequality of the theorem for small α. As already noted, the inequality of the theorem is trivial for large values, $\alpha \geq$ *const*. ○

Proof of Proposition 3.5.1. Define

$$n_k = \left[W^k\right], \; p_k = \mathbf{P}\left\{\max_{n_{k-1}<n\leq n_k} \frac{S_1^n}{\sqrt{n}(\ln\ln n)^\beta} \geq R\right\},$$

where $W > 1$ and square brackets refer to the integer part of a real number. To show that the upper limit of the proposition is finite, it suffices, by the Borel–Cantelli lemma, to demonstrate convergence of the series $\sum p_k$ for large enough values of R.

By the choice of the sequence n_k and the Lévy inequality, the estimate stipulated in the statement of the proposition yields the bound

$$\begin{aligned} p_k &\leq 2\mathbf{P}\left\{S_1^{n_k} \geq (1-\varepsilon)RW^{-1/2}(\ln k)^\beta\sqrt{n_k}\right\} \\ &\leq c_1 W^{-\gamma k} + \exp\left\{-(1-\varepsilon)^U R^U W^{-U/2}(\ln k)^{U\beta}\right\} \end{aligned}$$

for large values of k with arbitrarily small $\varepsilon > 0$. The series in the right-hand side converges if $\beta U \geq 1$ and the remaining parameters are chosen appropriately. This proves the proposition. ○

3.5.3 Inequalities for Moments

Suppose that the sequence of the rv's $S_{1,k}[X]$ is a submartingale with respect to the increasing sequence of σ-algebras $\mathcal{F}_k$. The main tool in this subsection is the following inequality due to D. Burkholder (it is cited without proof, see Appendix B for references).

Proposition 3.5.2 *The following inequality holds for all $p \geq 2$:*

$$\mathbf{E}\,(S_1^n\,[X])^p \leq c\left(\sum_{j=1}^n \mathbf{E}\,|X_j|^p + \mathbf{E}\left[\sum_{j=1}^n \mathbf{E}\left\{X_j^2\,|\mathcal{F}_j\,\right\}\right]^{p/2}\right).$$

The constant $c = c(p)$ depends on the order p only.

If $p \in (1, 2)$, then

$$\mathbf{E}\,(S_1^n\,[X])^p \leq c'(p)\mathbf{E}\left(\sum_{j=1}^n X_j^2\right)^{p/2} \leq c''(p)\sum_{j=1}^n \mathbf{E}\,|X_j|^p.$$

Let ξ_j be independent RV's with values in a real linear measurable space X and $|\cdot|$ a measurable seminorm on their range. As before, select some nonrandom vectors $a_j \in X$, $|a_j| < \infty$.

The following inequality for the moments of sums is a corollary of Proposition 3.5.2.

Theorem 3.5.2 *The assumptions of Theorem* 3.5.1 *yield for the moments of* $S_1^n[\xi]$ *the following bounds.*

a) *If* $p \in (1,\, 2)$, *then* $\mathbf{E}\,|S_1^n - \mathbf{E}S_1^n|^p \le c\sum_{j=1}^n \mathbf{E}\,|\xi_j - a_j|^p$.

b) *If* $p \ge 2$, *then*

$$\mathbf{E}\,|S_1^n - \mathbf{E}S_1^n|^p \le c\left(\sum_{j=1}^n \mathbf{E}\,|\xi_j - a_j|^p + \left[\sum_{j=1}^n \mathbf{E}\,|\xi_j - a_j|^2\right]^{p/2}\right).$$

If the sequence $\left(S_1^k\right)$ is a submartingale, the estimates of Theorem 3.5.2 can be used to obtain bounds on the moments of the maximal sum S_1^{n*}.

Proof of Theorem 3.5.2. Denote by $\mathcal{F}_{[1,k]}$ the "past" (3.1.1) of the sequence (ξ_j) and put

$$\mathbf{E}_k(\cdot) = \mathbf{E}\left\{\cdot\,|\mathcal{F}_{[1,k]}\right\}, \quad Z_k = |S_{1,n}[\xi] - (\xi_j - a_j)|.$$

By the triangle inequality, the martingale-differences of (3.3.1) admit the estimate

$$|X_k| = |\mathbf{E}_k\left(S_1^n - Z_k\right) - \mathbf{E}_{k-1}\left(S_1^n - Z_k\right)| \le |\xi_k - a_k| + \mathbf{E}\,|\xi_k - a_k|. \quad (3.5.7)$$

It follows by the Hölder inequality that if $p > 1$, then

$$\mathbf{E}\,|X_k|^p \le 2^{p-1}\left(\mathbf{E}\,|\xi_k - a_k|^p + [\mathbf{E}\,|\xi_k - a_k|]^p\right) \le 2^p\mathbf{E}\,|\xi_k - a_k|^p. \quad (3.5.8)$$

The bound of the theorem now follows from Proposition 3.5.2. ○

3.5.4 Estimates in Terms of Sums Without Repetitions

When the moments $\mathbf{E}\,|\xi_j - a_j|^p$ vary considerably with j, it may prove essential to evaluate probabilities of large deviations using more complicated combinations of moments than the sums considered in the preceding subsections.

Let $A = \left(A_{i(1)\cdots i(m)}\right)$ be an m-dimensional array (whose indices all have the range $i(\bullet) = \overline{1,n}$). The *ordered sum* of its elements is

$$\sum_{1}^{n}{}_{\uparrow}^{m}\,[A] = \sum_{1 \le i(1) < i(2) < \cdots < i(m) \le n} A_{i(1)\cdots i(m)}. \quad (3.5.9)$$

Denote the Kronecker power of a vector $b = (b_j) \in \mathbf{R}^n$ by $b^{\otimes m} = \left(b_{i(1)}\cdots b_{i(m)}\right)$. It is easy to see that for each $m \le n$

$$\sum_{1}^{n}{}_{\uparrow}^{m}\left[b^{\otimes m}\right] = \sum_{i=m}^{n} b_i \sum_{1}^{i-1}{}_{\uparrow}^{m-1}\left[b^{\otimes(m-1)}\right],$$

(3.5.10)

$$\sum\nolimits_{\uparrow 1}^{m\ n} \left[b^{\otimes m}\right] \leq \left(\sum\nolimits_{j=1}^{n} b_j\right)^m$$

if $b_j \geq 0$ for all j.

The next theorem provides an estimate in terms of ordered sums for the probabilities of large deviations of a sum of independent RV's. General assumptions about the RV's, the nonrandom vectors a_j, and the measurable seminorm are as in subsections 3.5.1–3.5.3.

Theorem 3.5.3 *Assume that* $|\xi_j - a_j| \leq L$ *a.s. for* $j = \overline{1,n}$.

The following inequality holds for each set of parameters $m \in \mathbf{N}$, $\gamma \in (0,\,1)$, *and* $t > 0$: *if* $x \geq 8(m-1)L/\gamma$, *then*

$$\mathbf{P}\left\{S_1^n \geq \mathbf{E}S_1^n + x\right\} \leq \exp\left\{-(1-\gamma)\min\left[A, B\right]\right\},$$

where

$$A = \frac{\gamma}{16} e^t x^2 \left(m! \sum\nolimits_{\uparrow 1}^{m\ n} \left[\{\mathbf{E}\,|\xi - a|\}^{\otimes m}\right]\right)^{-1/m},$$

$$B = \frac{x}{2L} \ln\left[\frac{\gamma}{8} x L^t \left(m! \sum\nolimits_{\uparrow 1}^{m\ n} \left[\{\mathbf{E}|\xi - a|^{1+t}\}^{\otimes m}\right]\right)^{-1/m}\right].$$

The proof of this theorem needs some preparations.

Lemma 3.5.2 *Assume that the sequence* (u_j) *obeys the restriction* $0 \leq u_k \leq U$. *Then for* $n \geq m$ *the ordered sums satisfy the inequality*

$$\sum\nolimits_{\uparrow 1}^{m\ n} \left[u^{\otimes m}\right] \geq \frac{1}{m!}\left[U_{1,n} - (m-1)U\right]_+^m, \quad U_{k,l} \stackrel{def}{=} \sum_{j=k}^{l} u_j.$$

Proof of Lemma 3.5.2 is an induction argument.

Its assertion is obvious for $m = 1$.

Assume that it is true for sums of order not exceeding m. Put $\widetilde{U}(x) = U_{1,k}$ if $x \in [U_{1,k},\, U_{1,k+1})$. It follows from (3.5.10) and the induction assumption that

$$\begin{aligned}\sum\nolimits_{\uparrow 1}^{m+1\ n} \left[u^{\otimes(m+1)}\right] &\geq \sum_{i=m+1}^{n} \frac{u_i}{m!}\left(U_{1,i-1} - (m-1)U\right)_+^m \\ &= \int_{U_{1,m}}^{U_{1,n}} \frac{1}{m!}\left(\widetilde{U}(x) - (m-1)U\right)_+^m dx.\end{aligned}$$

The restrictions of the lemma ensure that $\widetilde{U}(x) \geq x - U$ and $U_{1,m} \leq mU$. Consequently,

$$\sum\nolimits_{\uparrow 1}^{m+1\ n} \left[u^{\otimes(m+1)}\right] \geq \int_0^{U_{1,n}} \frac{(x - mU)_+^m}{m!}dx = \frac{(U_{1,n} - mU)_+^{m+1}}{(m+1)!}.$$

This is the estimate of the lemma for sums of order $m+1$, so the induction is completed. ○

One more ingredient for the proof of Theorem 3.5.3 is

Lemma 3.5.3 *Let* (X_k) *be the martingale differences of decomposition* (3.3.1), *and assume satisfied the conditions of Theorem* 3.5.3.

The following inequality holds for $h \geq 0$ *and* $t > 0$:

$$\phi_k(h) = \ln \mathbf{E}_{k-1} e^{hX_k} \leq \int_0^h [\breve{a}_k(h') + b_k(h')]\, dh',$$

where

$$\begin{aligned} \breve{a}_k(h) &= 4 \min\left\{L, he^t \mathbf{E}\,|\xi_k - a_k|^2\right\}, \\ b_k(h) &= 2 \min\left\{L, L^{-t}(e^{2hL} - 1)\mathbf{E}\,|\xi_k - a_k|^{1+t}\right\}. \end{aligned}$$

Proof of Lemma 3.5.3. By (3.5.7) the rv X_k is bounded. Hence the expectation $\mathbf{E}_{k-1} \exp\{hX_k\}$ is differentiable in h, and the order of differentiation and calculation of the conditional expectation can be reversed. Thus,

$$\phi_k(h) = \int_0^h \phi_k'(h')dh', \quad \phi_k'(h) = \exp\{-\phi_k(h)\}\, \mathbf{E}_{k-1} X_k \left(e^{hX_k} - 1\right),$$

since $\mathbf{E}_{k-1}X_k = 0$. By the Jensen inequality $\phi_k(h) \geq 0$ and $e^{\phi_k(h)} \geq 1$.

If $t, h > 0$, then

$$\phi_k'(h) = \hat{a} + \hat{b},$$

where

$$\begin{aligned} \hat{a} &= \exp\{-\phi_k(h)\}\, \mathbf{E}_{k-1} X_k \left(e^{hX_k} - 1\right) \mathbf{1}_{\{X_k \leq t/h\}}, \\ \hat{b} &= \exp\{-\phi_k(h)\}\, \mathbf{E}_{k-1} X_k \left(e^{hX_k} - 1\right) \mathbf{1}_{\{X_k > t/h\}}. \end{aligned}$$

Inequality (3.5.7) and the fact that X_k has zero conditional expectation imply the estimates

$$\begin{aligned} \left|\hat{a} + \hat{b}\right| &= \left|\frac{\mathbf{E}_{k-1} X_k e^{hX_k}}{\mathbf{E}_{k-1} e^{hX_k}}\right| \leq 2L, \\ |\hat{a}| &\leq 4L, \quad 0 \leq \hat{b} \leq e^{-\phi_k(h)} \mathbf{E}_{k-1} X_k e^{hX_k} \mathbf{1}_{\{X_k > t/h\}} \leq 2L. \end{aligned}$$

It is easy to see that $u\,(e^u - 1) \leq u^2 \exp\{u_+\}$ for each u. Hence (3.5.8) implies the estimate

$$\hat{a} \leq \frac{1}{h} \mathbf{E}_{k-1} \left[hX_k \left(e^{hX_k} - 1\right)\right] \mathbf{1}_{\{X_k \leq t/h\}} \leq he^t \mathbf{E}_{k-1} X_k^2 \leq 4he^t \mathbf{E}\,|\xi_k - a_k|^2$$

if $\hat{a} > 0$ (inequality $e^{\phi_k(h)} \geq 1$ is used once again).

The function $(e^u - 1)/u^t$ is nondecreasing for $u > t$, so it follows from (3.5.7) and (3.5.8) that for positive values of $\hat{b}$

$$\hat{b} \leq \mathbf{E} X_k \left(e^{hX_k} - 1\right) \mathbf{1}_{\{X_k > t/h\}} \leq 2L^{-t}(e^{2hL} - 1)\mathbf{E}\,|\xi_k - a_k|^{1+t}.$$

This proves the lemma: its estimates are trivially true for nonpositive values of $\hat{a}$ and $\hat{b}$. ○

Proof of Theorem 3.5.3. By the Chebyshev inequality and (3.5.1), the "tail" of S_1^n admits the bound

$$\mathbf{P} = \mathbf{P}\{S_1^n \geq \mathbf{E}S_1^n + x\} \leq e^{-hx}\mathbf{E}\exp\{h(X_1 + \cdots + X_n)\}.$$

When the conditional means E_k are calculated for $k = n, n-1, \ldots 1$ consecutively, one can apply Lemma 3.5.3. This yields the inequality

$$\mathbf{P} \leq \exp\{-(1-\gamma)xh + G(h)\}, \tag{3.5.11}$$

where the number $\gamma \in (0, 1)$ is arbitrary and

$$\begin{aligned} G(h) = G(h;\gamma) &= -\gamma xh + \textstyle\int_0^h [A(h') + B(h')]\, dh', \\ A(h) &\stackrel{def}{=} \textstyle\sum_{j=1}^n \breve{a}_j(h), \quad B(h) \stackrel{def}{=} \sum_{j=1}^n b_j(h). \end{aligned}$$

Both functions A and B are continuous and nondecreasing by Lemma 3.5.3, with $A(0) = B(0) = 0$. Consequently, the function G is convex, $G(0) = 0$, and $G'(0) = -\gamma x < 0$.

If $G(0) \leq 0$ for all $h > 0$, then the passage to the limit at $h \to \infty$ in (3.5.11) shows that $\mathbf{P} = 0$.

If $G(h)$ is positive for large values of h, this convex function has a unique minimum at some point $h_0 > 0$, with $G(h_0) < 0$. The point of minimum satisfies the necessary condition

$$\gamma x = A(h_0) + B(h_0),$$

so at least one of the two inequalities,

$$A(h_0) \geq \tfrac{1}{2}\gamma x, \quad B(h_0) \geq \tfrac{1}{2}\gamma x, \tag{3.5.12}$$

holds in this case.

If the first inequality in (3.5.12) is valid, then by Lemma 3.5.2

$$\begin{aligned} \tfrac{1}{2}\gamma x - (m-1)4L &\leq \left(m!\sum_1^n {}_{\uparrow}^{m} \left[\breve{a}^{\otimes m}\right]\right)^{1/m} \\ &\leq 4h_0 e^t \left(m!\sum_1^n {}_{\uparrow}^{m} \left[\mathbf{E}\,|\xi_\cdot - a_\cdot|^{\otimes m}\right]\right)^{1/m}. \end{aligned}$$

Hence

$$h^0 \geq \tfrac{1}{16}\gamma e^{-t} x \left(m!\sum_1^n {}_{\uparrow}^{m} \left[\mathbf{E}\,|\xi_\cdot - a_\cdot|^{\otimes m}\right]\right)^{-1/m}.$$

This inequality yields the bound of the theorem after it is substituted in (3.5.11).

If the second inequality in (3.5.12) holds, then by Lemma 3.5.2

$$\tfrac{1}{2}\gamma x - (m-1)2L \leq \left(m!\sum_1^n {}_{\uparrow}^{m} \left[b^{\otimes m}\right]\right)^{1/m}$$

$$\leq \frac{2e^{2h_0 L}}{L^t}\left(m!\sum_{1}^{n}{}^{m}_{\uparrow}\left[\left\{\mathbf{E}\,|\xi_{\cdot}-a_{\cdot}|^{1+t}\right\}^{\otimes m}\right]\right)^{1/m}.$$

It follows that

$$h_0 \geq \frac{1}{2L}\ln\left\{\frac{\gamma}{8}xL^t\left(m!\sum_{1}^{n}{}^{m}_{\uparrow}\left[\left\{E\,|\xi_{\cdot}-a_{\cdot}|^{1+t}\right\}^{\otimes m}\right]\right)^{-1/m}\right\}.$$

This latter inequality is substituted in (3.5.11) to complete the proof of the theorem. ○

3.6 The Cramér Transform

3.6.1 Definition and Inversion Formula

The results of this section refer to RV's with values in a separable real Banach space X. The norm in X and its conjugate X^* is, as usual, denoted by $|\cdot|$; for a functional $t \in X^*$ its value on the vector $x \in X$ is $\langle x, t\rangle$.

Consider a RV ξ with the distribution

$$F(B) = \mathbf{P}\{\xi \in B\}, \quad B \in \mathcal{B}(X). \tag{3.6.1}$$

In what follows, the RV ξ usually satisfies the *Cramér condition*:

$$\exists A_1 > 0 \;\; A_2 = \mathbf{E}e^{A_1|\xi|} < \infty. \tag{3.6.2}$$

In this case the *Laplace transform* of the distribution F,

$$\varphi(h) = \mathbf{E}e^{\langle\xi,h\rangle} = \int_X e^{\langle x,h\rangle}F(dx), \quad h \in X^*, \tag{3.6.3}$$

is well defined and finite at least for all $h \in X^*$ satisfying the restriction $|h| \leq A_1$. It can be extended to the "strip" $\{h' + ih'' : h', h'' \in X^*, |h''| \leq A_1\}$ in the complexification $X^* + iX^*$ of the conjugate space X^*. For the "purely imaginary" values of its argument, the Laplace transform $\varphi(ih)$, $h \in X^*$, of a distribution is evidently its CF.

Suppose that $\varphi(h)$ is finite for some $h \in X^*$. The corresponding *Cramér transform* of the distribution F is defined by the formula

$$F_h^*(B) = \frac{1}{\varphi(h)}\mathbf{E}e^{\langle\xi,h\rangle}\mathbf{1}_{\{\xi\in B\}} = \frac{1}{\varphi(h)}\int_B e^{\langle x,h\rangle}F(dx). \tag{3.6.4}$$

The distribution F_h^* is called the *Cramér conjugate* of F, and a RV with this distribution is sometimes also called the Cramér conjugate of ξ.

The Cramér transform is a common technique in the study of large deviations. Its applications to sums of independent RV's are based on the following inversion formula. Suppose that F_j, $j = \overline{1,n}$, are the distributions of independent RV's ξ_j. Assume that the Cramér transforms $F_{h,j}^*$ of these distributions are well defined for some value of $h \in X_*$, and that ξ_j^* are independent RV's with the conjugate distributions $F_{h,j}^*$.

Lemma 3.6.1 *If $f : X^n \to \mathbf{C}$ is a measurable complex-valued functional, then*

$$\mathbf{E} f(\xi_1, \dots, \xi_n) = \exp\left\{\sum_{j=1}^n \ln \varphi_j(h)\right\} \mathbf{E}\, e^{-\langle S_{1,n}[\xi^*], h\rangle} f(\xi_1^*, \dots, \xi_n^*),$$

where φ_j is the Laplace transform (3.6.3) *of the distribution of ξ_j. If either one of the expectations in the left- and right-hand side is finite, the other is also finite.*

Notation here is that of the preceding sections, e.g., $S_{1,n}[\xi^*] = \sum_{j=1}^n \xi_j^*$.

Proof of Lemma 3.6.1. To derive the equality of the lemma, it suffices to use the expression for the density of the joint distribution $\widetilde{F}_n$ of the "original" RV's ξ_j with respect to the joint distribution $\widetilde{F}_n^*$ of the conjugate ones ξ_j^*:

$$\frac{d\widetilde{F}_n}{d\widetilde{F}_n^*}(x_1, \dots, x_n) = \exp\left\{-\sum_{j=1}^n \langle x_j, h\rangle\right\} \prod_{j=1}^n \varphi_j(h). \ \bigcirc$$

The assumption that the Laplace transform is finite is actually a restriction on one-dimensional distributions of a RV, so it does not necessarily imply the existence of the expectations $\mathbf{E}\xi_j$ or $\mathbf{E}\xi_j^*$. In a situation where the "conjugate" RV's of Lemma 3.6.1 do have expectations (in either the Bochner or Pettis sense), the equality of Lemma 3.6.1 can be written in the form

$$\begin{aligned} \mathbf{E} f(\xi_1, \dots, \xi_n) &= \exp\left\{\sum_{j=1}^n \left(\ln \varphi_j(h) - \langle \mathbf{E}\xi_j^*, h\rangle\right)\right\} && (3.6.5) \\ &\times \ \mathbf{E} \exp\left\{-\langle S_{1,n}[\bar{\xi}], h\rangle\right\} f\left(\mathbf{E}\xi_1^* + \bar{\xi}_1, \dots, \mathbf{E}\xi_n^* + \bar{\xi}_n\right), \end{aligned}$$

where $S_{1,n}[\bar{\xi}]$ is the sum of the centered RV's $\bar{\xi}_j = \xi_j^* - \mathbf{E}\xi_j^*$.

Remark 3.6.1 The Cramér condition (3.6.2) and condition of Theorem 3.3.2 are essentially versions of the same restriction. Indeed, (3.6.2) implies the estimate

$$\mathbf{E}\,|\xi|^l \le \tfrac{1}{2} l! H^{l-2} \mathbf{E}\,|\xi|^2, \quad H = \sup_{l \ge 3} (2A_2)^{1/(l-2)} \left(A_1^l \mathbf{E}\,|\xi|^2\right)^{-1/(l-2)} < \infty.$$

To verify this inequality, note that $|x|^l \le l! e^{|x|}$ for all x.

Suppose that the condition of Theorem 3.3.2 is valid for $a_j = 0$. The power expansion of the exponential and this condition yield the inequality

$$\sum_{j=1}^n \mathbf{E} e^{h|\xi_j|} \le h \sum_{j=1}^n \mathbf{E}\,|\xi_j| + \frac{1}{2} \frac{B^2 h^2}{1 - hH}, \quad h < 1/H.$$

The Cramér condition is satisfied, e.g., by RV's with bounded norm: indeed, $\mathbf{E} e^{A|\xi|} \le e^{AL}$ with $A > 0$ and $\mathbf{E}\,|\xi|^l \le b^2 L^{l-2}$, $b^2 = \mathbf{E}\,|\xi|^2$ for all $l \ge 2$ if $|\xi| \le L$ a.s.

3.6.2 Moments of the Cramér Conjugate

The Laplace transform is a smooth function of its argument in the interior of the domain where it is defined.

Lemma 3.6.2 *If distribution* (3.6.1) *satisfies the Cramér condition* (3.6.2), *its Laplace transform* (3.6.3) *has weak differentials of arbitrary order*

$$\prod_{j=1}^{k}\langle g_j,\nabla\rangle\varphi(h) \equiv \left.\frac{\partial^k}{\partial t_1\cdots\partial t_k}\varphi(h+t_1g_1+\cdots+t_kg_k)\right|_{t_\bullet=0},$$

which are continuous in all their arguments. These differentials admit the representation

$$\prod_{j=1}^{k}\langle g_j,\nabla\rangle\varphi(h) = \mathbf{E}\mu_k(g_1,\dots,g_k;h),$$

where

$$\mu_k(g_1,\dots,g_k;h) \stackrel{def}{=} \prod_{j=1}^{k}\langle g_j,\nabla\rangle\exp\{\langle\xi,h\rangle\} = \exp\{\langle\xi,h\rangle\}\prod_{j=1}^{k}\langle\xi,g_j\rangle.$$

Remark 3.6.2 Notation for weak differentials deviates from the standard usage. Writing "directional derivatives" in this form is more common in physics. However, it seems more convenient in the probabilistic setting, where it is often necessary to take averages with respect to the distribution of "direction".

It follows from the existence of continuous weak derivatives that the Fréchet derivatives exist as well.

Formulae for derivatives of the Laplace transform similar to those of the lemma hold also for complex values of its argument. The fact is verified by analogy with the "real" case treated below.

Proof of Lemma 3.6.2. The exponential $\exp\{\langle\xi,h\rangle\}$ is evidently differentiable in h, so the task consists in verification of the possibility to interchange the order of differentiation and computing expectations in the right-hand side of (3.6.3).

The rv's μ_k are uniformly integrable for $|h|\le A_1-\varepsilon$ and $|g_j|\le G$ whatever the choice of $\varepsilon, G>0$. This follows by the Lebesgue theorem from the existence of an integrable majorant: if $\delta>0$ is sufficiently small, then

$$\begin{aligned}\mu_k^+ = e^{|\langle\xi,h\rangle|}\prod_{j=1}^{k}\langle\xi,g_j\rangle &\le c(\delta,k)\,e^{(|h|+k\delta)|\xi|}\prod_{j=1}^{k}|g_j| \\ &\le c(\delta,k)\exp\{A_1|\xi|\}\prod_{j=1}^{k}|g_j|.\end{aligned} \tag{3.6.6}$$

Hence the Lebesgue theorem is indeed applicable, and the order of integration and differentiation can be reversed. ○

If distribution (3.6.1) obeys the Cramér condition (3.6.2), the conjugate distribution (3.6.4) also satisfies it (with an exponent which may be smaller and satisfies the inequality $A_1^*\ge A_1-|h|$).

The moments of the conjugate distribution are expressed in terms of derivatives of the Laplace transform (3.6.3) φ. Thus,

$$\begin{aligned}\mathbf{E}\langle\xi^*,g\rangle &= \langle g,\nabla\rangle\ln\varphi(h), \\ \mathbf{cov}\left(\langle\xi^*,g_1\rangle,\langle\xi^*,g_2\rangle\right) &= \langle g_1,\nabla\rangle\langle g_2,\nabla\rangle\ln\varphi(h).\end{aligned} \tag{3.6.7}$$

Note that by the Jensen inequality and (3.6.7), in the domain of definition of the Laplace transform

$$\mathbf{E}\xi = 0 \implies \varphi(h) \geq 1; \quad |\varphi(h) - 1| \leq c|h|^2 \text{ if } |h| \leq c. \tag{3.6.8}$$

Hence for centered RV's and small values of h (viz., $|h| \leq c_0$)

$$\begin{aligned} |\mathbf{E}\xi^*| &\leq c_1 |h|, \\ |\mathbf{E}\xi^* - \mathcal{V}h| &= |\mathbf{E}\xi^* - \mathbf{E}\langle \xi, h\rangle \xi| \leq c_2 |h|^2, \end{aligned} \tag{3.6.9}$$

where $\mathcal{V}$ is the covariance operator of ξ. The values of the constants are determined by the distribution of ξ. They can be evaluated in terms of the constants in the Cramér condition.

Consider a symmetric bilinear form $\mathcal{A}$ on X. Define its norm by the usual agreement $|\mathcal{A}| = \sup_{|x|\leq 1, |y|\leq 1} \mathcal{A}(x, y)$. If $\mathcal{A}(\xi, \xi) \exp\{\langle \xi, h\rangle\}$ is summable, then

$$\mathbf{E}\mathcal{A}(\xi^*, \xi^*) = \frac{1}{\varphi(h)} \mathbf{E}\mathcal{A}(\xi, \xi) \exp\{\langle \xi, h\rangle\}.$$

For the centered RV $\bar{\xi} = \xi^* - \mathbf{E}\xi^*$ and $|h| \leq c_5$, this latter equality implies the bound

$$\begin{aligned} & |\mathbf{E}\mathcal{A}(\bar{\xi}, \bar{\xi}) - \mathbf{E}\mathcal{A}(\xi, \xi)| \\ = \; & |\mathbf{E}\mathcal{A}(\xi^*, \xi^*) - \mathcal{A}(\mathbf{E}\xi^*, \mathbf{E}\xi^*) - \mathbf{E}\mathcal{A}(\xi, \xi)| \leq c_4 |h|. \end{aligned} \tag{3.6.10}$$

A similar inequality is easily obtained for the averages of arbitrary tensor powers of the conjugate RV's: if $\mathcal{A}_k$ is a bounded k-linear form in X, then

$$\begin{aligned} & \left|\mathbf{E}\mathcal{A}_k\left([\xi^*]^{\otimes k}\right) - \mathbf{E}\mathcal{A}_k\left([\xi]^{\otimes k}\right)\right| \leq \frac{1}{\varphi(h)} \left|\mathbf{E}\mathcal{A}_k\left([\xi]^{\otimes k}\right)\left[e^{\langle \xi, h\rangle} - 1\right]\right| \\ + \; & \frac{|\varphi(h) - 1|}{\varphi(h)} \left|\mathbf{E}\mathcal{A}_k\left([\xi]^{\otimes k}\right)\right| \leq c_{5,k} |h|, \\ & \left|\mathbf{E}\mathcal{A}_k\left([\bar{\xi}]^{\otimes k}\right) - \mathbf{E}\mathcal{A}_k\left([\xi]^{\otimes k}\right)\right| \leq c_{5,k} |h|. \end{aligned} \tag{3.6.11}$$

The constants c_i in (3.6.9)-(3.6.11) are all determined by the values of the constants in the Cramér condition (3.6.2).

The following "collective" estimates for the moments of Cramér conjugates are useful in calculations with sums of independent, but not identically distributed, RV's.

Lemma 3.6.3 *Let ξ_j be RV's with values in a separable real Banach space X with zero expectations. Assume that for all natural numbers $l \geq 2$*

$$\sum_{j=1}^{n} \mathbf{E}|\xi_j|^l \leq \tfrac{1}{2} l! B_n^2 H^{l-2}.$$

Then the RV's ξ_j^* ***with Cramér conjugate distributions and the centered conjugate RV's*** $\bar{\xi}_j = \xi_j^* - \mathbf{E}\xi_j^*$ ***of Lemma*** 3.6.1 ***admit the following estimates: if*** $H\,|h| < c_3$,

$$\sum_{j=1}^{n} \left|\mathbf{E}\xi_j^* - \mathbf{E}\langle\xi_j, h\rangle\xi_j\right| \le c_1 |h|^2 B_n^2 H, \tag{3.6.12}$$

$$\sum_{j=1}^{n} \left|\mathbf{E}\langle\bar{\xi}_j, s\rangle\langle\bar{\xi}_j, t\rangle - \mathbf{E}\langle\xi_j, s\rangle\langle\xi_j, t\rangle\right| \le c_2 |h|\,|s|\,|t|\,B_n^2 H. \tag{3.6.13}$$

Moreover, if $u > 0$ ***and*** $(2u + |h|)\,H < 1$, ***then***

$$\sum_{j=1}^{n}\sum_{l=3}^{\infty} \frac{u^l}{l!}\mathbf{E}\left|\bar{\xi}_j\right|^l \le \frac{B_n^2 H\,(2u + |h|)^3}{1 - (2u + |h|)\,H}. \tag{3.6.14}$$

The constants in the lemma can be evaluated at the price of some additional calculations. For instance, the proof below furnishes values $c_1 = 6/(1 - 0.3)$ and $c_2 = 13.5/(1 - 0.8)$ if $c_3 = 0.1$.

The proof of this lemma makes use of the following elementary estimates:

$$\begin{aligned}
\sum_{m=0}^{k} \frac{k!}{(k-m)!m!} &= 2^k, \\
\sum_{l=1}^{\infty} \frac{(l+2)!}{l!}v^l &= \frac{d^2}{du^2}\frac{1}{1-u}\bigg|_{u=0}^{u=v} \le \frac{6v}{1-3v}, \\
\sum_{l=1}^{\infty} \frac{(l+3)!}{l!}v^l &= \frac{d^3}{du^3}\frac{1}{1-u}\bigg|_{u=0}^{u=v} \le \frac{24v}{1-4v}, \qquad 0 < v < 1.
\end{aligned} \tag{3.6.15}$$

Proof of Lemma 3.6.3. a) By (3.6.8)

$$\left|\mathbf{E}\xi_j^* - \mathbf{E}\langle\xi_j, h\rangle\xi_j\right| \le [\varphi(h) - 1]\,\mathbf{E}\left|\xi_j\right|^2 |h| + \mathbf{E}\left|\xi_j\right|^2 |h| \left(e^{|h||\xi_j|} - 1\right).$$

Using the Taylor expansion, the right-hand side of this inequality is estimated in terms of moments of the positive rv $|\xi_j|$. By the Hölder inequality

$$\mathbf{E}\left|\xi_j\right|^l \mathbf{E}\left|\xi_j\right|^m \le \mathbf{E}\left|\xi_j\right|^{l+m}$$

for each choice of $l \ge 0$ and $m \ge 0$. Inequality (3.6.12) follows from the bound obtained above and (3.6.15) after summing in j.

b) To derive (3.6.13), start from the following inequality (which is derived as a similar bound of part a):

$$\begin{aligned}
\kappa_j &= \left|\mathbf{E}\langle\xi_j^*, s\rangle\langle\xi_j^*, t\rangle - \mathbf{E}\langle\xi_j, s\rangle\langle\xi_j, t\rangle\right| \\
&\le |s|\,|t|\left([\varphi(h) - 1]\,\mathbf{E}\left|\xi_j\right|^2 + \mathbf{E}\left|\xi_j\right|^2\left[e^{|\xi_j||h|} - 1\right]\right) \\
&\le 2\,|s|\,|t| \sum_{l=1}^{\infty} \frac{|h|^l}{l!}\mathbf{E}\left|\xi_j\right|^{l+2}.
\end{aligned}$$

Combined with the assumption of the lemma and (3.6.15), this inequality yields the estimate

$$\sum_{j=1}^{n} \kappa_j \le \frac{6\,|s|\,|t|\,H\,|h|\,B_n^2}{1-3H\,|h|}. \tag{3.6.16}$$

The original RV's ξ_j are centered. Consequently, as also in part a),

$$\left|\langle \mathbf{E}\xi_j^*, s\rangle\right| \le |s|\,|h| \left(\mathbf{E}\,|\xi_j|^2 + 2\sum_{l=1}^{\infty} \frac{|h|^l}{l!}\mathbf{E}\,|\xi_j|^{l+2}\right). \tag{3.6.17}$$

One can get a less precise estimate for the same expectation directly from the definition of the conjugate distribution and the inequality in (3.6.8):

$$\left|\langle \mathbf{E}\xi_j^*, t\rangle\right| \le |t| \sum_{m=1}^{\infty} \frac{|h|^m}{m!}\mathbf{E}\,|\xi_j|^{m+1}.$$

This latter inequality and (3.6.17) imply the bound

$$\begin{aligned}\left|\langle \mathbf{E}\xi_j^*, s\rangle\langle \mathbf{E}\xi_j^*, t\rangle\right| &\le |s|\,|t|\,|h| \left(\sum_{m=1}^{\infty} \frac{|h|^m}{m!}\mathbf{E}\,|\xi_j|^{m+3}\right.\\ &+ \left. 2\sum_{l=1}^{\infty}\sum_{m=1}^{\infty} \frac{|h|^{l+m}}{l!m!}\mathbf{E}\,|\xi_j|^{l+m+3}\right).\end{aligned}$$

After summing in j and elementary algebraic calculations, the above estimates, (3.6.15), and the conditions of the lemma yield the inequality

$$\sum_{j=1}^{n} \left|\langle \mathbf{E}\xi_j^*, s\rangle\langle \mathbf{E}\xi_j^*, t\rangle\right| \le 60\,|s|\,|t|\,\frac{B_n^2\,|h|^2\,H^2}{1-8\,|h|\,H}.$$

Combined with (3.6.16), the last bound proves (3.6.13).

c) To derive the last estimate of the lemma one should but provide sufficiently precise estimates for the moments of $|\xi_j^*|$ since by the Hölder inequality

$$\mathbf{E}\,|\bar{\xi}_j|^m \le \mathbf{E}\left(|\xi_j^*| + \mathbf{E}\,|\xi_j^*|\right)^m \le 2^m\mathbf{E}\,|\xi_j^*|^m.$$

It follows from (3.6.8) and Lemma 3.6.2 that

$$\mathbf{E}\,|\xi_j^*|^m \le \sum_{l=0}^{\infty} \frac{|h|^l}{l!}\mathbf{E}\,|\xi_j|^{l+m}, \quad m \ge 3.$$

Consequently,

$$\begin{aligned}\sum_{j=1}^{n}\sum_{m=3}^{\infty} \frac{u^m}{m!}\mathbf{E}\,|\bar{\xi}_j|^m &\le \frac{1}{2}\sum_{m=3}^{\infty}\sum_{l=0}^{\infty} \frac{(l+m)!}{l!m!}(2u)^m\,|h|^l\,B_n^2 H^{l+m-2}\\ \le \frac{B_n^2}{2H^2}\sum_{k=3}^{\infty}\left[\sum_{l=0}^{k}\binom{k}{l}|h|^l\,(2u)^{k-l}\right]H^k &= \frac{1}{2}\frac{B_n^2 H\,(|h|+2u)^3}{1-H\,(|h|+2u)}.\end{aligned}$$

This inequality completes the proof of the lemma. ○

A slightly more convenient estimate can be derived for RV's with values in a separable real Hilbert space. As before, the norm is $|\cdot|$ also for RV's in Hilbert space X, and the restrictions of Lemma 3.6.3 are supposed to hold. Notation of preceding calculations is conserved below, and angle brackets refer to the natural autoduality of a Hilbert space.

Lemma 3.6.4 *Let $\mathcal{P}'X \to \mathcal{P}'X$ be an operator of orthogonal projection in the Hilbert space X.*

If $\bar{\xi}_j$ are the centered RV's defined as in Lemma 3.6.1, then the following estimate holds for $H\,|h| < c_3$:

$$\sum_{j=1}^{n}\left|\mathbf{E}\,|\mathcal{P}'\bar{\xi}_j|^2 - \mathbf{E}\,|\mathcal{P}'\xi_j|^2\right| \leq c_2\,|h|\,H.$$

If $(2u+|h|) < 1$ and $u > 0$, then

$$\sum_{j=1}^{n}\sum_{l=3}^{\infty}\frac{u^l}{l!}\mathbf{E}\,|\mathcal{P}'\bar{\xi}_j|^l \leq \frac{B_n^2 H\,(2u+|h|)^3}{1-(2u+|h|)\,H}.$$

Proof of Lemma 3.6.4 is a repetition of calculations from parts a) and b) of the proof of Lemma 3.6.3. It starts from the inequality $|\mathcal{P}'x| \leq |x|$, valid for all x, and the estimate

$$\left|\mathbf{E}\,|\mathcal{P}'\xi_j^*|^2 - \mathbf{E}\mathcal{P}'\,|\xi_j|^2\right| \leq \frac{(\varphi_j-1)}{\varphi_j(h)}\mathbf{E}\,|\xi_j|^2 + \frac{1}{\varphi_j(h)}\mathbf{E}\,|\xi_j|^2\left[e^{|h||\xi_j|}-1\right]. \ \bigcirc$$

3.7 Truncations

3.7.1 Inequalities for Probabilities

Truncation is a common technique for reducing problems about sums of unbounded RV's to simpler ones for sums of bounded RV's (see, e.g., subsection 3.5.2). This section contains some inequalities usually employed in the reduction procedure.

Notation of Section 3.1 is used without comment: S_1^n, V_1^n, etc. are the quantities defined in (3.1.2)–(3.1.6)). The values of the RV's considered are in a real linear measurable space X carrying some measurable seminorm $|\cdot|$ (e.g., separable Banach or Hilbert).

Let ξ be some RV. Its *truncation* at level $r > 0$ is defined by

$$\xi^r = \mathbf{1}_{\{|\xi|\leq r\}}\xi, \quad r > 0, \tag{3.7.1}$$

so $|\xi^r| \leq r$ a.s. (For $r = \infty$ the truncated RV $\xi^\infty = \xi$ is the original one.)

Truncation is usually applied to sequences of independent RV's ξ_j. It follows from the additivity of probability that for any RV's (possibly dependent) and

$r < R$ the truncations satisfy the inequality

$$\left|\mathbf{P}\left\{S_{1,n}\left[\xi^{r}\right]\in C\right\}-\mathbf{P}\left\{S_{1,n}\left[\xi^{R}\right]\in C\right\}\right| \begin{array}{l}\leq \sum_{j=1}^{n}\mathbf{P}\left\{\xi_j^r\neq\xi_j^R\right\}\\ \leq \sum_{j=1}^{n}\mathbf{P}\left\{r\leq|\xi_j|<R\right\}.\end{array} \tag{3.7.2}$$

This estimate holds for any measurable subset $C \subset X$ whatever the choice of $r < R \leq \infty$.

A more precise and special result of this kind is

Lemma 3.7.1 *Let ξ_j be independent RV's with values in a separable real Banach space X with norm $|\cdot|$ such that $\mathbf{E}|\xi_j|^p < \infty$, $j = \overline{1,n}$, for some $p \geq 1$. Put*

$$\Lambda = \left(\mathbf{E}\left|S_{1,n}[\xi^r]\right|^p\right)^{1/p} + \left(\mathbf{E}\left|S_{1,n}[\xi^R]\right|^p\right)^{1/p} + \sum_{j=1}^{n}\left(\left|\mathbf{E}\xi_j^r\right| + \left|\mathbf{E}\xi_j^R\right|\right).$$

There exist constants $c_i > 0$ such that the conditions

$$\widehat{\mathbf{p}} = \mathbf{P}\left\{\bigvee_1^n\xi \geq r\right\} \leq c_1, \quad \widehat{\mathbf{q}} = \sum_{j=1}^{n}\mathbf{P}\left\{r \leq |\xi_j| < R\right\} \leq c_2$$

imply the inequality

$$\left|\mathbf{P}\left\{S_{1,n}[\xi^r]\in C\right\}-\mathbf{P}\left\{S_{1,n}[\xi^R]\in C\right\}\right| \leq \frac{c_3}{\rho^p}\left(\widehat{\mathbf{q}}\Lambda^p + \sum_{j=1}^{n}\mathbf{E}|\xi_j|^p\mathbf{1}_{\{r\leq|\xi_j|<R\}}\right)$$

for each $\rho \geq 1$, $0 < r < R$, and measurable subset $C \subset \{x : |x| \geq \rho\}$.

As in later propositions of this section, the constants in Lemma 3.7.1 are determined by the choice of the exponent p.

A slightly more convenient inequality using only restrictions on the distributions of individual summands can be stated for independent RV's ξ_j taking values from a separable real Hilbert space with norm $|\cdot|$ and the inner product $(\cdot,\cdot)$.

Lemma 3.7.2 *Assume that the independent Hilbert-space-valued RV's ξ_j obey the moment restrictions $\mathbf{E}|\xi_j|^p < \infty$ and $\mathbf{E}\xi_j = 0$ with some $p \geq 2$. Put*

$$\begin{aligned}\widehat{\mathbf{q}} &= \sum_{j=1}^{n}\mathbf{P}\left\{r\leq|\xi_j|<R\right\},\\ \tilde{\Lambda} &= \left(\sum_{j=1}^{n}\mathbf{E}|\xi_j|^p\right)^{1/p} + \left(\sum_{j=1}^{n}\mathbf{E}|\xi_j|^2\right)^{1/2}\\ &\quad + \min_{2\leq p'\leq p} r^{-(p'-1)}\sum_{j=1}^{n}\mathbf{E}|\xi_j|^{p'}\mathbf{1}_{\{|\xi_j|\geq r\}}.\end{aligned}$$

There exist such positive constants $\bar{c}_i$ that, for $\rho \geq 1$ and $0 < r < R$, the inequality

$$\left|\mathbf{P}\left\{S_{1,n}\left[\xi^r\right]\in C\right\}-\mathbf{P}\left\{S_{1,n}\left[\xi^R\right]\in C\right\}\right| \leq \frac{\bar{c}_1}{\rho^p}\left(\widehat{\mathbf{q}}\tilde{\Lambda}^p + \sum_{j=1}^{n}\mathbf{E}|\xi_j|^p\mathbf{1}_{\{r\leq|\xi_j|<R\}}\right)$$

holds for each measurable subset $C \subseteq \{x : |x| \geq \rho\}$ provided that $\widehat{\mathbf{q}} \leq \bar{c}_2$.

Corollary 3.7.1 *Assume that the conditions of Lemma 3.7.2 are satisfied, and the RV's $\xi, \xi_1, \ldots$ are, moreover, identically distributed.*

There exist such positive constants $\widehat{c}_i$ that the inequality

$$n\mathbf{P}\left\{u \leq \tfrac{1}{\sqrt{n}} |\xi| < v\right\} \leq \widehat{c}_0$$

implies, for $0 < u < v$, $r = u\sqrt{n}$, and $R = v\sqrt{n} \geq 1$ the following bound:

$$\begin{aligned} &\left|\mathbf{P}\left\{\tfrac{1}{\sqrt{n}} S_{1,n}[\xi^r] \in C\right\} - \mathbf{P}\left\{\tfrac{1}{\sqrt{n}} S_{1,n}[\xi^R] \in C\right\}\right| \\ \leq\ & \frac{\widehat{c}_1}{v^p n^{p/2-1}} \left(\mathbf{E}|\xi|^p \mathbf{1}_{\{u \leq |\xi|/\sqrt{n} < v\}} + \widehat{\Lambda}^p\, \mathbf{P}\left\{u \leq \tfrac{1}{\sqrt{n}} |\xi| < v\right\}\right). \end{aligned}$$

In this inequality the measurable set $C \subseteq \{x : |x| \geq v\}$ is arbitrary, the order of moments satisfies the condition $p \geq 2$, and

$$\widehat{\Lambda}^p = \left[\mathbf{E}|\xi|^2\right]^{p/2} + \frac{\mathbf{E}|\xi|^p}{n^{p/2-1}} + \left(\frac{1}{u^{p-1} n^{p/2-1}} \mathbf{E}|\xi|^p \mathbf{1}_{\{|\xi| \geq u\sqrt{n}\}}\right)^p.$$

Proof of Corollary 3.7.1 consists in elementary algebraic transformations of the estimate of Lemma 3.7.2. ○

The proofs of Lemmas 3.7.1-3.7.2 are exposed in subsection 3.7.3 below. The necessary preparations are done in the next subsection.

3.7.2 Auxiliary Inequalities for Moments

The next lemma is applicable to arbitrary real linear measurable spaces. In what follows, ξ_j are independent RV's in such a space, and $|\cdot|$ is a measurable seminorm.

Lemma 3.7.3 *Fix arbitrary positive numbers p and q, $p \leq q$. There exist positive constants c_i such that if the maximal summand satisfies the inequality $\mathbf{P}\{\bigvee_1^n \xi \geq r\} \leq c_1$, then the moments of the truncated sum admit the bound*

$$\mathbf{E}|S_{1,n}[\xi^r]|^q \leq c_2 \left([\mathbf{E}(S_1^n[\xi])^p]^{q/p} + \frac{r^q}{r^p} \sum_{j=1}^n \mathbf{E}|\xi_j|^p\right).$$

The values of the constants c_i are determined by the choice of exponents p and q only.

Proof of Lemma 3.7.3. Put $L = [\mathbf{E}(S_1^n[\xi])^p]^{1/p}$. The original sum $S_{1,n}[\xi]$ coincides with the truncated one $S_{1,n}[\xi^r]$ if $\bigvee_1^n \xi \leq r$. For this reason the Chebyshev inequality yields the bound

$$\mathbf{P}\{S_1^n[\xi^r] \geq aL\} \leq \mathbf{P}\{S_1^n[\xi] \geq aL\} + \mathbf{P}\{\bigvee_1^n \xi \geq r\} \leq \frac{1}{a^p} + \mathbf{P}\{\bigvee_1^n \xi \geq r\}.$$

Under the restriction of the lemma on the value of r, the above bound shows that Theorem 3.1.3 can be applied to evaluate the moments of the truncated sum $S_{1,n}[\xi^r]$ if a is sufficiently large, $a = c_3$) (see Remark 3.1.3). It should also be taken into consideration that by (3.7.1)

$$(\mathsf{V}_1^n \xi^r)^q \le r^{q-p} (\mathsf{V}_1^n \xi)^p \le r^{q-p} \sum_{j=1}^n |\xi_j|^p .$$

Combined with inequality (3.1.13) of Remark 3.1.3, this estimate proves the lemma. ○

Recall that under the assumptions of Lemma 3.7.3

$$\mathbf{P}\{\mathsf{V}_1^n \xi \ge r\} \le \frac{1}{r^p} \sum_{j=1}^n \mathbf{E}|\xi_j|^p \mathbf{1}_{\{|\xi_j| \ge r\}}. \tag{3.7.3}$$

One more auxiliary result refers to more general sums.

Lemma 3.7.4 *Let ξ_j be independent RV's in a separable real Banach space with norm $|\cdot|$. If $p \ge 1$ and I is an arbitrary subset of the segment $[1,\, n] \bigcap \mathbf{N}$, then*

$$\mathbf{E}|S_I[\xi]|^p \le 2^{p/(p-1)} \left(\mathbf{E}(S_1^n[\xi])^p + \max_I |\mathbf{E}S_I[\xi]|^p \right),$$

where $S_I[\xi] = \sum_{j \in I} \xi_j$.

Proof of Lemma 3.7.4. It is obvious from the definition of S_I that

$$|S_I|^p \le [|S_I + \mathbf{E}S_{I^c}| + |\mathbf{E}S_{I^c}|]^p \le 2^{p/(p-1)} [|S_I + \mathbf{E}S_{I^c}|^p + |\mathbf{E}S_{I^c}|^p],$$

where $I^c = ([1,\, n] \bigcap \mathbf{Z}) \setminus I$ is the complement of I in $[1,\, n] \bigcap \mathbf{Z}$ and consequently $S_{1,n} = S_I + S_{I^c}$. To derive the bound of the lemma, one can use the Jensen inequality for the conditional expectations given S_I in the form

$$\mathbf{E}\{|S_I + S_{I^c}|^p \,|S_I\} \ge |S_I + \mathbf{E}S_{I^c}|^p .$$

This yields the bound of the lemma after some uncomplicated calculations. ○

3.7.3 Derivation of Truncation Inequalities

Proof of Lemma 3.7.1. Put $\alpha_j = \mathbf{P}\{r \le |\xi_j| < R\}$ and denote by

$$G_j(C) = \mathbf{P}\{\xi_j \in C \,| r \le |\xi_j| < R\}, \quad F_j^{(u)}(C) = \mathbf{P}\{\xi_j^u \in C\},$$

respectively, the distribution of a summand conditional on landing between two spheres and the distribution of ξ_j truncated at the level u. In addition, let $I : C \mapsto \mathbf{1}_C(0)$ be the degenerate distribution with unit mass at the origin; convolutions with it do not change measures, $F * I = F$.

For each summand the difference of distributions truncated at different levels admits the representation

$$F_j^{(R)} - F_j^{(r)} = \alpha_j (G_j - I), \quad r < R.$$

The following identity holds for convolutions of arbitrary measures:

$$\ast_{j=1}^{n} F_j - \ast_{j=1}^{n} H_j = \sum_{j=1}^{n} \left(\ast_{i=1}^{j-1} F_i\right) \ast (F_j - H_j) \ast \left(\ast_{k=j+1}^{n} H_k\right). \quad (3.7.4)$$

Let $\Phi_n^\rho = \mathcal{L}(S_{1,n}[\xi^\rho])$, $\rho = r, R$, be the distributions of truncated sums. For the difference of these distributions on an arbitrary set C, the above identity and the equality relating different level truncations of individual summands yield the inequality

$$\begin{aligned} \left|\Phi_n^R(C) - \Phi_n^r(C)\right| &= \left|\sum_{j=1}^{n} \alpha_j (G_j - I) \ast \widehat{H}_j(C)\right| \\ &\le \sum_{j=1}^{n} \alpha_j (G_j + I) \ast \widehat{H}_j(C), \end{aligned} \quad (3.7.5)$$

where $\widehat{H}_j = \left(\ast_{i<j} F_i^{(r)}\right) \ast \left(\ast_{k>j} F_k^{(R)}\right)$ is the distribution of the "combined" sum $S_j' = S_{1,j-1}[\xi^r] + S_{j+1,n}[\xi^R]$. Evidently, $G_j \ast H_j(C)$ is the distribution of $S_j' + \eta_j$, where the RV η_j is independent of S_j' and has distribution G_j.

It follows that the moments of the measures appearing as summands in the inequality for the difference of truncations admit the estimates

$$\begin{aligned} \int_X |x|^p \widehat{H}_j(dx) &= \mathbf{E}\left|S_{1,j-1}[\xi^r] + S_{j+1,n}[\xi^R]\right|^p \\ &\le c\left(\mathbf{E}\left|S_{1,j-1}[\xi^r]\right|^p + \mathbf{E}\left|S_{j+1,n}[\xi^R]\right|^p\right) \le c'\Lambda^p, \\ \int_X |x|^p\, G_j \ast \widehat{H}_j(dx) &= \mathbf{E}\left|S_{1,j-1}[\xi^r] + \eta_j + S_{j+1,n}[\xi^R]\right|^p \\ &\le \bar{c}\left(\mathbf{E}\left|S_{1,j-1}[\xi^r]\right|^p + \mathbf{E}\left|S_{j+1,n}[\xi^R]\right|^p + \mathbf{E}|\eta_j|^p\right) \\ &\le \bar{c}'\left(\Lambda^p + \mathbf{E}|\eta_j|^p\right). \end{aligned}$$

The deduction of these estimates employs Lemma 3.7.4 to evaluate moments of the incomplete sums and the Hölder inequality. Thus, for the measure considered

$$\int_X |x|^p \left|\Phi_n^R - \Phi_n^r\right|(dx) \le c\left(\sum_{j=1}^{n} \alpha_j \Lambda^p + \sum_{j=1}^{n} \alpha_j \mathbf{E}|\eta_j|^p\right).$$

This is the estimate of the lemma, since by the definition

$$\alpha_j \mathbf{E}|\eta_j|^p = \mathbf{E}|\xi_j|^p\, \mathbf{1}_{\{r \le |\xi_j| < R\}}. \ \bigcirc$$

Proof of Lemma 3.7.2. For centered RV's with values in a separable real Hilbert space $\mathbf{E}\left|S_{1,n}[\xi]\right|^2 = \sum_{j=1}^{n} \mathbf{E}|\xi_j|^2$. This equality is similar to the well-known additivity property of variance. Consequently, Remark 3.1.3 combined with the Chebyshev inequality yields the estimate

$$\mathbf{E}\left|S_{1,n}[\xi^R]\right|^p \le c\left(\sum_{j=1}^{n} \mathbf{E}|\xi_j|^p + \left[\sum_{j=1}^{n} \mathbf{E}|\xi_j|^2\right]^{p/2} + \left[\sum_{j=1}^{n} \left|\mathbf{E}\xi_j^R\right|\right]^p\right).$$

Moreover, the following estimate is available for centered RV's if $\tilde{p} > 1$: since

$\mathbf{E}\xi_j^r = -\mathbf{E}\mathbf{1}_{\{|\xi_j|\geq r\}}\xi_j,$

$$\left|\mathbf{E}\xi_j^r\right| = \left|-\mathbf{E}\mathbf{1}_{\{|\xi_j|\geq r\}}\xi_j\right| \leq \frac{1}{r^{\tilde{p}-1}}\mathbf{E}\left|\xi_j\right|^{\tilde{p}}\mathbf{1}_{\{|\xi_j|\geq r\}}.$$

It remains to apply Lemma 3.7.1 and use the above bounds for moments. ○

Chapter 4

Rough Asymptotics of Large Deviations

Theorems on large deviations (LD) for sums of i.i.d. RV's ($S_{1,n}[\xi]$ in notation of Chapter 3) deal with asymptotic behavior of the probabilities

$$\mathbf{p}_n(B, \varepsilon_n) = \mathbf{P}\{\varepsilon_n S_{1,n}[\xi] \in B\},$$

where the set B and the normalizing factor $\varepsilon_n > 0$ are so chosen that the probability is infinitesimal.

The term "rough theorems on large deviations" refers to theorems concerning the principal term of asymptotics of $\ln \mathbf{p}_n(B, \varepsilon_n)$. The theorems which describe asymptotics of the probability itself are called, by contrast, "fine".

Another direction of classification is with respect to the choice of the normalizing factor. Large deviations proper correspond to $\varepsilon_n = n^{-1}$. For moderately large deviations $\varepsilon_n = (r\sqrt{n})^{-1}$, with $r = r(n) \to \infty$ and $r = o(\sqrt{n})$.

In the range of moderately large deviations, the asymptotic behavior of probabilities is quite similar to that characteristic of Gaussian distributions. The domain of large deviations proper does not demonstrate this affinity.

The methods developed for analysis of large deviations are applicable to much more general objects. However, systematic treatment of sums of i.i.d. RV's seems useful because the scope of applications is wide enough, and some of the facts readily established in this special case are not easily reproduced in greater generality.

4.1 Deviation Functional and Deviation Rates

4.1.1 Definitions and Main Results

In this section X is a complete separable metrizable locally convex real linear topological space with the dual X', and the value of the functional $t \in X'$ at $x \in X$ is $\langle x, t\rangle$. Recall that the topolgy in X is generated by a sequence of seminorms. Each open neighborhood of a point $a \in X$ contains the "unit ball" $\{x \in X : |x - a| < 1\}$ for an appropriate seminorm on X.

The space X is considered as a linear measurable space. The class of its measurable subsets is the Borel σ-algebra $\mathcal{B}$ (X) (which is generated by the dual space X' as a cylindric σ-algebra). Distributions in X are Radon measures.

The RV's $\xi, \xi_1, \ldots \in X$ are i.i.d. with some common distribution $F = \mathcal{L}(\xi)$ (namely, $F(B) = \mathbf{P}\{\xi \in B\}$ on the Borel σ-algebra $\mathcal{B}(X)$). The notation $S_{1,n}[\xi] = \sum_{j=1}^{n} \xi_j$ of (3.1.2) may be abbreviated to $S_n \equiv S_{1,n}[\xi]$.

Put

$$\lambda_n(B) = \frac{1}{n} \ln 1/\mathbf{P}\left\{\frac{1}{n} S_{1,n}[\xi] \in B\right\} \geq 0, \quad B \in \mathcal{B}(X). \tag{4.1.1}$$

The asymptotic behavior of this sequence can be described in terms of the so-called *deviation functional* (DF) Λ of the distribution $F = \mathcal{L}(\xi)$. This latter is defined by the relation

$$\Lambda(a) = \sup_{h \in X'} [\langle x, h \rangle - \ln \varphi(h)], \ a \in X, \tag{4.1.2}$$

where $\varphi(h) = \int_X e^{\langle x,h \rangle} F(dx)$ is the Laplace transform (3.6.3) of F. (Note that $\Lambda(a)$ is nonnegative, but not necessarily finite.)

Theorem 4.1.1 a) *If $G \subset X$ is an open set, then*

$$\limsup_{n \to \infty} \left(-\frac{1}{n} \ln \mathbf{P}\left\{\frac{1}{n} S_{1,n}[\xi] \in G\right\}\right) \leq \inf_{x \in G} \Lambda(x).$$

b) *If $U \subset X$ is a convex open set, then*

$$\lim_{n \to \infty} \left(-\frac{1}{n} \ln \mathbf{P}\left\{\frac{1}{n} S_{1,n}[\xi] \in U\right\}\right) = \inf_{x \in U} \Lambda(x).$$

c) *If $K \subset X$ is a compact set, then*

$$\liminf_{n \to \infty} \left(-\frac{1}{n} \ln \mathbf{P}\left\{\frac{1}{n} S_{1,n}[\xi] \in K\right\}\right) \geq \inf_{x \in K} \Lambda(x).$$

Theorem 4.1.1 is a corollary of Theorem 4.1.2 below.

The study of large deviations employs a number of characteristics describing the rate at which the probability of hitting a set diminishes as the number of summands increases. Relatively easy to calculate for sets of simple structure, they can be used to describe large deviation asymptotics for more complex sets.

The rate at which the probability in (4.1.1) goes to zero for a given set B is characterized by the quantities

$$L^+(B) \equiv \limsup_{n \to \infty} \lambda_n(B), \quad L^-(B) \equiv \liminf_{n \to \infty} \lambda_n(B), \quad L^+(B) \geq L^-(B). \tag{4.1.3}$$

Both of them are nonnegative (though not necessarily finite) by (4.1.1), and both enjoy an evident monotonicity property:

$$B' \subseteq B'' \implies L^{\pm}(B') \geq L^{\pm}(B''). \tag{4.1.4}$$

For an open convex set U the two characteristics are equal: $L^+(U) = L^-(U) = L(U)$ (this is the assertion of Lemma 4.1.1 in the next subsection). The quantities $L^{\pm}(\cdot)$ serve to define two "local" deviation rates which play an important part in the study of large deviations.

Denote by

$$\mathcal{G}_a = \{G : a \in G, \ G \text{ is open}\} \tag{4.1.5}$$

the class of open subsets of X which contain a vector a. This class includes the smaller class of half-spaces covering a (notation below is that of (1.2.3)):

$$\mathcal{H}_a = \left\{H = H^c_{e,\rho} : a \in H, \, e \in X', \, \rho \in \mathbf{R}\right\}, \ \mathcal{H}_a \subset \mathcal{G}_a. \tag{4.1.6}$$

The "local" versions of deviation rate are defined by the relations

$$\Lambda_{\mathcal{G}}(a) = \sup\left\{L^{\pm}(G) : G \in \mathcal{G}_a\right\}, \tag{4.1.7}$$

$$\Lambda_{\mathcal{H}}(a) = \sup\left\{L(H) : H \in \mathcal{H}_a\right\}. \tag{4.1.8}$$

Both quantities are nonnegative and may assume the value $+\infty$.

By the monotonicity property (4.1.4), the class $\mathcal{G}_a$ in definition (4.1.7) can be replaced by the family of convex open subsets containing a since the space is locally convex. By the same monotonicity property

$$\Lambda_{\mathcal{G}}(a) \geq \Lambda_{\mathcal{H}}(a), \tag{4.1.9}$$

since a smaller open subset containing a can be selected in each half-space covering this point.

The main result of this section is

Theorem 4.1.2 *The following equalities hold for each $a \in X$:*

$$\Lambda_{\mathcal{H}}(a) = \Lambda_{\mathcal{G}}(a) = \Lambda(a).$$

Its proof is presented in subsection 4.1.4, as also the proof of Theorem 4.1.1.

4.1.2 The "Neighborhood" Deviation Rate

Lemma 4.1.1 *The limit*

$$L(U) \stackrel{def}{=} \lim_{n\to\infty} \lambda_n(U)$$

exists if the set U is open and convex.

The lemma shows that definition (4.1.7) is unambiguous: since the limit exists, in (4.1.3)

$$L^+(U) = L^-(U) = L(U).$$

The limit may be infinite.

The proof of the lemma makes use of the following technical proposition.

A nonnegative sequence $f_n \in [0, +\infty]$, $n \in \mathbf{N}$, whose terms may be infinite, is called ***subadditive*** if

$$\forall m, n \in \mathbf{N} \ \ f_{m+n} \leq f_m + f_n. \tag{4.1.10}$$

Lemma 4.1.2 *If the sequence f_n is subadditive and its values are finite for all n large enough, $n \geq n_0$, then it has a finite limit:*

$$\lim_{n\to\infty} f_n/n = \varphi_- \stackrel{def}{=} \inf_{n\geq n_0} [f_n/n] .$$

Proof of Lemma 4.1.2. Fix an arbitrary value of $\varepsilon > 0$ and choose a natural number $\tilde{n} \geq n_0$ so as to approach the lower bound: $f_{\tilde{n}}/\tilde{n} \leq \varphi_- + \varepsilon$.

Each natural number $n \geq n_0$ admits the representation

$$\begin{aligned} n &= m\tilde{n} + k, \\ m &= [(n-n_0)/\tilde{n}] = (n/\tilde{n}) \cdot (1 + o(1)), \; n \to \infty, \\ n_0 &\leq k = k(n) \equiv n - m\tilde{n} \leq n_0 + \tilde{n}. \end{aligned} \tag{4.1.11}$$

(Here $[\cdot]$ is the integer part of a number). This decomposition and (4.1.10) lead to the inequality

$$f_n \leq m f_{\tilde{n}} + \max_{n_0 \leq k \leq n_0 + \tilde{n}} f_k .$$

It follows by the choice of $\tilde{n}$ and φ_- that

$$\varphi_- = \inf_{n\geq n_0} f_n/n \leq \liminf_{n\to\infty} f_n/n \leq \limsup_{n\to\infty} f_n/n \leq \varphi_- + \varepsilon .$$

This proves the lemma since ε can be taken arbitrarily small. ○

Proof of Lemma 4.1.1. a) The sequence $f_n = -\ln \mathbf{P}\left\{\frac{1}{n} S_n \in U\right\}$ is evidently nonnegative. It is also subadditive. Indeed, convexity of the set U implies the inclusion of events

$$\left\{\tfrac{1}{m+n} S_{m+n} \in U\right\} \supseteq \left\{\tfrac{1}{m} S_m \in U\right\} \bigcap \left\{\tfrac{1}{n} S_{m+1,n} \in U\right\} .$$

The RV's S_m and $S_{m+1,n} = S_{m+n} - S_m$ in the right-hand side are independent. Since all summands have identical distributions, $\mathcal{L}(S_{m+1,n}) = \mathcal{L}(S_n)$, i.e., $S_{m+1,n}$ has the distribution of S_n. Consequently, the logarithms of probabilities in (4.1.1) obey the relations

$$f_{m+n} \leq f_m + f_n, \quad m, n \in \mathbf{N}.$$

b) If $f_n = +\infty$ for all $n \in \mathbf{N}$, existence of infinite limit $\lim_{n\to\infty} f_n/n$ is obvious.

c) Suppose that $f_{\tilde{n}} < \infty$ for some $\tilde{n}$. In the rest of the proof the number $\tilde{n}$ is fixed.

By Lemma 4.1.2, the sequence f_n/n has a finite limit if there exists a number n_0 such that $f_n < \infty$ for $n \geq n_0$.

The distribution of $\tilde{n}^{-1} S_{\tilde{n}}$ is Radon. For this reason there is a convex compactum and a continuous seminorm $|\cdot|$ in X such that

$$\mathbf{P}\left\{\tilde{n}^{-1} S_{\tilde{n}} \in K\right\} > 0, \quad U \supset K + 2V, \quad V = \{|x| < 1\} . \tag{4.1.12}$$

(The possibility to choose the seminorm exists because U is open, K is compact, and the space is locally convex).

Each number n admits the decomposition

$$n = m\tilde{n} + k,\ 0 \le k < \tilde{n},\quad m = [n/\tilde{n}] > 0 \text{ if } n \ge \tilde{n}.$$

The corresponding decomposition of the sum of RV's is

$$\frac{1}{n}S_n = \frac{\tilde{n}}{n}\sum_{j=1}^{m}\frac{1}{\tilde{n}}S_j' + \frac{1}{n}S_m'',\quad S_j' = S_{(j-1)\tilde{n}+1,j\tilde{n}},\ S_m'' = S_{m\tilde{n}+1,n},$$

where the distributions of the summands are $\mathcal{L}(S_j') = \mathcal{L}(S_{\tilde{n}})$ and $\mathcal{L}(S_m'') = \mathcal{L}(S_k)$ because all these are sums of i.i.d. RV's.

For large enough values of n, $n \ge n'$,

$$m\tilde{n}/n = 1 + o(\tilde{n}/n) \implies (1 - m\tilde{n}/n)\, K \subset V$$

since the set K is compact (cf. (4.1.12)).

For the original open convex set, convexity of the compactum K in (4.1.12) leads to the inclusion

$$\begin{aligned}\left\{\frac{1}{n}S_n \in U\right\} &\supset \left\{\frac{\tilde{n}}{n}\left[\frac{1}{\tilde{n}}S_1' + \cdots + \frac{1}{\tilde{n}}S_m'\right] \in V + K\right\} \bigcap \left\{\frac{1}{n}S_m'' \in V\right\} \\ &\supset \left\{\frac{1}{n}S_m'' \in V\right\} \bigcap \bigcap_{j=1}^{m}\left\{\frac{1}{\tilde{n}}S_j' \in K\right\}.\end{aligned}$$

The corresponding bound on probabilities is, for large values of n,

$$\mathbf{P}\left\{\tfrac{1}{n}S_n \in U\right\} \ge \mathbf{P}\left\{\tfrac{1}{n}S_k \in V\right\}\left(\mathbf{P}\left\{\tfrac{1}{\tilde{n}}S_{\tilde{n}} \in K\right\}\right)^m,$$

where, since V is an open neighborhood of the origin and $\tilde{n}$ a fixed number,

$$\liminf_{n\to\infty}\ \min_{0\le k\le \tilde{n}} \mathbf{P}\left\{\tfrac{1}{n}S_k \in V\right\} = 1.$$

The above implies that there exists a natural number n_0 such that for $n \ge n_0$

$$\mathbf{P}\left\{\tfrac{1}{n}S_n \in U\right\} > 0,\quad f_n = -\ln \mathbf{P}\left\{\tfrac{1}{n}S_n \in U\right\} < +\infty.$$

The assertion of the lemma now follows from Lemma 4.1.2. ○

Lemma 4.1.1 shows that the functional $\Lambda_{\mathcal{G}}$ is well defined by (4.1.7).

Lemma 4.1.3 *The functional $\Lambda_{\mathcal{G}}$ is convex and lower semicontinuous:*

$$\liminf_{x\to a} \Lambda_{\mathcal{G}}(x) \ge \Lambda_{\mathcal{G}}(a).$$

Proof of Lemma 4.1.3. a) To show that $\Lambda_{\mathcal{G}}(x)$ is semicontinuous, take an open set $G \ni a$ and select a convex neighborhood of zero U so small that $G \supset a + 2U$. Then for all $x \in a + U$ there is inclusion $G \supset x + U$, and it follows from (4.1.7) that

$$\Lambda_{\mathcal{G}}(x) \ge L(x + U) \ge L^+(G).$$

Hence

$$\liminf_{x\to a} \Lambda_{\mathcal{G}}(x) \ge L^+(G).$$

In the above, the open set $G \in \mathcal{G}$ was arbitrary, so the inequality remains valid also for the supremum in (4.1.7). This establishes semicontinuity of $\Lambda_{\mathcal{G}}$.

b) To verify convexity, fix two points $a, b \in X$ and a number $\mu \in (0,\ 1)$. Put $x = \mu a + (1-\mu)\, b$.

Dependence of $L^{\pm}(\cdot)$ on the set-valued argument is monotone, and the space X is locally convex and metrizable. Hence it is possible to choose a sequence of continuous seminorms $|\cdot|_m$ in such a way that the corresponding neighborhoods of zero $U_m = \{x\ :\ |x|_m < 1\}$ satisfy the relation

$$L\,(x + U_m) \searrow \Lambda_{\mathcal{G}}(x).$$

Consider a sequence of natural numbers $n \mapsto k = k(n) \in \mathbf{N}$ such that

$$\mu_n \stackrel{def}{=} k(n)/n = \mu + o(1), \quad n \to \infty.$$

By the choice of μ, its values satisfy the inequality $k < n$ for large n. Thus, the sum $S_{1,n}\,[\xi]$ can be decomposed into a pair of independent RV's $S_{1,k}\,[\xi]$ and $S_{k+1,n}\,[\xi]$ with the distributions of S_k and S_{n-k} respectively. Hence the convexity of U_m implies the inclusions

$$\begin{aligned} &\left\{\left|\tfrac{1}{n}S_n - x\right|_m < 1\right\} \\ \supset\ &\left\{\left|\mu_n \tfrac{1}{k}S_{1,k} - \mu_n a + (1-\mu_n)\tfrac{1}{n-k}S_{k+1,n} - (1-\mu_n)\, b\right|_m < 1-\delta\right\} \\ \supset\ &\left\{\left|\tfrac{1}{k}S_{1,k} - a\right|_m < 1-\delta\right\} \bigcap \left\{\left|\tfrac{1}{n-k}S_{k+1,n} - b\right|_m < 1-\delta\right\} \end{aligned}$$

which hold for an arbitrary value of δ if n is sufficiently large. The corresponding inequality for probabilities is

$$\begin{aligned} \mathbf{P}\left\{\left|\tfrac{1}{n}S_n - x\right|_m < 1\right\} \ \geq\ & \mathbf{P}\left\{\left|\tfrac{1}{k}S_k - a\right|_m < 1-\delta\right\} \\ \times\ & \mathbf{P}\left\{\left|\tfrac{1}{n-k}S_{n-k} - b\right|_m < 1-\delta\right\}. \end{aligned}$$

This inequality and Lemma 4.1.1 lead to the estimate

$$\begin{aligned} L\,(x + U_m) \ &\leq\ \mu L\,(a + [1-\delta]\, U_m) \ +\ (1-\mu)\, L\,(b + [1-\delta]\, U_m) \\ &\leq\ \mu\Lambda_{\mathcal{G}}(a) \ +\ \Lambda_{\mathcal{G}}(b) \end{aligned}$$

which establishes convexity of $\Lambda_{\mathcal{G}}$ after the passage to the limit in $m \to \infty$. ○

4.1.3 DF and "Half-Space" Deviation Rate

The "half-space" rate $\Lambda_{\mathcal{H}}$ of (4.1.7) can be identified with the DF (4.1.11). The calculations are based upon the following corollary of Proposition A.3.1. Notation for half-spaces used below is that of (1.2.3).

Lemma 4.1.4 *The limit*

$$L(H) \ =\ \lim_{n\to\infty} \left[-\frac{1}{n}\ln \mathbf{P}\left\{\frac{1}{n}S_{1,n}\,[\xi] \in H\right\}\right]$$

exists for each open half-space $H = H^c_{e,\rho}$.

If $\mathbf{P}\{\langle\xi, e\rangle \le \rho\} < 1$, *then* $L(H) = \sup_{\vartheta\ge 0}[\rho\vartheta - \ln \mathbf{E}\exp\{\vartheta\langle\xi, e\rangle\}]$.

Proof of Lemma 4.1.4 is an application of Proposition A.3.1, as the rv $Y = \langle\xi, e\rangle$ and its independent copies Y_j satisfy its conditions. ○

Lemma 4.1.5 *The DF* Λ *is convex and semicontinuous, and*

$$\forall\, a \in X \quad \Lambda_{\mathcal{H}}(a) = \Lambda(a).$$

Proof of Lemma 4.1.5. a) Convexity of the DF follows immediately from its definition. To show its semicontinuity, note that for each $h \in X'$

$$\liminf_{x\to a} \Lambda(x) \ge \lim_{x\to a}[\langle x, h\rangle - \ln\varphi(h)] = \langle a, h\rangle - \ln\varphi(h).$$

The inequality remains valid if the right-hand side is replaced by its supremum.

b) By the Chebyshev inequality

$$\forall\, e \in X' \;\; \mathbf{P}\left\{\langle n^{-1}S_n, e\rangle \ge \rho\right\} \le \exp\{n[-\rho + \ln\varphi(e)]\}.$$

The half-space $H^c_{e,\rho}$ contains a if $\rho < \langle a, e\rangle$. Hence for each a the values of functionals (4.1.8) and (4.1.2) satisfy the inequality

$$\Lambda_{\mathcal{H}}(a) = \sup_{H\in\mathcal{H}} L(H) \ge \Lambda(a).$$

To arrive at the conclusion of the lemma, it remains to explore a few special cases.

c1) If $\Lambda_{\mathcal{H}}(a) = 0$, the equality of the lemma follows from the estimate of part b) since the DF is nonnegative.

c2) Suppose that $\mathbf{P}\{\langle\xi, e\rangle \le \langle a, e\rangle\} < 1$ for all $e \in X'$, so $\mathbf{P}\{\langle\xi, e\rangle \le \rho\} < 1$ if $\rho < \langle a, e\rangle$. Then Lemma 4.1.4 can be used to establish the existence of the limit and inequality in the relation

$$L\left(H^c_{e,\rho}\right) = \sup_{\vartheta\ge 0}[\rho\vartheta - \ln\varphi(\vartheta e)] \le \sup_{\vartheta\ge 0}[\langle a, \vartheta e\rangle - \ln\varphi(\vartheta e)] \le \Lambda(a).$$

In this formula, the half-space containing a is arbitrary. Hence $\Lambda_{\mathcal{H}}(a) \le \Lambda(a)$, and the equality of the lemma follows from the converse bound of part b).

c3) Suppose that for some $e \in X'$

$$\mathbf{P}\{\langle\xi, e\rangle \le \langle a, x\rangle\} = 1, \;\; \mathbf{P}\{\langle\xi, e\rangle = \langle a, e\rangle\} > 0.$$

Then $\mathbf{P}\{\langle\xi, e\rangle \le \rho\} < 1$ for each $\rho < \langle a, e\rangle$, and one can reason as in part c1) using Lemma 4.1.4 to establish the following inequalities:

$$L\left(H^c_{e,\rho}\right) \le \sup_{\vartheta\ge 0}[\langle a, \vartheta e\rangle - \ln\varphi(\vartheta e)] \le \Lambda(a).$$

c4) If for some $e \in X'$ all the mass of the one-dimensional distribution $\mathcal{L}(\langle\xi, e\rangle)$ lies strictly to the left of $\langle a, e\rangle$, viz., $\mathbf{P}\{\langle\xi, e\rangle < \langle a, e\rangle\} = 1$, then

$$\langle a, \vartheta e\rangle - \ln \mathbf{E}\varphi(\vartheta e) = -\ln \mathbf{E}\exp\{\vartheta\langle\xi - a, e\rangle\} \nearrow +\infty \text{ as } \vartheta \nearrow +\infty.$$

Thus, the equality of the lemma is valid also in this case: $\Lambda(a) = \infty$ and by part b) $\Lambda_{\mathcal{H}}(a) = \infty$. ○

4.1.4 Identity of DF and Local Rates

Consider the limit $L(H)$ of Lemma 4.1.4 for an open half-space (1.2.3) $H = H^c_{e,\rho}$.

Lemma 4.1.6 *Assume that for some $\gamma \geq 0$ the open half-space H has no common points with the set $B_\gamma = \{x : \Lambda_{\mathcal{G}}(x) \leq \gamma\}$. Then $L(H) \geq \gamma$.*

Proof of Lemma 4.1.6. a) The assertion of the lemma is obvious if $L(H) = +\infty$.

b) If $L(H) < \infty$, then there is a smaller convex precompact set C, also separated from B_γ, for which $L^-(C)$ can be only slightly greater than $L(H)$. The following construction can be used to verify its existence.

Fix an arbitrary value of $\varepsilon > 0$ and choose a number $\tilde{n} = \tilde{n}(\varepsilon)$ in such a way that

$$\lambda_{\tilde{n}}(H) < L(H) + \varepsilon.$$

The distribution of $S_{\tilde{n}}$ is a σ-additive measure. Hence there exists a smaller half-space $H' \subset H$ with the same normal to the boundary, $H' = \{\langle x, e\rangle > r\}$, $r > \rho$, satisfying the condition

$$\lambda_{\tilde{n}}(H') < L(H) + 2\varepsilon.$$

The closure of H' is evidently a subset of H.

Further, the distribution of $S_{\tilde{n}}$ is a tight measure, so there exists a convex compact set K for which

$$\lambda_{\tilde{n}}(C) < L(H) + 3\varepsilon, \quad C = H' \bigcap K.$$

The set C is convex as well, so (cf. Proof of Lemma 4.1.1) for $m \in \mathbf{N}$

$$\mathbf{P}\left\{\frac{1}{m\tilde{n}} S_{1,m\tilde{n}} \in C\right\} \geq \prod_{j=1}^{m} \mathbf{P}\left\{\frac{1}{\tilde{n}} S_{j\tilde{n}-\tilde{n},j\tilde{n}} \in C\right\} = \left[\mathbf{P}\left\{\frac{1}{\tilde{n}} S_{1,\tilde{n}} \in C\right\}\right]^m.$$

(All the independent sums $S_{(j-1)\tilde{n},j\tilde{n}}$ have same distribution $\mathcal{L}(S_{\tilde{n}})$ because they are made of i.i.d. RV's.) It follows that

$$L^-(C) \leq \lambda_{\tilde{n}}(C) < L(H) + 3\varepsilon. \tag{4.1.13}$$

c) The quantity $L^-(C)$ can be evaluated in terms of γ from below.

Recall that C is the intersection of a convex compact set with a half-space whose closure is contained in the interior of H. Hence its closure $\overline{C}$ is a compact subset of the open half-space H, and each point $y \in \overline{C}$ has a convex open neighborhood U_y which contains no points of the set B_γ. The neighborhood U_y can be chosen to satisfy also the restriction

$$L(U_y) > \gamma - \varepsilon \tag{4.1.14}$$

(see (4.1.7) and Lemma 4.1.1). The collection of neighborhoods $\{U_y, y \in \overline{C}\}$ is a covering of the compact set $\overline{C}$, so it contains a finite subcovering $\{U_i = U_{y(i)}\}$, $i = \overline{1, N}$. By the construction, for each n

$$\mathbf{P}\left\{\tfrac{1}{n} S_{1,n} \in \overline{C}\right\} \le \textstyle\sum_{j=1}^{N} \mathbf{P}\left\{\tfrac{1}{n} S_{1,n} \in U_j\right\}.$$

Hence, Lemma 4.1.1 yields the bound

$$L^-(C) = \liminf_{n\to\infty} \lambda_n(C) \ge \liminf_{n\to\infty} \left[-\frac{1}{n}\ln N + \min_{1\le j\le N} \lambda_n(U_j)\right] \ge \gamma - \varepsilon. \quad (4.1.15)$$

It follows from (4.1.13-4.1.15) that

$$L(H) + 3\varepsilon \ge L^-(C) \ge \gamma - \varepsilon, \quad (4.1.16)$$

and a passage to the limit in $\varepsilon \to 0$ yields the inequality of the lemma. ○

It is possible now to prove the main results of this section.

Proof of Theorem 4.1.2. The functional $\Lambda_{\mathcal{G}}$ is convex and lower semicontinuous. Hence its level sets $B_\gamma = \{x : \Lambda_{\mathcal{G}}(x) \le \gamma\}$ are convex and closed for each γ.

Fix a point $a \in X$ and a number γ such that $\gamma < \Lambda_{\mathcal{G}}(a)$ (note that the inequality is strict). Then $a \notin B_\gamma$, and it follows from the Hahn–Banach theorem that there exists an open half-space $H = H^c_{e,p}$ which contains a, but has no common points with B_γ.

By Lemma 4.1.1 and (4.1.8)

$$\Lambda_{\mathcal{H}}(a) \ge L(H) \ge \gamma.$$

After a passage to the limit in $\gamma \nearrow \Lambda_{\mathcal{G}}(a)$ this inequality yields the estimate

$$\Lambda_{\mathcal{H}}(a) \ge \Lambda_{\mathcal{G}}(a).$$

Combined with the converse inequality (4.1.9), this estimate shows that $\Lambda_{\mathcal{H}}(a) = \Lambda_{\mathcal{G}}(a)$. The assertion of the theorem now follows from Lemma 4.1.5 which identifies $\Lambda_{\mathcal{H}}$ and the DF. ○

Proof of Theorem 4.1.1. a1) Existence of the limit and the estimate

$$L(U) \le \inf\nolimits_{x\in U} \Lambda(x) \quad (4.1.17)$$

follow from Lemma 4.1.1, definition (4.1.7) of the quantity $\Lambda_{\mathcal{G}}$, and Theorem 4.1.2.

a2) To derive the lower bound on $L(U)$, choose a number γ such that $\gamma < \inf_{x\in U} \Lambda(x)$ (with strict inequality). By its choice, the set $C_\gamma = \{x : \Lambda(x) \le \gamma\}$ has no common points with U.

The set C_γ is convex and closed (see Lemma 4.1.5). Hence by the Hahn–Banach theorem there exists an open half-space $H = H^c_{e,p}$ which contains U and has no common points with C_γ. Lemma 4.1.4 shows the existence of the limit $L(H)$ satisfying the inequalities

$$\gamma \le L(H) \le L(U).$$

After the passage to the limit in $\gamma \nearrow \inf_{x\in U} \Lambda(x)$, the latter bound leads to the inequality

$$L(U) \geq \inf_{x\in U} \Lambda(x).$$

In conjunction with (4.1.17), it proves assertion a) of the theorem.

b) Fix an arbitrary point $x \in G$ and its convex neighborhood U, $U \subset G$. Evidently,

$$\mathbf{P}\left\{\tfrac{1}{n}S_{1,n} \in G\right\} \geq \mathbf{P}\left\{\tfrac{1}{n}S_{1,n} \in U\right\}.$$

Combined with Lemma 4.1.1, definition (4.1.7), and Theorem 4.1.2 this inequality yields the estimate

$$\limsup_{n\to\infty} \lambda_n(G) \leq L(U) \leq \Lambda(x).$$

This proves assertion b) of the theorem, since the point x is arbitrary, and one can replace the right-hand side by its lower bound.

c) Fix an arbitrary small number $\varepsilon > 0$. For each point $x \in K$ choose a convex neighborhood U_x such that

$$L(U_x) > \min\left\{\tfrac{1}{\varepsilon},\ \Lambda(x) - \varepsilon\right\}.$$

The family of neighborhoods $(U_x, x \in K)$ is evidently an open covering of K. This covering contains a finite subcovering $\left(U_{x(j)}, j = \overline{1,N}\right)$ because the set K is compact.

It follows from the additivity property of probability that

$$\begin{aligned}
&\limsup_{n\to\infty}\left[-\tfrac{1}{n}\ln \mathbf{P}\left\{\tfrac{1}{n}S_{1,n} \in K\right\}\right] \\
\geq\ &\limsup_{n\to\infty}\left[-\tfrac{1}{n}\ln \textstyle\sum_{j=1}^{N} \mathbf{P}\left\{\tfrac{1}{n}S_{1,n} \in U_{x(j)}\right\}\right] \\
\geq\ &\min_{j=\overline{1,N}} L\left(U_{x(j)}\right) \geq \min\left\{\tfrac{1}{\varepsilon},\ \inf_{x\in K}\Lambda(x) - \varepsilon\right\},
\end{aligned}$$

so assertion c) follows from the passage to the limit in $\varepsilon \to 0$. ○

4.2 Information Distance

4.2.1 Definition of the Information Distance

Below X is a separable Banach space with norm $|\cdot|$.

Large deviation asymptotics are related to an important concept of information theory, the so-called ***Kullback–Leibler information distance*** (or ***relative entropy***).

For two distributions F and G of arbitrary nature, the Kullback–Leibler information is defined by the equality

$$I_F(G) = \begin{cases} \int_X g(x)\ln g(x)\, F(dx) &, \quad G \ll F, \\ +\infty &, \quad G \not\ll F. \end{cases} \tag{4.2.1}$$

where notation $G \ll F$ implies absolute continuity and $g = dG/dF$ is the density of G with respect to F.

Consider a distribution F in the real Banach space X. Assume that it satisfies a very strong Cramér-type condition:

$$\forall\, t > 0 \;\; A(t) \;=\; \int_X e^{t|x|} F(dx) \;<\; \infty. \tag{4.2.2}$$

Under this restriction, Cramér transform (3.6.4) $F_h^*(dx)$ is well defined for any value of the vector parameter $h \in X^*$. For this new distribution, the norm has finite moments of arbitrary order (see Section 3.6). The expectation of F_h^* is the Fréchet derivative of the Laplace transform of F:

$$F_h^*(dx) \;=\; \frac{\exp\{\langle x, h\rangle\}\, F(dx)}{\varphi(h)}, \quad \overline{X}_h \overset{def}{=} \int_X x F_h^*(dx) \;=\; \nabla \ln \varphi(h). \tag{4.2.3}$$

The Kullback–Leibler information is easily calculated for the pair of Cramér-conjugate distributions (F, F_h^*):

$$I_F(F_h^*) \;=\; \langle \overline{X}_h, h\rangle - \ln \varphi(h). \tag{4.2.4}$$

For each $h \in X^*$, the expectation $\overline{X}_h$ and Laplace transform φ can be used to define the convex functional $U_h(t)$ over the conjugate space X^*:

$$U_h(t) \overset{def}{=} \langle \overline{X}_h, t\rangle - \ln \varphi(t) \;=\; \langle \nabla\varphi(h), t\rangle - \ln \varphi(t), \quad t \in X^*. \tag{4.2.5}$$

Suppose that the distribution F satisfies the nondegeneracy condition

$$\forall\, s \in X^* \quad \int_X \langle x - \overline{X}_0, s\rangle^2 F(dx) \;=\; \langle s, \nabla\rangle^2 \varphi(0) \;>\; 0. \tag{4.2.6}$$

(Notation of weak differentials in X^* is that of Lemma 3.6.2.) This condition stipulates that all one-dimensional projections of a RV with distribution F have nonzero variances, and it is easy to see that it remains valid for distributions obtained from F by the Cramér transform.

Under restriction (4.2.6), the functional $t \mapsto U_h(t)$ attains its unique maximum at $t = h$ (cf. (4.2.5)):

$$\max_{t\in X^*} U_h(t) \;=\; \langle \overline{X}_h, h\rangle - \ln \varphi(h).$$

Indeed, $U_h(t)$ is a smooth function in its argument t (see Section 3.6), and it satisfies the sufficient condition for the existence of an extremum at $t = h$:

$$\nabla_t U_h(t)\big|_{t=h} \;=\; 0, \quad \langle s, \nabla_t\rangle^2 U_h(t)\big|_{t=h} \;=\; -\,\langle s, \nabla_t\rangle^2 \ln \varphi(t)\big|_{t=h} \;<\; 0, \quad s \in X^*.$$

The maximum itself is the value of DF (4.1.2) at $\overline{X}_h$, and by (4.2.4) it equals the Kullback–Leibler information for (F, F_h^*):

$$\Lambda\left(\overline{X}_h\right) \overset{def}{=} \max_{t\in X^*} \left[\langle \overline{X}_h, t\rangle - \ln \varphi(t)\right] \;=\; \max_{t\in X^*} U_h(t) \;=\; U_h(h) \;=\; I_F(F_h^*).$$

It will be shown below that the value of the deviation functional at an arbitrary point a admits a similar interpretation in terms of the information distance between F and the class of distributions with expectation a. This distance is defined by the equality

$$\hat{I}(a) = \hat{I}(a;F) \stackrel{def}{=} \inf_{G\in\mathcal{G}_a} I_F(G), \quad \mathcal{G}_a = \left\{G : \int_X xG(dx) = a\right\}. \qquad (4.2.7)$$

The rest of this section is dedicated to the study of the information theoretic concepts introduced above in more detail. The following result will be essential for the applications to large deviations considered below.

Theorem 4.2.1 *The functional $\hat{I}(a)$ is nonnegative in X and vanishes only for distributions with expectation a:*

$$\hat{I}(a) = 0 \iff \int_X xF(dx) = a.$$

The proof of Theorem 4.2.1 is placed in subsection 4.2.2 .

4.2.2 An Extremal Property of Information Distance

Consider the class of Borel-measurable strictly positive functions on X

$$\mathcal{U}(X) = \{u : X \to \mathbf{R}_+ \mid \exists \alpha,\beta > 0\ \forall x \in X\ \alpha \le u(x) \le \beta\}. \qquad (4.2.8)$$

Denote by $\mathcal{U}_C$ its subclass of bounded continuous functions with positive lower bound:

$$\mathcal{U}_C(X) = \mathbf{C}\bigcap\mathcal{U}. \qquad (4.2.9)$$

Lemma 4.2.1 *The following equality holds for arbitrary distributions F and G on the Borel σ-algebra $\mathcal{B}(X)$:*

$$I_F(G) = \sup_{u\in\mathcal{U}} \left[\int_X \ln u(x)\, G(dx) - \ln\left(\int_X u(x)F(dx)\right)\right].$$

The right-hand side of this equality is finite if and only if the distribution G is absolutely continuous with respect to F and its density $g(x) = (dG/dF)(x)$ satisfies the condition

$$\int_X g(x) \ln |g(x)|\ F(dx) < \infty. \qquad (4.2.10)$$

The upper bound of the lemma does not change if $\mathcal{U}$ be replaced by the class of continuous functions $\mathcal{U}_C$.

Proof of Lemma 4.2.1. a) Assume satisfied condition (4.2.10). It follows from the Jensen inequality that for each function from $\mathcal{U}$

$$\begin{aligned}
&\int_X \ln u(x)\, G(dx) - \ln\left(\int_X u(x)F(dx)\right)\\
&= \int_X g(x)\ln g(x)\, F(dx) + \int_X g(x)\ln\frac{u(x)}{g(x)}\, F(dx) - \ln\left(\int_X u(x)F(dx)\right)\\
&\le \int_X g(x)\ln g(x)\, F(dx) = I_F(G).
\end{aligned}$$

To show that $I_F(G)$ is indeed the supremum, consider the functions

$$u_n(x) = \begin{cases} g(x), & 1/n < g(x) < n, \\ 1/n, & g(x) \le 1/n, \\ n, & g(x) \ge n. \end{cases}$$

They are clearly in $\mathcal{U}$ and for all $x \in X$

$$0 \le u_n(x) \le g(x), \quad g(x) |\ln u_n(x)| \le g(x) |\ln g(x)|,$$

while $u_n(x) \to g(x)$ as $n \to \infty$. Consequently by the Lebesgue theorem

$$\lim_{n\to\infty} \left[\int_X \ln u_n(x) G(dx) - \ln \left(\int_X u_n(x) F(dx) \right) \right]$$
$$= \int_X g(x) \ln g(x)\, F(dx) - \ln \left(\int_X g(x) F(dx) \right) = I_F(G),$$

since $\int_X g(x)F(dx) = \int_X G(dx) = 1$. The above formulae prove the equality of the lemma.

b) To avoid a special argument for the case where G is not absolutely continuous with respect to F, each distribution G is ascribed a density $g(x) = (dG/dF)(x)$ which assumes the value $g(x) = +\infty$ on the support of its singular component.

The integrand in (4.2.10) is bounded on the set where $0 < g(x) < 1$ because

$$\forall\, g \in (0,\, 1] \quad g\, |\ln g| \le 1/e. \tag{4.2.11}$$

Hence the integral in (4.2.10) converges if and only if it converges with the integration domain restricted to $\{x : g(x) \ge 1\}$, and it converges simultaneously with the corresponding integral for the function $g \ln(1+g)$. Indeed, if $g(x) \ge 1$, then

$$0 \le \ln g(x) \le \ln(g(x)+1) \le \ln[2g(x)],$$

so

$$\int_{\{g<1\}} g(x) |\ln g(x)| F(dx) \le 1/e,$$
$$\int_{\{g\ge 1\}} g(x) \ln g(x) F(dx) \le \int_{\{g\ge 1\}} g(x) \ln[g(x)+1]\, F(dx) \tag{4.2.12}$$
$$\le \int_{\{g\ge 1\}} g(x)[\ln 2 + \ln g(x)]\, F(dx) \le \ln 2 + \int_{\{g\ge 1\}} g(x) \ln g(x) F(dx).$$

Consider the function $\hat{u}_n(x) = 1 + \min\{g(x), n\}$. Obviously, $\hat{u}_n(x) \nearrow 1 + g(x)$ and $0 \le \ln \hat{u}_n(x) \nearrow \ln[1+g(x)]$ as $n \to \infty$.

With respect to F the set $\{g(x) = \infty\}$ has zero measure, so by the monotone convergence theorem

$$1 \le \lim_{n\to\infty} \int_X \hat{u}_n(x) F(dx) = \int_{\{g<\infty\}} [1 + g(x)] F(dx) = 1 + G(\{g(x) < \infty\}) \le 2.$$

Applied to the integral with respect to G, the monotone convergence theorem and the estimate just obtained yield the relation

$$\lim_{n\to\infty}\left[\int_X \ln \hat{u}_n(x)G(dx) - \ln\left(\int_X \hat{u}_n(x)F(dx)\right)\right]$$

$$= \int_X \ln\left[1+g(x)\right]G(dx) - \ln\left(1+G\{x : g(x)<\infty\}\right).$$

Thus the supremum in the statement of the lemma is infinite if condition (4.2.10) is violated, i.e., $I_F(G) = +\infty$.

c) To ascertain that the upper bound in the right-hand side of the equality of the lemma is not changed if the class $\mathcal{U}$ is replaced by its subclass of continuous functions, note that one can approximate the functions from $\mathcal{U}$ by sequences of continuous functions belonging to $\mathcal{U}_C$ which converge in measure for the dominating measure $F+G$. ○

Lemma 4.2.2 *The Kullback–Leibler information $I_F(G)$ is nonnegative. This functional is convex in its argument G if the distribution F is fixed.*

The functional $I_F(G)$ is lower semicontinuous in both its arguments in the following sense:

$$\liminf_{n\to\infty} I_{F_n}(G_n) \geq I_F(G)$$

if the sequence of pairs of distributions (F_n, G_n) is so chosen that there is weak convergence $F_n \overset{weak}{\longrightarrow} F$ and $G_n \overset{weak}{\longrightarrow} G$.

Proof of Lemma 4.2.2. The mapping $u \mapsto u\ln u$ is convex on the half-line $u>0$. Hence the fact that $I_F(G)$ is nonnegative and convex is established using a few forms of the Jensen inequality. Thus, for $G \ll F$

$$\int_X g(x)\ln g(x)F(dx) \geq \overline{G}\ln\overline{G} = 0, \quad \overline{G} = \int_X g(x)F(dx) = \int_X G(dx) = 1.$$

If there is weak convergence $F_n \overset{weak}{\to} F$ and $G_n \overset{weak}{\to} G$, then it follows from Lemma 4.2.1 that each continuous function from $\mathcal{U}_C$ satisfies the inequality

$$\liminf_{n\to\infty} I_{F_n}(G_n) \geq \int_X \ln u(x)G(dx) - \ln\left(\int_X u(x)F(dx)\right).$$

This proves the lemma, since $u \in \mathcal{U}_C$ is arbitrary. ○

Lemma 4.2.3 *Consider a tight family of measures $\{F_n, n\in \mathbf{N}\}$. If*

$$I = \sup_n I_{F_n}(G_n) < \infty,$$

then the family of measures $\{G_n, n\in\mathbf{N}\}$ is also tight.

Proof of Lemma 4.2.3. Fix an arbitrary value of $\varepsilon>0$ and choose a compact set $K\subset X$ such that $F_n(X\setminus K)<\varepsilon$ for all n. Denote by $g_n = (dG_n/dF_n)(x)$ the density of G_n with respect to F_n.

If $u \geq 10$, it follows from the condition of the lemma that

$$G_n(X \setminus K) \leq uF_n(X \setminus K) + \frac{1}{\ln u} \int_{\{g_n(x) \geq u\}} g_n(x) \ln g_n(x) F_n(dx)$$

for all n. The integral in the right-hand side does not exceed the quantity

$$\mathbf{q} = \int_X g_n(x) \ln g_n(x) F(dx) + 2 \int_{\{g_n(x)<1\}} g_n(x) \left|\ln g_n(x)\right| F_n(dx) \leq l + 2/e.$$

The latter inequality holds by (4.2.11). Hence for all n the measure G_n admits the estimate

$$G_n(X \setminus K) \leq \delta(\varepsilon), \;\; \delta(\varepsilon) = \min_{u \geq 10} \left[u\varepsilon + (l + 2/e)/\ln u\right] \to 0$$

as $\varepsilon \to 0$. The estimate holds for an arbitrary $\varepsilon > 0$, so the family $\{G_n\}$ is tight. ○

Lemma 4.2.4 *If condition* (4.2.2) *is fulfilled, then the inequality*

$$\int_B |x| \, G(dx) \leq \sqrt{A(2t)F(B)} + \frac{1}{t} \left(I_F(G) + 2/e\right)$$

holds for each $t > 1$ *and measurable set* $B \subset X$.

Proof of Lemma 4.2.4. The functions $u \mapsto u \ln u$ and $v \mapsto e^v - 1$ are Legendre transforms of each other. Hence the Young inequality implies the following estimate valid for all $t > 1$ and $x \in X$:

$$|x| \, g(x) \leq \exp\{t\,|x| - 1\} + \frac{g(x)}{t} \ln \frac{g(x)}{t} \leq \exp\{t\,|x| - 1\} + \frac{g(x)\left|\ln g(x)\right|}{t},$$

where, as usual, $g(x) = (dG/dF)(x)$.

By the Cauchy inequality

$$\int_B e^{t|x|-1} F(dx) \leq \frac{1}{e} \sqrt{F(B)A(2t)}.$$

It follows (see (4.2.11) and the proof of Lemma 4.2.3) that

$$\begin{aligned} &\int_X g(x) \left|\ln g(x)\right| F(dx) \\ \leq \;& \int_X g(x) \ln g(x) F(dx) + 2 \int_{\{g(x)<1\}} g(x) \left|\ln g(x)\right| F(dx) \quad (4.2.13) \\ \leq \;& I_F(G) + 2/e. \end{aligned}$$

This proves the lemma. ○

Consider sequences of distributions F_n and G_n defined on $\mathcal{B}(X)$.

Lemma 4.2.5 *Assume that* F_n *is weakly compact,*

$$\forall t > 0, A(t) = \sup_{n \in \mathbf{N}} \int_X e^{t|x|} F_n(dx) < \infty,$$

and

$$I_{F_n}(G_n) \le l < \infty.$$

If $G_n \overset{weak}{\to} \overline{G}$, *then the mean values of these measures also converge:*

$$\lim_{n\to\infty} \int_X x G_n(dx) = \int_X x\overline{G}(dx).$$

Proof of Lemma 4.2.5. It is easy to see that, the family F_n being tight,

$$\lim_{u\to\infty} \sup_{n\in\mathbf{N}} F_n\{x: |x| \ge u\} = 0.$$

Hence the following relations hold by Lemma 4.2.4:

$$\forall\, t > 1 \qquad \limsup_{u\to\infty} \sup_{n\in\mathbf{N}} \int_{\{|x|\ge u\}} |x|\, G_n(dx) \le \frac{1}{t}(l + 2/e),$$

$$\lim_{u\to\infty} \sup_{n\in\mathbf{N}} \int_{\{|x|\ge u\}} |x|\, G_n(dx) = 0.$$

Convergence of the mean values now follows from the weak convergence of G_n stipulated in the statement of the lemma. ○

Consider the class of distributions with uniformly bounded Kullback–Leibler information with respect to a given measure F:

$$\mathcal{G}^*{}_l = \{G: I_F(G) \le l\}, \quad l \in \mathbf{R}_+.$$

Denote by $\overline{X}_l$ the set of expectations with respect to distributions from this class:

$$\overline{X}_l = \left\{ y = \int_X xG(dx): \ G \in \mathcal{G}^*{}_l \right\}.$$

Lemma 4.2.6 *For each* $l < \infty$ *the set* $\overline{X}_l$ *is a convex compact subset of the space* X.

Proof of Lemma 4.2.6. Convexity of $\overline{X}_l$ is a consequence of Lemma 4.2.2.

For each point in $\overline{X}_l$ there is a distribution from $\mathcal{G}^*{}_l$ whose mean coincides with this point. The family of distributions $\mathcal{G}^*{}_l$ is tight by Lemma 4.2.3. Hence each sequence of distributions G_n belonging to this class contains a weakly convergent subsequence $G_{n'} \overset{weak}{\to} \overline{G}$, $n' = n'(k) \nearrow \infty$. By Lemma 4.2.3 the limit $\overline{G}$ is also a distribution from $\mathcal{G}^*{}_l$, and the mean values converge to the mean of the limit distribution:

$$\lim_{n'\to\infty} \int_X xG_{n'}(dx) = \int_X x\overline{G}(dx) \in \overline{X}_l.$$

Thus, each sequence of points in $\overline{X}_l$ contains a convergent subsequence. This proves the lemma. ○

The preceding results suffice for the

Proof of Theorem 4.2.1. The fact that the information is nonnegative follows from Lemma 4.2.2. It remains to verify its last assertion.

If $\int_X xF(dx) = a$, then the distribution F belongs to $\mathcal{G}_a$, so the minimal information vanishes: $\hat{I}(a) = 0$.

For a distribution F with $\hat{I}(a; F) = 0$ there exists a sequence of distributions G_n, $G_n \ll F$, whose mean is a and the Kullback–Leibler information is infinitesimal:

$$\int_X xG_n(dx) = a, \quad \lim_{n\to\infty} I_F(G_n) = 0.$$

By Lemma 4.2.3 and Lemma 4.2.5, this sequence contains a weakly convergent subsequence whose mean values also converge:

$$G_{n'} \overset{weak}{\to} \overline{G}, \quad n' = n'(k) \nearrow \infty, \quad \int_X xG_{n'}(dx) \to \int_X x\overline{G}(dx).$$

Hence $\int_X x\overline{G}(dx) = a$, and by Lemma 4.2.2 $I_F(\overline{G}) = 0$.

The above argument shows that the class $\mathcal{G}_a$ contains a distribution with zero information distance from F. By definition (4.2.1)

$$I_F(\overline{G}) = 0 \implies G \ll F, \ (d\overline{G}/dF)(x) = 1 \pmod{F}.$$

In other words, $\overline{G} = F$ if $I_F(\overline{G}) = 0$. Hence $\overline{G}$ coincides with F and a is the mean value of this distribution. This proves the theorem. ○

4.3 Information Distance and DF

4.3.1 Identity of DF and Minimal Information Distance

The purpose of the section is to study in more detail the properties of DF (4.1.2) and its relation to the minimal information distance (4.2.7).

Below X is a separable real Banach space, and notations of norm, duality between X and its conjugate X^*, etc. are standard.

Consider a distribution F in X and a RV $\xi \in X$ with this distribution: $\mathbf{P}\{\xi \in B\} = F(B)$. The distribution F is supposed to satisfy the Cramér-type condition (4.2.2), so its Laplace transform (3.6.3) $\varphi(h)$ and DF (4.1.2) $\Lambda(a) = \Lambda_F(a)$ are well defined for all values of arguments, as well as the information distance (4.2.7) to the class of distributions with given expectation, $\hat{I}(a) = \hat{I}(a; F)$. Below the distribution F is fixed, so it is not specified in notation.

An information-theoretical interpretation of the DF is provided by

Theorem 4.3.1 *If the distribution obeys Cramér-type condition* (4.2.2), *then its DF satisfies the equality*

$$\forall\, a \in X \quad \Lambda(a) = \hat{I}(a).$$

In the setting of this section, the DF Λ is a smooth function of its argument. It grows faster than a linear function of $|x|$ under restriction (4.2.2): indeed, in this case

$$\forall\, t > 0 \quad \Lambda(x) \geq t\,|x| - \ln A(t), \quad \liminf_{|x|\to\infty} \Lambda(x)/|x| = +\infty. \tag{4.3.1}$$

The following proposition provides more information about Λ.

Theorem 4.3.2 *Under condition* (4.2.2), *DF* (4.1.2) *is nonnegative, convex, and lower semicontinuous. All its level sets*

$$C_u = \{x : \Lambda(x) \leq u\}, \quad u \in [0, \infty), \tag{4.3.2}$$

are compact in X.

The proofs of Theorem 4.3.1 and Theorem 4.3.2 are exposed in the next subsection.

4.3.2 Identification of DF and Minimal Information Distance

This subsection identifies DF $\Lambda(a) = \Lambda_F(a)$ with the minimal information distance $\hat{I}(a) = \hat{I}(a; F)$.

Proof of Theorem 4.3.1 follows directly from Lemma 4.3.1 and Lemma 4.3.2 below. ○

Lemma 4.3.1 a) *If the minimal information* $\hat{I}(a)$ *is finite, then there is a distribution* $G \in \mathcal{G}_a$ *such that* $I_F(G) = \hat{I}(a)$.

b) *The inequality* $\hat{I}(a) \geq \Lambda(a)$ *holds for each* $a \in X$.

Proof of Lemma 4.3.1. a) Assume that $\hat{I}(a)$ is finite. Select some sequence of distributions $G_n \in \mathcal{G}_a$ such that $I_F(G_n) \searrow \hat{I}(a)$. By Lemma 4.2.3 this sequence contains a weakly convergent subsequence $G_{n'} \xrightarrow{weak} \overline{G}$, and $\overline{G} \in \mathcal{G}_a$ by Lemma 4.2.5. At the same time, Lemma 4.2.2 shows that

$$\hat{I}(a) = \lim_{n' \to \infty} I_F(G_{n'}) \geq I_F(\overline{G}).$$

Hence $I_F(\overline{G}) = \hat{I}(a)$ by definition (4.2.7).

b) The inequality of part b) is trivial if $\hat{I}(a) = +\infty$.

Assume that $\hat{I}(a) < \infty$ and select a distribution $G \in \mathcal{G}_a$ such that $I_F(G) < \infty$. Then by Lemma 4.2.1 the following inequality holds for each bounded positive function which is bounded from below by a positive constant (viz., $u \in \mathcal{U}$):

$$I_F(G) \geq \int_X \ln u(x) G(dx) - \ln \left(\int_X u(x) F(dx) \right).$$

If the Laplace transform $\varphi(h)$ of F is finite for a value of h, then one can choose a sequence of functions $u_n \in \mathcal{U}$ and a constant $c > 0$ so that a.s. with respect to F

$$u_n(x) \to e^{\langle x,h \rangle}, \quad |\ln u_n(x)| \leq c\,[1 + |x|], \quad |u_n(x)| \leq 1 + e^{\langle x,h \rangle}.$$

It follows from Lemma 4.2.4 that the norm $|x|$ is summable with respect to G. Hence the dominated convergence theorem of Lebesgue implies that

$$\int_X u_n(x) F(dx) \to \varphi(h), \quad \int_X \ln u_n(x) G(dx) \to \int_X \langle x, h \rangle \, G(dx).$$

Thus, by the choice of the distribution G

$$I_F(G) \geq \int_X \langle x, h\rangle\, G(dx) - \varphi(h) = \langle a, h\rangle - \varphi(h).$$

This inequality remains valid if the right-hand side is replaced by its upper bound. The lemma is proved. ○

To prove the lower estimate, which is the converse to that of Lemma 4.3.1, it is necessary to have recourse to its special case for one-dimensional distributions, Proposition A.3.2.

Lemma 4.3.2 *If the distribution F satisfies the exponential integrability condition* (4.2.2), *then its DF* (4.1.2) *and the minimal information* (4.2.7) *are equal for all values of the argument:*

$$\forall\, a \in X \quad \hat{I}(a) = \Lambda(a).$$

Proof of Lemma 4.3.2. By Lemma 4.3.1, only the inequality $\hat{I}(a) \leq \Lambda(a)$ has to be verified. The proof proceeds by the rule of contraries. For the sake of convenience it is subdivided into a number of steps.

a) Suppose that l is some number satisfying the inequalities $\hat{I}(a) > l > \Lambda(a)$. By Lemma 4.2.6, the set $\overline{X}_l$ of expected values is a convex compact subset of X for the class $\mathcal{G}^*{}_l$ of distributions with bounded information. By the choice of l, the point a does not belong to $\overline{X}_l$, so it follows from the Hahn–Banach theorem that there exists a hyperplane which separates a from this set:

$$\exists e \in X^* \ \rho \in \mathbf{R} \ : \ \forall y \in \overline{X}_l \quad \langle y, e\rangle > \rho > \langle a, t\rangle\,. \tag{4.3.3}$$

b) Consider the one-dimensional projection of the measure F defined by the relation

$$\lambda^F(B) = F\{x : \langle x, e\rangle \in B\}, \quad B \in \mathcal{B}(\mathbf{R})\,.$$

If a measure is absolutely continuous with respect to another, $G \ll F$, its one-dimensional projections enjoy the same property: $\lambda^G \ll \lambda^F$. The one-dimensional density is the conditional expectation

$$\frac{d\lambda^G}{d\lambda^F}(u) = \mathbf{E}\left\{ \frac{dG}{dF}(\xi) \middle| \langle \xi, e\rangle = u \right\}, \quad \mathcal{L}(\xi) = F.$$

For this reason the Kullback–Leibler information does not increase if distributions are replaced with their one-dimensional projections:

$$I_{\lambda^F}\left(\lambda^G\right) \leq I_F\,(G)\,. \tag{4.3.4}$$

On the contrary, given a one-dimensional distribution μ, $\mu \ll \lambda^F$, one can construct a distribution in X with same Kullback–Leibler information with respect to F — a distribution with the desired property is, e.g.,

$$G_\mu(B) \stackrel{def}{=} \int_B \frac{d\mu}{d\lambda^F}\left(\langle x, e\rangle\right) F\,(dx)\,, \quad I_{\lambda^F}(\mu) = I_F\,(G_\mu)\,. \tag{4.3.5}$$

The mean values of distributions and their projections to the real line are related through the identities

$$\int_{\mathbf{R}} u\lambda^G(du) = \left\langle \int_X xG(dx), t\right\rangle, \quad \left\langle \int_X xG_\mu(dx), e\right\rangle = \int_{\mathbf{R}} u\mu(du). \qquad (4.3.6)$$

c) It follows from (4.3.5) that a one-dimensional distribution μ obeying the restriction $I_{\lambda^F}(\mu) \le l$ generates through (4.3.6) the infinite-dimensional one $G_\mu \in \mathcal{G}^*{}_l$ such that

$$\left\langle \int_X xG_\mu(dx), e\right\rangle \ge \rho.$$

Hence by (4.3.3) the number $\langle a, e\rangle$ is not an expectation of any one-dimensional distribution whose Kullback–Leibler information with respect to λ^F does not exceed l:

$$\langle a, e\rangle \notin \overline{R}_l = \left\{\int_{\mathbf{R}} u\mu(du) : \mu \in \overline{\Gamma}_l\right\}, \quad \overline{\Gamma}_l = \{\mu : I_{\lambda^F}(\mu) \le l\}.$$

It follows that the set $\overline{\Gamma}_l$ has no common points with

$$\Gamma_{(a,e)} = \left\{\mu : \int_{\mathbf{R}} u\mu(du) = \langle a, u\rangle\right\}.$$

Thus, there is inequality

$$\hat{I}^0(\langle a, e\rangle) \overset{def}{=} \inf\left\{I_{\lambda^F} : \mu \in \Gamma_{(a,e)}\right\} \ge l. \qquad (4.3.7)$$

At the same time, Proposition A.3.2 implies the relations

$$\begin{aligned} \hat{I}^0(\langle a, e\rangle) &= \sup\nolimits_{v\in\mathbf{R}} \left[v\langle a, e\rangle - \ln \int_{\mathbf{R}} e^{vu}\lambda^F(du)\right] \\ &= \sup\nolimits_{v\in\mathbf{R}} \left[v\langle a, e\rangle - \ln \int_X e^{\langle x, ve\rangle} F(dx)\right] \le \Lambda(a). \end{aligned}$$

This is the contradiction that proves the lemma. ○

Proof of Theorem 4.3.2. Fix a value of u and consider a sequence of points a_n belonging to tle level set C_u. By the already proved Theorem 4.2.1 and Theorem 4.3.1, the deviation function $\Lambda(a)$ coincides with the minimal Kullback–Leibler information in the class of distributions with mean a, $\hat{I}(a)$. Hence there is a sequence of measures G_n, $G_n \ll F$, such that (cf. (4.2.7))

$$a_n = \int_X xG_n(dx), \quad I_F(G_n) \le u + 1.$$

Using Lemma 4.2.3 and Lemma 4.2.5 one can select from this sequence a weakly convergent subsequence whose mean values also converge:

$$G_{n'} \overset{weak}{\longrightarrow}, \quad n' = n'(k) \nearrow \infty, \quad a_{n'} = \int_X xG_{n'}(dx) \to \overline{a} = \int_X x\overline{G}(dx).$$

The limit value $\overline{a}$ also belongs to the level set C_u because the DF is lower semicontinuous, and, consequently, this set is closed. This shows that C_u is compact.

The remaining assertions of Theorem 4.3.2 are easily verified. To do so one can either start directly from definition (4.1.2) or reformulate them in terms of the minimal Kullback–Leibler information (4.2.7) and apply Theorems 4.2.1 and 4.3.1. ○

4.4 LD Asymptotics for Sums in Banach Space

4.4.1 LD Asymptotics and Deviation Functional

In this section X is a separable real Banach space. Notation for its norm, conjugate, etc. are those of the preceding sections.

The RV's $\xi, \xi_1, \ldots \in X$ are i.i.d. with the distribution $F = \mathcal{L}(\xi)$ satisfying the Cramér-type condition (4.2.2).

As in (3.1.2), $S_{1,n}[\xi] = \sum_{j=1}^{n} \xi_j$, $S_1^n = |S_{1,n}|$, etc. and (see Section 4.1)

$$\lambda_n(B) = -\frac{1}{n} \ln \mathbf{P}\left\{\frac{1}{n} S_{1,n}[\xi] \in B\right\}, \quad B \in \mathcal{B}(X).$$

The theorems collected in Sections 4.2 and 4.3 make it possible to describe the main term of asymptotics of λ_n using DF (4.1.2). The first result of this section refers to probabilities.

Theorem 4.4.1 a) *If the set $G \subset X$ is open, then*

$$\limsup_{n \to \infty} \lambda_n(G) \le \inf_{x \in G} \Lambda(x).$$

b) *If $F \subset X$ is a closed set, then*

$$\liminf_{n \to \infty} \lambda_n(F) \ge \inf_{x \in F} \Lambda(x).$$

The LD behavior of sums can also be described in terms of exponential moments $\mathbf{E} \exp\left\{ng\left(\frac{1}{n} S_{1,n}\right)\right\}$ for suitable functionals g.

Assume that the functional $g : X \to \mathbf{R}$ is continuous and grows no faster than a linear function of $|x|$ as $|x| \to \infty$: there exist constants $\gamma_i > 0$ such that

$$\forall x \in X \quad g(x) \le \gamma_1 + \gamma_2 |x|. \tag{4.4.1}$$

In this case by (4.3.1)

$$\gamma^* \overset{def}{=} \sup_{x \in X} [g(x) - \Lambda(x)] < \infty, \tag{4.4.2}$$

and it follows from Theorem 4.1.1 that this quantity is finite. Indeed, consider some convex compactum $K \subset X$ such that $\mathbf{p} \overset{def}{=} \mathbf{P}\{\xi \in K\} > 0$. By convexity $\mathbf{P}\left\{\frac{1}{n} S_{1,n} \in K\right\} \ge \mathbf{p}^n$, so $\inf_{x \in K} \Lambda(x) \le \ln 1/\mathbf{p}$. Thus, there are points where $\Lambda(x)$ is finite, so by (4.4.2) $\gamma^* > -\infty$.

Theorem 4.4.2 *If the Cramér-type condition* (4.2.2) *and* (4.4.1) *are satisfied, then*

$$\lim_{n \to \infty} \frac{1}{n} \ln \mathbf{E} \exp\left\{ ng\left(\tfrac{1}{n} S_{1,n}[\xi]\right) \right\} = \gamma^*.$$

4.4.2 Calculation of Probability Asymptotics

The proof of Theorem 4.4.1 exposed below makes use of truncation (cf. Section 3.7).

Select a sequence of convex compact sets $K_m \subset X$ which satisfies the condition

$$\lim_{m\to\infty} \delta(m) = 0, \quad \delta(m) = F(X \setminus K_m). \tag{4.4.3}$$

This is possible since distributions in X are tight.

One can decompose the RV's $\xi, \xi_1, \ldots$ and sums of these RV's in the following manner (cf. (3.7.1)):

$$\begin{aligned} \xi &= \xi^0 + \hat{\xi}, \\ \xi^0 &= \xi^0(m) = \mathbf{1}_{K_m}(\xi)\xi, \ \hat{\xi} = \hat{\xi}(m) = [1 - \mathbf{1}_{K_m}(\xi)]\,\xi, \\ S_{1,n}[\xi] &= S_n^0 + \hat{S}_n, \end{aligned} \tag{4.4.4}$$

where $S_n^0 = S_{1,n}\left[\xi^0(m)\right] = \sum_{j=1}^n \xi_j^0(m)$ and $\hat{S}_n = S_{1,n}\left[\hat{\xi}(m)\right] = \sum_{j=1}^n \hat{\xi}_j(m)$ are sums of RV's in decomposition (4.4.4) for independent copies ξ_j, $\xi_j^0(m)$, and $\hat{\xi}_j(m)$. By analogy with the distribution $F = \mathcal{L}(\xi)$ and its Laplace transform $\varphi(h)$, the corresponding notation for truncated RV's $\xi^0(m)$ is F_m and $\varphi_m(h) = \int_X e^{(x,h)} dF_m(x)$.

The "residue term" $\hat{\xi}$ in (4.4.4) is small in a sense specified by

Lemma 4.4.1 *For each* $\varepsilon > 0$

$$\hat{L}_\varepsilon(m) \stackrel{def}{=} \liminf_{n\to\infty}\left[-\frac{1}{n}\ln \mathbf{P}\left\{\frac{1}{n}S_1^n[\hat{\xi}(m)] \geq \varepsilon\right\}\right] \to +\infty \quad \textit{as } m \to \infty.$$

Proof of Lemma 4.4.1. To derive the assertion of the lemma from (4.4.3), one can use Theorem 3.3.1 or a similar "one-dimensional" bound for the sums of independent rv's $\left|\hat{\xi}_j\right|$: for each $t > 0$

$$\begin{aligned} \mathbf{p} = \ & \mathbf{P}\left\{\tfrac{1}{n}S_1^n[\hat{\xi}(m)] \geq \varepsilon\right\} \\ \leq \ & \exp\left\{-nt\left(\varepsilon - \tfrac{1}{n}\mathbf{E}S_1^n[\hat{\xi}]\right) + nt\mathbf{E}\left(e^{t|\hat{\xi}|} - 1 - t\left|\hat{\xi}\right|\right)\right\}. \end{aligned}$$

Evidently, $\mathbf{E}S_1^n[\hat{\xi}] \leq n\mathbf{E}|\hat{\xi}|$, so $\mathbf{p} \leq \exp\left\{-nt\varepsilon + \mathbf{E}\left(e^{t|\hat{\xi}|} - 1\right)\right\}$.

The Cramér-type condition (4.2.2), Cauchy inequality, and (4.4.3) yield the bound

$$\mathbf{E}\exp\left\{t\left|\hat{\xi}(m)\right|\right\} - 1 \leq \mathbf{E}e^{t|\xi|}\mathbf{1}_{X\setminus K_m}(\xi) \leq \sqrt{\delta(m)A(2t)}, \quad t > 0, \tag{4.4.5}$$

where $\delta(m) \to 0$ as $m \to \infty$. Hence for each $t > 0$ and $m \in \mathbf{N}$

$$\mathbf{p} \leq \exp\left\{-n\left(t\varepsilon - \sqrt{\delta(m)A(2t)}\right)\right\}, \quad \hat{L}_\varepsilon(m) \geq t\varepsilon - \sqrt{\delta(m)A(2t)}.$$

The left-hand side in the last inequality does not depend on the arbitrarily chosen $t > 0$. To obtain the assertion of the lemma, it remains to pass to the limit first in $m \to \infty$ with t fixed to eliminate the term containing $\delta(m)$, and then in $t \nearrow \infty$. ○

The "principal" truncated part $\xi^0(m)$ in decomposition (4.4.4) has DF

$$\Lambda_m(x) \stackrel{def}{=} \sup_{h \in X^*} [\langle x, h \rangle - \ln \varphi_m(h)].$$

It approximates the original DF $\Lambda(x) = \sup_h [\langle x, h \rangle - \ln \varphi(h)]$ closely enough for large values of m in the following sense.

For a closed set $C \subset X$, denote by C^ε, $\varepsilon > 0$, its ε-neighborhood, by $\overline{C^\varepsilon}$ the closure of this neighborhood, and put

$$\Lambda_*(C) \stackrel{def}{=} \liminf_{\varepsilon \to 0} \liminf_{m \to \infty} \inf_{x \in \overline{C^\varepsilon}} \Lambda_m(x).$$

Lemma 4.4.2 *For each convex closed set* C

$$\Lambda_*(C) \geq \inf_{x \in C} \Lambda(x).$$

Proof of Lemma 4.4.2. Select sequences

$$m(k) \nearrow \infty, \quad \varepsilon(k) \searrow 0, \quad a_k \in \overline{C^{\varepsilon(k)}} \bigcap K_{m(k)}$$

in such a way that $\lim_{k \to \infty} \Lambda_{m(k)}(a_k) = \Lambda_*(C)$. (The compact sets above are those of (4.4.3).)

The lower bounds in the definition of $\Lambda_*(C)$ are attainable by Theorem 4.1.2 and Lemma 4.3.1. Hence one can choose some distributions G_k for which

$$I_{F_{m(k)}}(G_k) \to \Lambda_*(C), \quad \int_X x G_k(dx) = a_k,$$

where F_m is the truncated distribution of ξ^0 in (4.4.4).

By the construction there is weak convergence $F_{m(k)} \stackrel{weak}{\longrightarrow} F$. Hence by Lemma 4.2.3 it is possible to select a subsequence $k' = k'(l)$ for which also the sequence $G_{k'}$ is weakly convergent to some measure $\overline{G}$. Besides, the mean values of $G_{k'}$ also converge by Lemma 4.2.5. Thus, as $k' = k'(l)$ goes to infinity,

$$G_{k'} \stackrel{weak}{\longrightarrow} \overline{G}, \quad F_{m(k')} \stackrel{weak}{\longrightarrow} F, \quad a_{k'} = \int_X x G_{k'}(dx) \to \overline{a} = \int_X x \overline{G}(dx) \in C.$$

The last inclusion holds because C is closed and $a_{k'} \in \overline{C^{\varepsilon(k')}}$ by the construction.

By Lemma 4.3.1 and (4.2.7) $\Lambda(a) \leq I_F(\overline{G})$, while by the construction and Lemma 4.2.2

$$I_F(\overline{G}) \leq \liminf_{k' \to \infty} I_{F_{m(k')}}(G_{k'}) = \Lambda_*(C).$$

This proves the lemma. ○

Proof of Theorem 4.4.1. Assertion a) is a special case of Theorem 4.1.2.

To prove assertion b), consider the truncated sums $S_n^0 = S_n^0(m)$ and the residues $\hat{S}_n = \hat{S}_n(m)$ from decomposition (4.4.4). The following inequality is evident: for each $\varepsilon > 0$

$$\mathbf{P}\left\{\tfrac{1}{n} S_{1,n}[\xi] \in C\right\} \leq \mathbf{P}\left\{\left|\tfrac{1}{n}\hat{S}_n\right| \geq \varepsilon\right\} + \mathbf{P}\left\{\tfrac{1}{n} S_n^0 \in \overline{C^\varepsilon}\right\}.$$

Hence, in notation of (4.1.1),

$$\liminf_{n\to\infty} \lambda_n(C) \geq \min\{A_1(\varepsilon,m), A_2(\varepsilon,m)\}$$

where

$$A_1(\varepsilon,m) = \liminf_{n\to\infty}\left(-\tfrac{1}{n}\ln \mathbf{P}\left\{\left|\tfrac{1}{n}\hat{S}_n(m)\right| \geq \varepsilon\right\}\right),$$

$$A_2(\varepsilon,m) = \liminf_{n\to\infty}\left(-\tfrac{1}{n}\ln \mathbf{P}\left\{\tfrac{1}{n}S_n^0(m) \in \overline{C^\varepsilon}\right\}\right).$$

It follows from Lemma 4.4.1 that $A_1(\varepsilon,m)\to 0$ for a fixed value of $\varepsilon > 0$ as $m\to\infty$. Hence for each $\varepsilon > 0$

$$\liminf_{n\to\infty} \lambda_n(C) \geq \liminf_{m\to\infty} A_a(\varepsilon,m),$$

and by Lemma 4.4.2 also

$$\liminf_{n\to\infty} \lambda_n(C) \geq \inf_{x\in C} \Lambda(x).$$

This completes the proof of the theorem. ○

4.4.3 Asymptotics of Exponential Moments

Proof of Theorem 4.4.2. a) Fix an arbitrary value of $\varepsilon > 0$ and choose a point $a\in X$ with its convex neighborhood U in such a way that

$$g(a)-\Lambda(a) \geq \gamma^* - \varepsilon, \quad \sup_{x\in U}|g(x)-g(a)| \leq \varepsilon.$$

Lemma 4.1.1 and (4.1.7) imply then the inequality

$$L(U) = \lim_{n\to\infty} -\frac{1}{n}\ln \mathbf{P}\left\{\tfrac{1}{n}S_{1,n}[\xi]\in U\right\} \leq \Lambda(a),$$

which yields the estimates

$$\begin{aligned}&\liminf_{n\to\infty}\frac{1}{n}\ln \mathbf{E}\exp\left\{ng\left(\tfrac{1}{n}S_{1,n}[\xi]\right)\right\}\\ \geq\ & g(a)-\varepsilon+\lim_{n\to\infty}\frac{1}{n}\ln \mathbf{P}\left\{\tfrac{1}{n}S_{1,n}[\xi]\in U\right\}\\ \geq\ & g(a)-\varepsilon-\Lambda(a) \geq \gamma^*-2\varepsilon.\end{aligned}$$

Consequently,

$$\liminf_{n\to\infty}\frac{1}{n}\ln \mathbf{E}\exp\left\{ng\left(\tfrac{1}{n}S_{1,n}[\xi]\right)\right\} \geq \gamma^*.$$

b) Suppose that $r > 0$. Evidently,

$$\mathbf{E}\exp\left\{ng\left(\tfrac{1}{n}S_{1,n}[\xi]\right)\right\} \leq \mathbf{E}\exp\left\{ng\left(S_{1,n}[\xi]\right)\right\}\left(\mathbf{1}^{(0)}+\mathbf{1}^{(1)}\right),$$

where $\mathbf{1}^{(0)}$ and $\mathbf{1}^{(1)}$ are the indicators

$$\mathbf{1}^{(0)} = \mathbf{1}\left\{\left|\tfrac{1}{n}S_{1,n}[\xi]\right| \leq r\right\},\quad \mathbf{1}^{(1)} = \mathbf{1}\left\{\left|\tfrac{1}{n}S_{1,n}[\xi]\right| > r\right\}.$$

By the Cauchy inequality, the Cramér-type condition (4.2.2), and growth condition (4.4.1), there is inequality

$$\mathbf{E}\exp\left\{ng\left(\tfrac{1}{n}S_{1,n}[\xi]\right)\right\}\mathbf{1}^{(1)} \le \mathbf{E}\exp\left\{n\gamma_1+\gamma_2\sum_{j=1}^{n}|\xi_j|\right\}\mathbf{1}^{(1)}$$

$$\le \exp\left\{n\left(\gamma_1+\tfrac{1}{2}\ln A(2\gamma_2)\right)\right\}\sqrt{\mathbf{P}\left\{\left|\tfrac{1}{n}S_{1,n}[\xi]\right|\ge r\right\}},$$

where the constants γ_i are those from (4.4.1). Combined with Theorem 4.4.1, this inequality leads to the relation

$$\limsup_{n\to\infty}\frac{1}{n}\ln\mathbf{E}\exp\left\{ng\left(\tfrac{1}{n}S_{1,n}[\xi]\right)\right\}\mathbf{1}^{(1)} \le \gamma_1+\tfrac{1}{2}\ln A(2\gamma_2)-\tfrac{1}{2}\Lambda_r, \tag{4.4.6}$$

where $\Lambda_r \stackrel{def}{=} \inf_{|x|\ge r}\Lambda(x)$.

On the ball $B_r=\{x:\ |x|\le r\}$ the functional g is bounded:

$$g_r \stackrel{def}{=} \sup_{|x|\le r} g(x) \le \gamma_1+\gamma_2 r.$$

The intersection of this ball with a level set of DF (4.1.2),

$$K_\varepsilon \stackrel{def}{=} \{x:\ \Lambda(x)\le g_r+|\gamma^*|+\varepsilon\}\bigcap\{x:\ |x|\le r\},$$

is compact for whatever choice of $\varepsilon>0$ by Theorem 4.3.2. This makes it possible to select, for a fixed $\varepsilon>0$, a finite collection of points $a_i\in K_\varepsilon$ with convex neighborhoods $U_i\ni a_i$ so that (see Lemma 4.1.1)

$$K_\varepsilon\subset\bigcup_{i=1}^{N(\varepsilon)}U_i,\quad L(U_i)\ge\Lambda(a_i)-\varepsilon.$$

It is easy to see that

$$\mathbf{E}\exp\left\{ng\left(\tfrac{1}{n}S_{1,n}[\xi]\right)\right\}\mathbf{1}_{\bigcup U_i}\left(S_{1,n}[\xi]\right)$$

$$\le N(\varepsilon)\max_{i=\overline{1,N(\varepsilon)}}\left[\exp\{n[g(a_i)+\varepsilon]\}\mathbf{P}\left\{\tfrac{1}{n}S_{1,n}[\xi]\in U_i\right\}\right].$$

Hence

$$\limsup_{n\to\infty}\frac{1}{n}\ln\mathbf{E}\exp\left\{ng\left(\tfrac{1}{n}S_{1,n}[\xi]\right)\right\}\mathbf{1}_{B_r\cap\cup U_i}\left(\tfrac{1}{n}S_{1,n}[\xi]\right)$$

$$\le \max_{i=\overline{1,N}}\left[g(a_i)-\Lambda(a_i)\right]+2\varepsilon. \tag{4.4.7}$$

The complementary set $C_{r,\varepsilon}\stackrel{def}{=}B_r\setminus\bigcup_{i=1}^{N(\varepsilon)}U_i$ is closed, so Theorem 4.4.1 provides the bound

$$\limsup_{n\to\infty}\frac{1}{n}\ln\mathbf{E}\exp\left\{ng\left(\tfrac{1}{n}S_{1,n}[\xi]\right)\right\}\mathbf{1}_{C_{r,\varepsilon}}\left(\tfrac{1}{n}S_{1,n}[\xi]\right)$$

$$\le g_r-\inf_{x\in C_{r,\varepsilon}}\Lambda(x)\le -|\gamma^*|-\varepsilon. \tag{4.4.8}$$

It follows from (4.4.6)–(4.4.7) that

$$\begin{aligned}&\limsup_{n\to\infty}\frac{1}{n}\ln \mathbf{E}\exp\left\{ng\left(\tfrac{1}{n}S_{1,n}\left[\xi\right]\right)\right\}\\ \le\ &\max\left\{\gamma_1+\tfrac{1}{2}\ln A\left(2\gamma_2\right)-\tfrac{1}{2}\Lambda_r, 2\varepsilon+\max_i\left[g\left(a_i\right)-\Lambda\left(a_i\right)\right], -\left[\varepsilon+\left|\gamma^*\right|\right]\right\}.\end{aligned}$$

By (4.3.1), the DF satisfies the relation $\Lambda_r \to +\infty$ as $r\to\infty$. Consequently,

$$\limsup_{n\to\infty}\frac{1}{n}\ln \mathbf{E}\exp\left\{ng\left(\tfrac{1}{n}S_{1,n}\left[\xi\right]\right)\right\} \le \sup_{a\in X}\left[g(a)-\Lambda(a)\right] = \gamma^*.$$

Combined with the estimate of part a), the latter inequality proves the theorem. ○

4.5 Rough Gaussian LD Asymptotics

4.5.1 Asymptotic Expressions for LD Probabilities

LD asymptotics for Gaussian distributions can be used as a pattern for similar formulae for sums in the range of moderately large deviations.

Below X is a complete separable metrizable locally convex real linear topological space (cf. Section 4.1). Notations are standard: $\langle x,t\rangle$ is the value of linear functional $t\in X'$ at $x\in X$, etc. The family of measurable sets is the Borel σ-algebra $\mathcal{B}(X)$. Recall that topology of X can be reconstructed using as neighborhoods of points the "solid spheres" $U=\{|x-a|_U<r\}$ for appropriate continuous seminorms in X.

In the sequel, $\eta\in X$ is a centered Gaussian RV with the CF

$$\forall\, t\in X' \ \ \mathbf{E}e^{i\langle\eta,t\rangle} = \exp\left\{\ -\tfrac{1}{2}\mathbf{E}\left\langle\eta,t\right\rangle^2\ \right\}. \tag{4.5.1}$$

In this case DF (4.1.2) is

$$\Lambda_0(x) = \sup_{t\in X'}\left[\langle x,t\rangle-\tfrac{1}{2}\mathbf{E}\left\langle\eta,t\right\rangle^2\right]. \tag{4.5.2}$$

If η were a centered Gaussian RV from $\mathbf{R}^k$ with nondegenerate covariance operator $\mathcal{V}$, the formula for its DF would be more explicit: $\Lambda_0(x)=\frac{1}{2}|\mathcal{V}^{-1/2}x|^2$.

The section is dedicated to the study of asymptotic behavior at $r\to\infty$ of the quantities

$$g_r\left(B\right) \stackrel{def}{=} \frac{1}{r^2}\ln 1/\mathbf{P}\left\{\frac{1}{r}\eta\in B\right\},\quad B\in\mathcal{B}(X). \tag{4.5.3}$$

Lemma 4.5.1 *If $U\subset X$ is an convex open set, then $g_r(U)$ converges to some limit $g_\infty(U)$ as $r\to\infty$, and this limit satisfies the equality*

$$g_\infty(U) = \inf_{r>0} g_r(U).$$

Lemma 4.5.1 is proved in the next subsection.

For an arbitrary subset of X, put

$$\Lambda(B) \stackrel{def}{=} \inf_{x \in B} \Lambda_0(x). \tag{4.5.4}$$

Theorem 4.5.1 a) *If $G \subset X$ is an open set, then*

$$g_\infty^+(G) \stackrel{def}{=} \limsup_{r \to \infty} g_r(G) \leq \Lambda(G).$$

b) *If the set $K \subset X$ is compact, then*

$$g_\infty^-(K) \stackrel{def}{=} \liminf_{r \to \infty} g_r(K) \geq \Lambda(K).$$

c) *If U is a convex open subset of X, then $g_r(U)$ converges to some limit $g_\infty(U)$ as $r \to \infty$, and*

$$g_\infty(U) = g_\infty^\pm(U) = \Lambda(U).$$

Proof of Theorem 4.5.1 repeats that of Theorem 4.1.1 after Lemma 4.5.1 is proved. ○

It follows from Theorem 4.5.1 and Lemma 4.5.1 that for each convex open set U and $\varepsilon > 0$ there is a convex compact set K such that

$$K \subset U, \quad \Lambda(K) < \Lambda(U) + \varepsilon. \tag{4.5.5}$$

Indeed, by part c) of Theorem 4.5.1 one can choose $r > 0$ so that $g_r(U) < \Lambda(U) + \frac{1}{2}\varepsilon$ and, since $\mathcal{L}(\eta)$ is tight, there is a convex compactum K such that $g_r(K) < g_r(U) + \frac{1}{2}\varepsilon$, so $\Lambda(K) \leq g_r(K)$ satisfies the desired inequality.

Suppose that X is a separable real Banach space with norm $|\cdot|$, and the RV $\eta \in X$ is centered Gaussian. In this special case the result of Theorem 4.5.1 can be made slightly more precise.

It was shown in Section 2.1 that

$$\mathbf{E} \exp\left\{\alpha \left|\eta\right|^2\right\} < \infty$$

provided that $\alpha > 0$ is sufficiently small. Consequently, the distribution of η obeys the Cramér-type condition (4.2.2), and the results of Sections 4.2–4.4 can be applied to study LD asymptotics of Gaussian Banach-space-valued RV's. In particular, DF (4.5.2) has compact level sets, and Theorem 4.5.1 can be complemented with

Theorem 4.5.2 *If C is an arbitrary closed subset of separable real Banach space X, then*

$$g_\infty^-(C) \geq \Lambda(C).$$

Theorem 4.5.2 is proved in the next subsection. It is possible to derive from Theorem 4.5.1 and Theorem 4.5.2 asymptotic formulae for expectations $\mathbf{E} \exp\left\{r^2 f(\eta)\right\}$ similar to that of Theorem 4.4.2. However, these are not treated here, since calculations are fairly similar to those already performed.

4.5.2 Reduction to LD Asymptotics for Sums

The study of LD asymptotics for centered Gaussian RV's relies upon the characteristic property of Gaussian laws:

If η, η_j, $j \in \mathbf{N}$, are i.i.d. centered Gaussian RV's, k any natural number, and a_j arbitrary real coefficients, then the distributions of η and $\sum_{j=1}^k a_j\eta_j$ differ only in scale:

$$\mathcal{L}\left(\sum_{j=1}^k a_j\eta_j\right) = \mathcal{L}(\bar{a}\eta), \ \bar{a} = \sqrt{a_1^2 + \cdots + a_k^2}. \tag{4.5.6}$$

To make reading easier, it is convenient to expose separately the "geometric" part of the construction employed to prove Lemma 4.5.1. Notation of algebraic operations with subsets of X used below is that of subsection A.1.1. Note that $A + B$ is convex if both sets in the sum are convex, and open if at least one of them is open. Besides, $(1+u)A \supset A$ for $u \geq 0$ if $0 \in A$ and A is convex. If V is convex and contains zero, then

$$A + mV - x \supset A + (m-1)V$$

for each $x \in V$ and $m \in \mathbf{N}$. Indeed, consider a vector $a + (m-1)y$ with $a \in A$ and $y \in V$. Obviously, $mx, my \in mV$, and by convexity

$$x + (m-1)y = \tfrac{1}{m}(mx) + (1 - \tfrac{1}{m})(my) \in mV,$$

which means that

$$a + (m-1)y = (a + x + (m-1)y) - x \in A + mV - x.$$

Lemma 4.5.2 *Let U be an open set and $K \subset U$ its compact subset. There is a continuous seminorm $|\cdot|_*$ such that*

$$K + 3V \subset U, \ \ V = \{x \in X : \ |x|_* < 1\}.$$

If both K and U are convex, there is a number $\varepsilon > 0$ such that

$$(1+u)(K + 2V) \supset K + V \ \ \text{if } 0 \leq u \leq \varepsilon.$$

Proof of Lemma 4.5.2. The space X being locally convex, each point $y \in K$ has a neighborhood $\{x : |x - y|_y < 1\} \subset U$ defined using some continuous seminorm. The collection of sets $V_y = \left\{x : |x - y|_y < \frac{1}{8}\right\}$, $y \in K$, is an open covering of K, so it contains a finite open covering $(V_{y(j)})$ of neighborhoods numbered by $j = \overline{1, N}$. Put $|x|_* = 8 \max_{j=\overline{1,N}} |x|_{y(j)}$ and $V = \{x : \ |x|_* < 1\}$. This seminorm and neighborhood have the stipulated properties.

Indeed, assume that $x \in K + 3V$, i.e., $|x - y|_* < 1$ for some $y \in K$. The point y is covered by one of sets $V_{y(j)}$, so

$$|x - y(j)|_{y(j)} \leq \tfrac{1}{8} \ |x - y|_* + |y - y(j)|_{y(j)} < \tfrac{1}{2} \ ,$$

and thus $x \in U$.

To prove the second assertion of the lemma assuming that $K \neq \emptyset$, take some point $a \in K$ so that $K - a + 2V$ is a convex set containing the origin. Then

$$\begin{aligned}(1+u)(K+2V) = &\ a + ua + (1+u)([K-a]+2V) \\ \supset &\ a + ua + [K-a] + 2V \supset K + V\end{aligned}$$

if $0 \le u < 1/|a|_*$ (and, consequently, $\pm ua \in V$). If $K = \emptyset$, the inclusion is obvious. ○

Proof of Lemma 4.5.1. (i) Consider first the case of $\boldsymbol{\mu} \stackrel{def}{=} \inf_{\rho>0} g_\rho(U) < \infty$.

Fix an arbitrary pair of numbers $\varepsilon, \rho > 0$ and choose a convex compact set K so that

$$K \subset U, \quad g_\rho(K) < g_\rho(U) + \varepsilon. \tag{4.5.7}$$

This construction is possible because $\mathcal{L}(\eta)$ is a Radon measure. Choose, moreover some continuous seminorm $|\cdot|_*$ so as to satisfy the inclusion of Lemma 4.5.2.

Consider the identity

$$r^2 = n\rho^2 + \tau^2, \quad n = [r^2/\rho^2], \quad 0 \le \tau^2 < \rho^2, \tag{4.5.8}$$

with $[\cdot]$ the integer part of a number. Evidently, as $r \to \infty$,

$$n \to \infty, \quad n\rho^2/r^2 \to 1.$$

One can associate with (4.5.8) a representation of the distribution $\mathcal{L}(\eta)$ as that of a sum of scaled independent copies of η:

$$\mathcal{L}(\eta) = \mathcal{L}\left(\frac{\rho}{r} S_{1,n}[\eta_j] + \frac{\tau}{r}\eta_{n+1}\right), \quad S_{1,n}[\eta] = \sum_{j=1}^{n} \eta_j.$$

Consequently,

$$\begin{aligned}\mathbf{P}\left\{\tfrac{1}{r}\eta \in U\right\} &\ge \mathbf{P}\left\{\frac{\rho}{r^2} S_{1,n}[\eta] + \frac{\tau}{r^2}\eta_{n+1} \in K + 3V\right\} \\ &\ge \mathbf{P}\left\{\frac{\tau}{r^2}|\eta_{n+1}|_* < 1\right\} \mathbf{P}\left\{\frac{1}{(1+u)\rho}\left(\tfrac{1}{n}S_{1,n}[\eta]\right) \in K + 2V\right\},\end{aligned} \tag{4.5.9}$$

where $u = r^2/(\rho^2 n) - 1 > 0$, while $|\cdot|_*$ and V are the seminorm and neighborhood of Lemma 4.5.2 for the compactum and open convex set of (4.5.7).

The first factor in the right-hand side converges to one as $r \to \infty$: indeed,

$$\mathbf{P}\left\{|\eta|_* < r^2/\tau\right\} \le \mathbf{P}\left\{|\eta|_* < r^2/\rho\right\}$$

for $\tau < \rho$, so

$$\lim_{r\to\infty} \min_{0 \le \tau < \rho} \mathbf{P}\left\{|\eta|_* < r^2/\tau\right\} = 1.$$

Hence

$$\begin{aligned}&\limsup_{r\to\infty} \frac{\ln 1/\mathbf{P}\left\{\frac{1}{r}\eta \in U\right\}}{[r^2/\rho^2]} \\ &\le \limsup_{n\to\infty} \frac{1}{n} \ln 1/\mathbf{P}\left\{\frac{1}{\rho}\left(\tfrac{1}{n}S_{1,n}[\eta]\right) \in (1+u)(K+2V)\right\}.\end{aligned}$$

For large $r > 0$ (or, equivalently, large n and small $u > 0$) by Lemma 4.5.2 there is inclusion $(1+u)(K+2V) \supset K+V$. Consequently,

$$\limsup_{n\to\infty} \frac{1}{n} \ln 1/\mathbf{P}\left\{\frac{1}{\rho}\ \tfrac{1}{n}S_{1,n}[\eta] \in (1+u)(K+2V)\ \right\}$$
$$\leq\ \limsup_{n\to\infty} \frac{1}{n} \ln 1/\mathbf{P}\left\{\ \tfrac{1}{n}S_{1,n}[\eta] \in \rho(K+V)\ \right\},$$

where the set $K+V$ is convex and, as well as V, open. Thus it follows from (4.5.10) that

$$\limsup_{r\to\infty} \frac{1}{r^2} \ln 1/\mathbf{P}\left\{\tfrac{1}{r}\eta \in U\right\} = \limsup_{r\to\infty} \frac{1}{\rho^2 n(r)} \ln 1/\mathbf{P}\left\{\tfrac{1}{r}\eta \in U\right\}$$
$$\leq \frac{1}{\rho^2} \limsup_{n\to\infty} \ln 1/\mathbf{P}\left\{\tfrac{1}{n}S_{1,n}[\eta] \in \rho(K+V)\right\}.$$

Using Lemma 4.1.2 (see also (4.5.7) and the proof of Lemma 4.1.1), this estimate can be written as

$$g_\infty^+(U) \leq \inf_{n\in\mathbf{N}} g_{\rho\sqrt{n}}(K+V) \leq g_\rho(K+V) \leq g_\rho(K) \leq g_\rho(U) + \varepsilon. \quad (4.5.10)$$

Since both $\varepsilon > 0$ and $\rho > 0$ are arbitrary, this leads to the bounds

$$g_\infty^+(U) \leq g_\rho(U), \quad g_\infty^+(U) \leq \boldsymbol{\mu} = \inf_{\rho>0} g_\rho(U). \quad (4.5.11)$$

At the same time, it is a direct consequence of the definition of $\boldsymbol{\mu}$ that $g_\infty^-(U) \geq \boldsymbol{\mu}$. Hence the upper and lower limits $g_\infty^\pm(U)$ coincide, and the limit of the lemma does indeed exist.

The remaining possibility, $\boldsymbol{\mu} = +\infty$, implies equality $g_\rho(U) = \infty$ for all ρ, so existence of an infinite limit is obvious.

(ii) It remains to calculate the limit $g_\infty(U)$, which exists by the part of the lemma already proved. Thus it suffices to consider the limit along the subsequence $r_n = \sqrt{n}$. Theorem 4.1.1 and (4.5.6) show that for this subsequence

$$g_\infty(U) = \lim_{n\to\infty} \frac{1}{n} \ln 1/\mathbf{P}\left\{\ \tfrac{1}{n}S_{1,n}[\eta] \in U\ \right\} = \boldsymbol{\Lambda}(U).$$

This completes the proof of the lemma. ○

The argument that proves the refinement of Theorem 4.5.2 for Banach-space-valued RV's is as follows.

Proof of Theorem 4.5.2. As all distributions in a Banach space, the law $\mathcal{L}(\eta)$ is tight. Hence for each number $q \in (0, \frac{1}{2})$ there is a compact subset $K = K_q$ such that $\mathbf{P}\{\eta \notin K_q\} \leq q$. This compactum can be chosen convex and symmetrical: $K = \{x : |x|_+ \leq 1\}$, where $|\cdot|_+$ is a seminorm in X (stronger than the original norm $|\cdot|$). For this seminorm $\mathbf{E}\exp\left\{\alpha\,|\eta|_+^2\right\} < \infty$ if $\alpha > 0$ is sufficiently small (see Theorem 2.1.2 or the proof of Theorem 2.1.1). Consequently, for each $a > 0$ there is inequality

$$\mathbf{P}\left\{|\eta|_+ > ra\right\} \leq c\exp\left\{-\alpha a^2 r^2\right\}. \quad (4.5.12)$$

By the construction, the set aK is compact for each value of $a > 0$.

Evidently,

$$\mathbf{P}\left\{\tfrac{1}{r}\eta \in C\right\} \le \mathbf{P}\left\{|\eta|_+ > ra\right\} + \mathbf{P}\left\{\tfrac{1}{r}\eta \in C \cap aK\right\}.$$

Hence (4.5.12) and the fact that K is a compactum imply the estimates

$$\begin{aligned} g_\infty^-(C) &\ge \min\left\{g_\infty^-(C \cap aK), g_\infty^-(X \setminus aK)\right\} \\ &\ge \min\left\{\Lambda(C \cap aK), \alpha a^2\right\} \ge \Lambda(C) \end{aligned}$$

if a is large enough. This proves the theorem. ○

4.6 Moderately Large Deviations for Sums

4.6.1 Asymptotic Formulae

In this section X is a complete separable metrizable locally convex topological linear space. Notation of preceding section and Chapter 3 is used without additional comment.

The RV's $\xi, \xi_1, \ldots \in X$ are i.i.d.

It is assumed that one-dimensional projections of ξ obey the Cramér condition

$$\forall t \in X' \; \exists A_1 = A_1(t) > 0 : \; A_2(t) \stackrel{def}{=} \mathbf{E}\exp\left\{A_1 \left|\langle \xi, t\rangle\right|\right\} < \infty. \tag{4.6.1}$$

They are, moreover, centered:

$$\forall t \in X' \quad \mathbf{E}\langle \xi, t\rangle = 0. \tag{4.6.2}$$

It is general knowledge that the two conditions above imply asymptotic normality of all one-dimensional projections $\langle \zeta_n, t\rangle$ of the normalized sums

$$\zeta_n \stackrel{def}{=} \tfrac{1}{\sqrt{n}} S_{1,n}[\xi] = \tfrac{1}{\sqrt{n}}\left(\xi_1 + \cdots + \xi_n\right). \tag{4.6.3}$$

In what follows, sums (4.6.3) are assumed to be asymptotically normal in a stronger sense: the central limit theorem is applicable to the sequence $\xi_1, \xi_2, \ldots$ in its infinite-dimensional form. Namely, there exists a centered Gaussian RV $\eta \in X$ such that

$$\forall t \in X' \; \mathbf{E}\exp\left\{i\langle \eta, t\rangle\right\} = \exp\left\{-\tfrac{1}{2}\mathbf{E}\langle \xi, t\rangle^2\right\}, \tag{4.6.4}$$

and there is weak convergence of distributions in X:

$$\mathcal{L}(\zeta_n) \stackrel{weak}{\to} \mathcal{L}(\eta) \quad \text{as} \quad n \to \infty. \tag{4.6.5}$$

The weak convergence condition (4.6.4)–(4.6.5) is in the general case an independent restriction imposed upon the distribution of ξ. It is equivalent to the assumption that

$$\liminf_{n\to\infty} \mathbf{P}\left\{\zeta_n \in G\right\} \ge \mathbf{P}\left\{\eta \in G\right\} \tag{4.6.6}$$

for each open set $G \subset X$. Under special assumptions about the properties of topology in X, the "collective" condition (4.6.5) can be replaced by more explicit restrictions on the distribution of a generic summand ξ. Thus, if X is a separable Banach space of type 2, condition (4.6.5) follows from the conditions $\mathbf{E}\,|\xi|^2 < \infty$ and $\mathbf{E}\xi = 0$.

This section is a study of asymptotics of moderately large deviations for the sum $S_{1,n}[\xi]$. Asymptotic expressions are derived for the quantities

$$l_n(B) \stackrel{def}{=} \frac{1}{r_n^2} \ln 1/\mathbf{P}\left\{\frac{1}{r_n}\zeta_n \in B\right\}, \quad B \in \mathcal{B}(X), \tag{4.6.7}$$

where the positive real sequence r_n satisfies the condition

$$\lim_{n\to\infty} r_n = +\infty, \quad \lim_{n\to\infty} r_n/\sqrt{n} = 0. \tag{4.6.8}$$

Under assumptions of a fairly general nature, the rough LD asymptotics in the range of moderately large deviations is quite similar to that of Gaussian distributions. It can be described using DF (4.5.2) of the Gaussian limit distribution.

Theorem 4.6.1 *Assume conditions* (4.6.1)–(4.6.2), (4.6.4)–(4.6.5), *and* (4.6.8) *to be satisfied. Then the following assertions hold for sequence* (4.6.7).

a) *If the set $G \subset X$ is open, then*

$$l_\infty^+ \stackrel{def}{=} \limsup_{n\to\infty} l_n(G) \le \Lambda(G).$$

b) If the set $K \subset X$ is compact, then

$$l_\infty^-(K) \stackrel{def}{=} \liminf_{n\to\infty} l_n(K) \ge \Lambda(K).$$

c) *The sequence $l_n(U)$ converges to some limit $l_\infty(U)$ as $n \to \infty$ if the set $U \subset X$ is open and convex, and*

$$l_\infty(U) = l_\infty^\pm(U) = \Lambda(U).$$

In the special case of RV's with values from a separable real Banach space one can obtain a slightly more precise result (see subsection 4.6.3 below).

The proof of Theorem 4.6.1 is exposed in subsection 4.6.3.

4.6.2 An Auxiliary Result

The following technical lemma is important in the proof of Theorem 4.6.1. It is used to reduce the order of deviations in later calculations.

Lemma 4.6.1 *Consider a real positive sequence r_n, $n \in \mathbf{N}$, satisfying condition* (4.6.8). *Assume that another sequence ρ_n is nondecreasing and fulfills the following assumptions:*

$$\rho_n \ge 1, \quad \lim_{n\to\infty} \rho_n = +\infty, \quad \lim_{n\to\infty} \rho_n/r_n = 0.$$

Then there exists a sequence of natural numbers m_n such that

$$\lim_{n\to\infty} m_n = +\infty, \quad \lim_{n\to\infty} m_n/n = 0,$$

and, moreover,

$$\rho_{m_n}/\sqrt{m_n} \le r_n/\sqrt{n}, \quad \rho_{m_n}/\sqrt{m_n} = [1+o(1)]\, r_n/\sqrt{n}.$$

Proof of Lemma 4.6.1. Put

$$m_n = \min\left\{m \in \mathbf{N} : m/\rho_m^2 \ge n/r_n^2 - 10\right\}.$$

It follows from the restrictions imposed upon the sequence ρ_m that the inequality $m_n \le n$ holds for large enough values of n; at the same time, $m_n \to \infty$ since $n/r_n^2 \to \infty$ as $n \to \infty$.

The quantity m/ρ_m^2 cannot increase by more than 1 as m is changed to $m+1$. Hence there is inequality

$$n/r_n^2 - 10 \le m_n/\rho_{m_n}^2 \le n/r_n^2,$$

so the last two restrictions stipulated in the lemma are fulfilled. Finally, by the assumption $\rho_n^2/r_n^2 \to 0$ as $n \to \infty$, and it is easily seen that

$$m_n/n \le \rho_{m_n}^2/r_n^2 \le \rho_n^2/r_n^2.$$

This completes the proof of the lemma. ○

4.6.3 Calculation of Asymptotics

It is convenient to summarize the steps of proof of Theorem 4.6.1 in separate lemmas.

Lemma 4.6.2 *Under the conditions of Theorem* 4.6.1, *the inequality*

$$l_\infty^-(U) \ge \Lambda(U)$$

holds for each open convex set U.

Proof of Lemma 4.6.2. Fix an arbitrary number $\lambda < \Lambda(U)$ and consider the corresponding level set $B = \{x \in X : \Lambda_0(x) \le \lambda\}$ of the deviation functional. The set B is convex and closed (see Lemma 4.1.5), so there is an open half-space $H = H_{e,\rho}^c = \{\langle x, e\rangle > \rho\}$ such that $H \supset U$ and $B \bigcap H = \emptyset$. It follows that

$$\mathbf{P}\left\{\tfrac{1}{r_n}\zeta_n \in U\right\} \le \mathbf{P}\left\{\tfrac{1}{r_n}\langle\zeta_n, e\rangle > \rho\right\}.$$

The right-hand side can be evaluated using S. N. Bernstein's bound of Proposition 3.3.1:

$$l_n(U) \ge \tfrac{1}{2}\rho^2 \mathbf{E}\langle\xi, e\rangle^2 [1+\kappa_n], \quad \kappa_n = O(r_n/\sqrt{n}).$$

Thus, $l_\infty^-(U) \ge \frac{1}{2}\rho^2 \mathbf{E}\langle\xi, e\rangle^2$.

On the other hand, Lemma 4.5.1 combined with (4.1.3) implies the relations

$$\tfrac{1}{2}\rho^2 \mathbf{E}\,\langle \xi, e\rangle^2 \;=\; \lim_{r\to\infty} \frac{1}{r^2} \ln 1/\mathbf{P}\left\{\tfrac{1}{r}\langle \eta, e\rangle > \rho\right\} \;=\; \inf_{x\in H} \Lambda_0(x) \;\geq\; \lambda,$$

where η is a centered Gaussian RV with same covariances, so $l^-_\infty(U) \;\geq\; \lambda$ for each $\lambda < \Lambda(U)$. The lemma is proved. ○

In the calculations to follow it will often be convenient to use integer-valued sequences r_n^2 in order to avoid additional work verifying that the rounding errors are small. The necessary preparations are done in Lemma 4.6.3 below.

The assumptions concerning the range of deviations in Lemma 4.6.3 are much more strict than those of Theorem 4.6.1. In the proof of the latter, they will be relaxed using a reduction to the "convenient" range of deviations with the help of Lemma 4.6.1.

Lemma 4.6.3 *Assume that the distribution of ξ satisfies the asymptotic normality condition* (4.6.4)–(4.6.6), *and $U \subset X$ is an open set.*

a) *There exists a nondecreasing sequence $\rho_n^2 \nearrow \infty$ such that the inequality*

$$\limsup_{n\to\infty} \widehat{\rho}_n^{\,-2} \ln 1/\mathbf{P}\left\{\widehat{\rho}_n^{\,-1}\zeta_n \in U\right\} \;\leq\; \Lambda(U)$$

is satisfied whenever $\widehat{\rho}_n^{\,2}$ is an integer sequence obeying the restrictions

$$\widehat{\rho}_n^{\,2} \;\leq\; \rho_n^2, \quad n \geq n_0, \quad \lim_{n\to\infty} \widehat{\rho}_n^{\,2} \;=\; +\infty.$$

b) *If the original sequence r_n is such that* $\lim_{n\to\infty} r_n = \infty$, *then the sequence ρ_n can be selected so as to fulfill the additional requirement*

$$\lim_{n\to\infty} \rho_n / r_n = 0.$$

Proof of Lemma 4.6.3. It follows from the asymptotic normality assumption (see (4.6.6)) that for each given $m \in \mathbf{N}$ there exists a natural number $N_0(m)$ such that for $k = \overline{1, m}$ and $n \geq N_0(m)$

$$\frac{1}{k} \ln 1/\mathbf{P}\left\{ \tfrac{1}{\sqrt{k}}\zeta_n \in U \right\} \;\leq\; \frac{1}{k} \ln 1/\mathbf{P}\left\{ \tfrac{1}{\sqrt{k}}\eta \in U \right\} + \frac{1}{m}.$$

The sequence $N_0(m)$ can be chosen nondecreasing, and its growth with $m \to \infty$ can be made arbitrarily fast. The function $N_0 : \mathbf{N} \to \mathbf{N}$ has a nondecreasing inverse on the set $N_0(\mathbf{N})$ which can be extended to a nondecreasing function M_0 on all the set of natural numbers using piecewise constant interpolations. Obviously, $M_0\left(N_0(m)\right) \equiv m$.

The inequality stipulated in the lemma holds for $k \;=\; \overline{1, M_0(n)}$ by the construction, so the sequence $\rho_n \;=\; \sqrt{M_0(n)}$ satisfies the requirements of the lemma. ○

Lemma 4.6.4 *Assume all conditions of Theorem* 4.6.1 *to be satisfied.*

If $U \subset X$ is a convex open set, then $l^+_\infty(U) \;\leq\; \Lambda(U)$.

Proof of Lemma 4.6.4. Fix a number $\varepsilon > 0$. By (4.5.5) there is a convex compactum K and a continuous seminorm $|\cdot|_*$ such that

$$K + 3V \subset U, \quad \Lambda(K+V) \le \Lambda(K) < \Lambda(U) + \varepsilon,$$

where $V = \{x : |x|_* < 1\}$ is the open "unit ball" for the seminorm considered.

Given the sequence r_n and the convex open set $K + V$, choose a sequence ρ_n for which

$$\limsup_{n\to\infty} \widehat{\rho}_n^{-2} \ln 1/\mathbf{P}\left\{\widehat{\rho}_n^{-1}\zeta_n \in K+V\right\} \le \Lambda(K+V)$$

if $\widehat{\rho}_n$ obeys the conditions of Lemma 4.6.3. Use the sequence ρ_n to construct the natural sequence m_n of Lemma 4.6.1, and put

$$\begin{aligned} l(n) &= [n/m(n)] - 1, \quad q(n) = n - l(n)m(n), \\ \kappa(n) &\stackrel{def}{=} l(n)\left(\rho_{m(n)}/r_n\right)\sqrt{m(n)/n} \end{aligned}$$

(with $[\cdot]$, as usual, the integer part of a number.)

By the construction,

$$\forall n \;\; \kappa(n) \le 1, \quad \kappa(n) \to 1, \quad m(n)/n \to 0, \quad m(n) \le q(n) < 2m(n).$$

(Below $\rho_{m(n)}$, $\kappa(n)$, $m(n)$, $q(n)$, etc., are abbreviated to $\rho = \rho_m$, κ, m, q, etc. whenever there is no risk of ambiguity.)

Evidently, $n = lm + q$, and the corresponding decomposition of the sum ζ_n into independent summands is

$$\frac{1}{r_n}\zeta_n = \frac{\kappa}{l\rho_m}\left(\tfrac{1}{l}S_{1,l}[\zeta_m]\right) + \frac{1}{r_n}\sqrt{q/n}\,\zeta_q',$$

where the sum $S_{1,l}[\zeta_m]$ consists of independent copies of ζ_m, and the "residue" ζ_q' has the distribution of ζ_q.

Reasoning as in the proof of Lemma 4.5.1, one can conclude that

$$\begin{aligned} &\mathbf{P}\left\{\tfrac{1}{r}\zeta_n \in U\right\} \ge \mathbf{P}\left\{\tfrac{1}{r}\zeta_n \in K+3V\right\} \\ \ge\;& \mathbf{P}\left\{|\zeta_q'|_* < r\sqrt{n/q}\right\}\mathbf{P}\left\{\frac{\kappa}{\rho}\,\tfrac{1}{l}S_{1,l}[\zeta_m] \in (K+2V)\right\} \\ \ge\;& \mathbf{P}\left\{|\zeta_q'|_* < r\sqrt{n/q}\right\}\mathbf{P}\left\{\frac{1}{\rho}\,\tfrac{1}{l}S_{1,l}[\zeta_m] \in K+V\right\} \end{aligned}$$

because $1 \le 1/\kappa(n) = 1 + o(1)$ as $n \to \infty$. In the right-hand side, the first factor goes to one by the assumption of asymptotic normality as $r_n\sqrt{n/q(n)} \to \infty$. By convexity

$$\mathbf{P}\left\{\frac{1}{\rho}\left(\tfrac{1}{l}S_{1,l}[\zeta_m]\right) \in K+V\right\} \ge \left[\mathbf{P}\left\{\tfrac{1}{\rho}\zeta_m \in K+V\right\}\right]^l.$$

Hence,

$$\frac{1}{r^2}\ln 1/\mathbf{P}\left\{\frac{1}{r}\zeta_n \in U\right\} \le \frac{1}{r^2}\ln 1/\mathbf{P}\left\{|\zeta_q'|_* < r\sqrt{n/q}\right\}$$

$$+ \quad \frac{l\rho^2}{r^2}\frac{1}{\rho^2}\ln 1/\mathbf{P}\left\{\zeta_m \in \rho(K+V)\right\}.$$

As $l\rho^2/r^2 = 1 + o(1)$, this yields the estimate

$$l^+_\infty(U) \le \Lambda(K+V) < \Lambda(U) + \varepsilon.$$

The last inequality and a passage to the limit in $\varepsilon \searrow 0$ prove the lemma. ○

Proof of Theorem 4.6.1. Lemma 4.6.2 and Lemma 4.6.4 ensure existence of the limit $l_\infty(U)$. To prove all assertions of the theorem, it remains to repeat (with obvious modifications) the proof of Theorem 4.5.1. ○

4.6.4 Moderately Large Deviations in Banach Space

This subsection deals with sums of RV's with values in a separable real Banach space X. Notations below are standard: thus, $\langle\cdot,\cdot\rangle$ and $|\cdot|$ denote, respectively, the duality relation between the spaces X, X^*, and the norms in these spaces.

As in the preceding subsections, ξ, ξ_1, ... are i.i.d. RV's which satisfy conditions (4.6.2) and (4.6.4)–(4.6.5), the latter with X' replaced by X^*. The assumption of exponential summability of one-dimensional projections (4.6.1) is replaced by the much more restrictive Cramér type condition for the distribution of the norm

$$\exists\, A_1 > 0 : \; A_2 = \mathbf{E}\exp\left\{A_1 |\xi|\right\} < \infty. \qquad (4.6.9)$$

The main result proved in this subsection is

Theorem 4.6.2 *Assume conditions* (4.6.2), (4.6.4)–(4.6.5), (4.6.8), *and* (4.6.9). *Then the estimate*

$$l^-_\infty(C) \ge \Lambda(C)$$

holds for each closed subset C of the separable real Banach space X.

The proof of Theorem 4.6.2 is based upon a special finite-dimensional approximation for the distribution of a summand.

Lemma 4.6.5 *Let the RV $\xi \in X$ obey the condition $\mathbf{E}|\xi|^2 < \infty$ and be centered,* $\mathbf{E}\xi = 0$.

For each number $\delta > 0$ there exists a Borel measurable mapping $\tau_\delta : X \to X$ which satisfies the inequality

$$\mathbf{E}\left|\xi - \tau_\delta(\xi)\right|^2 \le \delta^2 \qquad (4.6.10)$$

and has the following properties:

a) *the range of the mapping $\tau_\delta(X)$ is some finite-dimensional subspace of X;*

b) *there exists a constant $c = c_\delta > 0$ such that almost surely $\tau_\delta(\xi) \le c$;*

c) $\mathbf{E}\tau_\delta(\xi) = 0$;

d) *the covariances of $\tau_\delta(\xi)$ are dominated by those of ξ:*

$$\mathbf{E}\left\langle\tau_\delta(\xi), f\right\rangle^2 \le \mathbf{E}\left\langle\xi, f\right\rangle^2, \quad f \in X^*.$$

Proof of Lemma 4.6.5. To construct the mapping of the lemma, fix $\delta > 0$ and a number $\varepsilon > 0$ (whose value will later be chosen in accordance with that of δ).

Distributions in X are tight, so it is possible to select a compact set $K \subset X$ with arbitrarily small measure $\mathbf{P}\{\xi \notin K\}$. By the well-known absolute continuity property of the integral, the compactum can be chosen to satisfy the inequality

$$\mathbf{P}\{\xi \notin K\} + \mathbf{E}\,|\xi|^2\,\mathbf{1}_{\{X\setminus K\}}(\xi) \le \varepsilon^2. \tag{4.6.11}$$

The compact set K has a finite covering by balls of radius ε: for some $N = N(\varepsilon)$ and x_j, $j = \overline{1, N}$

$$K \subset \bigcup_{j=1}^{N} B_j^{(\varepsilon)}, \quad B_j^{\varepsilon} = x_j + \varepsilon B, \quad B = \{x : |x| < 1\}.$$

The sets

$$K \bigcap \left(B_j^{(\varepsilon)} \setminus \textstyle\bigcup_{j'<j} B_{j'}^{(\varepsilon)}\right), \quad j = \overline{1, N},$$

also cover K and are, moreover, disjoint. With the set $F_0 = X \setminus K$ added, they provide a partition of the space X into a union of disjoint sets:

$$X = \bigcup_{j=0}^{N} F_j. \tag{4.6.12}$$

This partition can be used to introduce a new RV which is the conditional expectation of ξ with respect to the σ-algebra $\mathcal{A}$ generated by the events $\{\xi \in F_j\}$: namely,

$$\tau_\delta(\xi) \stackrel{def}{=} \mathbf{E}\{\xi\,|\mathcal{A}\} = \sum_{j=0}^{N} \mathbf{E}\{\xi\,|\xi \in F_j\}\,\mathbf{1}_{F_j}(x)\big|_{x=\xi}\,. \tag{4.6.13}$$

The mapping in (4.6.13) has the properties specified by the lemma provided that the number $\varepsilon = \varepsilon(\delta)$ is small.

Indeed, it is evident that the values of $\tau_\delta(x)$ belong to a finite dimensional subspace, and the RV $\tau_\delta(\xi)$ is centered.

The inequality for the covariances follows from the Jensen inequality:

$$\mathbf{E}\,\langle \xi, f\rangle^2 \ge \mathbf{E}\,(\mathbf{E}\{\langle \xi, f\rangle\,|\mathcal{A}\})^2.$$

Some calculations are necessary to evaluate the precision of approximation. By the choice of the sets in (4.6.12), $|\xi - x_j| < \varepsilon$ if $\xi \in F_j$, so for $x \in F_j$

$$|\tau_\delta(x) - x_j| \le \int_{|x'-x_j|<\varepsilon} |x' - x_j|\,\mathbf{Pr}\{\xi \in dx'\,|\xi \in F_j\} \le \varepsilon, \quad j = \overline{1, N}.$$

Consequently,

$$\mathbf{E}\,|\xi - \tau_\delta(\xi)|^2\,\mathbf{1}_{F_j}(\xi) \le \varepsilon^2\mathbf{P}\{\xi \in F_j\}, \quad j \ge 1. \tag{4.6.14}$$

A similar bound for $j = 0$ is derived in a slightly different way. Combined with the Cauchy inequality, (4.6.11) yields the estimates

$$\mathbf{E}\,|\xi - \tau_\delta(\xi)|^2\,\mathbf{1}_{F_0}(\xi) \;\le\; 2\Big[\mathbf{E}\,|\xi|^2\,\mathbf{1}_{F_0}(\xi) + \mathbf{E}\,|\mathbf{E}\,\{\xi\,|\,\xi \in F_0\}|^2\,\mathbf{1}_{F_0}(\xi)\Big] \;\le\; 4\varepsilon^2. \tag{4.6.15}$$

The above estimates show that it suffices to take $\varepsilon < c'\delta$ to obtain the inequality of the lemma for $\mathbf{E}\,|\tau_\delta(\xi) - \xi|^2$. This completes the proof. ○

For a given value of $\delta > 0$, the mapping $\tau = \tau_\delta$ of Lemma 4.6.5 yields the decomposition

$$\xi = \xi^0 + \xi^+, \quad \xi^0 = \tau_\delta(\xi)\,.$$

The corresponding decomposition of the normalized sum is

$$\zeta_n = \tfrac{1}{\sqrt{n}} S_{1,n}[\xi] = \zeta_n^0 + \zeta_n^+, \tag{4.6.16}$$

with

$$\zeta_n^0 = \tfrac{1}{\sqrt{n}} S_{1,n}[\xi^0], \quad \zeta_n^+ = \tfrac{1}{\sqrt{n}} S_{1,n}[\xi^+]\,.$$

Lemma 4.6.6 *Under the restrictions of Theorem* 4.6.2 *the inequality*

$$l_\infty^{0-}(C) \stackrel{def}{=} \liminf_{n\to\infty} \frac{1}{r_n^2} \ln 1/\mathbf{P}\left\{\frac{1}{r_n}\zeta_n^0 \in C\right\} \ge \Lambda(C)$$

holds for each closed set $C \subseteq X$.

Proof of Lemma 4.6.6. Denote by E^0 the finite-dimensional subspace which contains all values of ξ^0, and by $\overline{B}$ the closure of the unit ball in X.

The norms $|\xi_j^0|$ of the summands in the sum ζ_n^0 are a.s. bounded, and the summands are centered. In a finite-dimensional space, it is possible to use the classical bound of S.N.Bernstein (see Proposition 3.3.1) to derive an estimate for large deviation probabilities:

$$\mathbf{P}\{\zeta_n^0 \notin \alpha r_n B\} \le c_1 \exp\left\{-c_2\alpha^2 r_n^2\left[1 + c_3\alpha r_n/\sqrt{n}\right]^{-1}\right\},$$

where the values of $c_i > 0$ depend only upon the number of dimensions of E^0 and the upper bound of the values of $|\xi^0|$ (see Lemma 4.6.5, part b).

By the choice of the subspace E^0,

$$\left\{\zeta_n^0 \in C \bigcap (\alpha r_n B)\right\} = \left\{\zeta_n^0 \in r_n \widehat{K}_\alpha\right\},$$

where the set $\widehat{K}_\alpha = (\alpha r_n \overline{B}) \bigcap C \bigcap E^0$ is bounded and, since E^0 is finite-dimensional, compact. This permits one to apply the bound of Theorem 4.6.1,

$$l_\infty^{0-}\left(\widehat{K}_\alpha\right) \ge \Lambda^0\left(\widehat{K}_\alpha\right), \quad \Lambda^0(K) = \inf_{x\in K} \Lambda_0^0(K) \ge \Lambda(K),$$

where by Lemma 4.6.5, d) the "finite-dimensional" DF of the accompanying

Gaussian law for ξ^0 satisfies the inequality

$$\begin{aligned}\Lambda_0^0(x) &= \sup_{h\in X^*}\left[\langle x,h\rangle - \tfrac{1}{2}\mathbf{E}\langle \xi^0,h\rangle^2\right] \\ &\geq \Lambda_0(x) = \sup_{h\in X^*}\left[\langle x,h\rangle - \tfrac{1}{2}\mathbf{E}\langle \xi,h\rangle^2\right].\end{aligned}$$

Combining the above bounds, it is easily seen that

$$l_\infty^{0-}(C) \geq \min\left\{\Lambda(C), c_2\alpha^2\right\},$$

where the number $\alpha > 0$ is arbitrary. The proof of the lemma is completed by the passage to the limit in $\alpha \to \infty$. ○

The next lemma deals with the infinite-dimensional residue in decomposition (4.6.16).

Lemma 4.6.7 *Let r_n be an arbitrary sequence satisfying condition* (4.6.8). *Fix $\delta > 0$, and use the mapping τ_δ of Lemma* 4.6.5 *to define decomposition* (4.6.16). *The following inequality holds for each number $\varepsilon > 0$:*

$$\liminf_{n\to\infty} \frac{1}{r_n^2} \ln 1/\mathbf{P}\left\{\left|\zeta_n^+\right| \geq \varepsilon r_n\right\} \geq \tfrac{1}{2}(\varepsilon/\delta)^2.$$

Proof of Lemma 4.6.7. By the assumption of asymptotic normality (4.6.4)–(4.6.6), the sequence of distributions of ζ_n is tight. Hence (see Remark 3.1.3) there exists a constant $c' > 0$ such that

$$\mathbf{E}\left|\zeta_n\right| \leq c', \quad n \in \mathbf{N}.$$

The laws of the finite-dimensional RV's $\zeta_n^0 \in E^0$ are also asymptotically normal whatever the choice of δ in (4.6.16) provided that this parameter is fixed. Hence also for these sums of a.s. bounded RV's there is inequality

$$\mathbf{E}\left|\zeta_n^0\right| \leq c'' < \infty, \quad n \in \mathbf{N},$$

with some positive constant c''.

Thus, a similar estimate holds for the infinite-dimensional residual term in decomposition (4.6.16):

$$\mathbf{E}\left|\zeta_n^+\right| \leq c' + c'',$$

and one can use the estimate of Theorem 3.3.1. Indeed, the i.i.d. summands satisfy the inequality

$$\mathbf{E}\exp\left\{h\left|\xi^+\right|\right\} \leq \frac{\mathbf{E}\exp\{2h|\xi^0|\} + \mathbf{E}\exp\{2h|\xi|\}}{2},$$

so its conditions are fulfilled. The estimate of the lemma now follows from the inequality of Theorem 3.3.1 and (4.6.10). ○

Proof of Theorem 4.6.2. Fix a value of $\delta > 0$ and denote by τ the corresponding mapping of Lemma 4.6.5. Decompose the sum ζ_n as explained before Lemma 4.6.6. (The asymptotic behavior of LD probabilities of both summands in this decomposition is described in the two preceding lemmas.) Evidently, for each $\varepsilon > 0$

$$\mathbf{P}\{\zeta_n \in C\} \leq \mathbf{P}\{|\zeta_n^+| \geq \varepsilon\} + \mathbf{P}\{\zeta_n^0 \in \overline{C^\varepsilon}\},$$

where $\overline{C^\varepsilon}$ is the closure of the ε-neighborhood of the set C.

It follows from Lemma 4.6.6 and Lemma 4.6.7 that

$$l_\infty^-(C) \geq \min\left\{\Lambda\left(\overline{C^\varepsilon}\right), \tfrac{1}{2}(\varepsilon/\delta)^2\right\}.$$

The left-hand side of this inequality is independent of the choice of ε or δ, and neither of these quantities is subject to any restrictions. The passage to the limit in $\delta \to \infty$ yields the bound

$$l_\infty^-(C) \geq \Lambda\left(\overline{C^\varepsilon}\right), \quad \varepsilon > 0.$$

To derive from the latter inequality the assertion of the theorem, it remains to substantiate the possibility of the passage to the limit in $\varepsilon \to \infty$.

If a number l satisfies the inequality $l < \Lambda(C)$, then the set $\{x : \Lambda_0(x) \leq l\}$ has no common points with the set C^ε provided that $\varepsilon > 0$ is sufficiently small.

Indeed, by Theorem 4.3.2 the level set of the Gaussian deviation functional Λ_0 is compact, and Λ_0 is lower semicontinuous. It is easy to arrive at a contradiction assuming that C_l and $\overline{C^\varepsilon}$ have common points for all $\varepsilon > 0$. Indeed, if there were common points, it would be possible to choose a convergent sequence $x_n \to x_\infty$ with $x_n \in C_l$ and $x_\infty \in \overline{C^\varepsilon}$. Such a sequence could be used to establish the inequality

$$l \geq \liminf_{n\to\infty} \Lambda_0(x_n) \geq \Lambda_0(x_\infty)$$

which is contradictory with the choice of l. Thus, C_l and $\overline{C^\varepsilon}$ have no common points for sufficiently small $\varepsilon > 0$.

It follows that

$$\liminf_{\varepsilon\to 0} \Lambda\left(\overline{C^\varepsilon}\right) \geq l \text{ if } l < \Lambda(C).$$

Since the choice of l is not subject to other restrictions, the last inequality proves the theorem. ○

One can use methods similar to those applied in the above proof to establish versions of Theorem 4.6.1 and Theorem 4.6.2 in a less comprehensive range of deviations under less restrictive conditions of type (4.6.9).

Chapter 5

Gaussian and Related Approximations for Distributions of Sums

The aim of this chapter is to expose a technique of approximating distributions of sums of independent RV's. The situation considered is one where summand's distributions concentrate around its mean, so that it is possible, e.g., to define their higher-order moments. In this setting, the best known approximation to the law of a sum is given by the central limit theorem (CLT). It is the accompanying Gaussian law which has same mean and covariances as the sum. One can construct more complicated approximations, Edgeworth expansions, using higher-order moments.

The problem of approximating sums' distributions is considered here from a rather special viewpoint. Attention is concentrated on some aspects of the construction which are of interest mainly in infinite dimensions, and no attempt is made to include all known facts.

In one or several dimensions Gaussian laws have extremely regular densities. Edgeworth expansions provide corrections to these densities, which improve the accuracy of Gaussian approximation. The resulting approximation to the distribution of a sum also has a very regular density. It is not necessarily a probability (the approximation can be a signed measure), but its values are relatively easy to calculate. Corrections to Gaussian density are obtained by applying to it certain differential polynomials whose coefficients are described in terms of variances and higher moments of summands. To obtain approximations and estimates for their errors, the adequate apparatus is the Fourier transform, and a major part of work is the study and transformation of power expansions of summands' CF's near the origin.

The approach described above fails in infinite dimensions, as there is no Lebesgue measure and, consequently, no densities to approximate.

A natural way to circumvent this difficulty is to concentrate on images of sums' distributions under regular mappings to one or several dimensions. Much of recent progress in studies of convergence rates for infinite-dimensional CLT was reached in this direction. This is the approach adopted below.

Approximation of RV distributions includes two related problems. One is the construction proper of an asymptotic expansion. The other is evaluation of its error. Most attention below is paid to the former and, presumably, simpler of the two.

The method employed to construct approximations is an adaptation of the classical operator approach of J. W. Lindeberg (reflected, e.g., in the classical book of W. Feller) to smooth mappings from a Banach or Hilbert space to finite dimensions. Fourier transforms appear at a later stage when one considers superpositions of such mappings with less regular functions on the real line or in finite dimensions.

The problem of estimating approximation error is considered in less detail. The chapter demonstrates technical means used to this end, but makes no attempt at exhaustive treatment. Some estimates of accuracy of higher-order Edgeworth expansions are included in the last section without proof. The reason for this omission is that known results of this kind are technically extremely demanding, and the restrictions they use seem to be far from their ultimate form.

5.1 Operator Approach to Limit Theorems

5.1.1 Translations and Derivatives

In this section, X is a separable real Banach space with the conjugate X^*. The norms in both are denoted $|\cdot|$, as also the norms of linear operators acting in these spaces or their Cartesian powers, and the absolute values of complex or real numbers. $\langle x, t\rangle$ is the value of a linear functional $t \in X^*$ on a vector $x \in X$.

If this does not cause misunderstanding, all mappings $f : X \to Y$ are called functions on X, whatever the nature of their values.

For reasons of convenience, notation for derivatives, differentials, etc. deviates from the standard usage and follows patterns more common in physics or engineering. Weak differentials of all orders are denoted

$$\prod_{j=1}^{m} \langle a_j, \nabla\rangle^{k_j} f(x) = \frac{\partial^{k_1+\cdots+k_m}}{\partial t_1^{k_1}\cdots\partial t_m^{k_m}} f\left(x+\textstyle\sum_{j=1}^m t_j a_j \right)\Big|_{\mathbf{t}=0}, \tag{5.1.1}$$

where $\mathbf{t} = \left(t_j, j=\overline{1,m}\right) \in \mathbf{R}^m$. Fréchet (or weak) derivatives are $\nabla^{\otimes k} f$, and $\nabla^{\otimes 0} f \stackrel{def}{=} f$ denotes the function itself when this is convenient. This notation aims at stressing the fact that the value of a higher-order strong derivative of a real-valued function can be considered as an element of an appropriate tensor power of the dual space:

$$\left\langle \otimes_{j=1}^m a_j, \nabla^{\otimes m} f(x)\right\rangle = \textstyle\prod_{j=1}^m \langle a_j, \nabla\rangle f(x).$$

To make formulae easier to read, no attempt is made to discern between the operator of multiplication by a constant number and this number, identity operator and the number "1", etc.

It is convenient to first restrict attention to a very small class of complex-valued functions $\mathcal{M}(X \to \mathbf{C})$ which is similar to the class of entire functions of exponential type (playing an important part in the theory of approximation). A function from this class has bounded continuous Fréchet derivatives of arbitrary orders in all the space:

$$\mathcal{M}(X \to \mathbf{C}) = \{f : \langle f \rangle \stackrel{def}{=} \limsup_{k\to\infty} |\nabla^{\otimes k} f|_\infty^{1/k} < \infty\}, \tag{5.1.2}$$

where $|\nabla^{\otimes k} f|_\infty = \sup_{x\in X} |\nabla^{\otimes k} f(x)|$. It is evident that in $\mathcal{M}$ weak differentiation defines a bounded operator $f \mapsto \langle a, \nabla \rangle f$.

In the infinite-dimensional setting of this section, the class $\mathcal{M}$ is too small for most applications. Nevertheless, it is convenient to elaborate some formulae first for functions from $\mathcal{M}$ to have a pattern for the uglier calculations applying to more realistic situations.

The connection between translations and differentiation in X is stressed by special notation for translations operators:

$$\exp\{\langle a, \nabla \rangle\} f(x) \stackrel{def}{=} f(x+a), \quad x, a \in X.$$

If $f \in \mathcal{M}$, $a \in X$, and $n \in \mathbf{N}$, then the Taylor formula can be written as

$$\begin{aligned} e^{\langle a,\nabla\rangle} f(x) &= \sum_{k=0}^{n} \frac{1}{k!} \langle a, \nabla \rangle^k f(x) \\ &+ \int_0^1 \frac{1}{n!}(1-u)^n \langle a, \nabla \rangle^{n+1} f(x+ua)du \\ &= \left[\sum_{k=0}^{n} \langle a, \nabla \rangle^k + \int_0^1 \frac{(1-u)^n}{n!} \langle a, \nabla \rangle^{n+1} e^{u\langle a,\nabla\rangle} du\right] f(x). \end{aligned} \tag{5.1.3}$$

In identities for differential operators and translations, it will often be convenient to drop the function to which these are applied. For instance, the infinite-series version of Taylor's formula and (5.1.3) can be written as

$$e^{\langle a,\nabla\rangle} = \sum_{k=0}^{\infty} \frac{1}{k!} \langle a, \nabla \rangle^k,$$

$$e^{\langle a,\nabla\rangle} = \sum_{k=0}^{n} \frac{1}{k!} \langle a, \nabla \rangle^k + \int_0^1 \frac{1}{n!}(1-u)^n \langle a, \nabla \rangle^{n+1} e^{u\langle a,\nabla\rangle} du.$$

The infinite-series representation of translation is obtained by passage to the limit (for functions from $\mathcal{M}$) from the second one. It could be used to extend $f \in \mathcal{M}$ to the complexification of the original space $X + iX$.

5.1.2 Average Translations and Their Expansions

Each RV $\xi \in X$ is associated with the operators (later called *average translations*)

$$\mathbf{E}\exp\{u\langle\xi,\nabla\rangle\} f(x) \stackrel{def}{=} \mathbf{E}\exp\{\langle u\xi,\nabla\rangle\} f(x) = \mathbf{E} f(x+u\xi), \qquad (5.1.4)$$

where $u \in \mathbf{R}$ and $\mathbf{E}\exp\{0\langle\xi,\nabla\rangle\} = 1$. The domain of definition for operators (5.1.4) is determined by natural integrability assumptions. For a function with bounded Fréchet derivatives $\nabla^{\otimes j} f$, $j \leq k$, the average translation operator is related to differentiation by the formulae

$$\begin{aligned} \mathbf{E}\langle\xi,\nabla\rangle^k f(x) &= \frac{d^k}{du^k}\mathbf{E}e^{u\langle\xi,\nabla\rangle} f(x)\bigg|_{u=0}, \\ \nabla^{\otimes k}\mathbf{E}e^{\langle\xi,\nabla\rangle} f(x) &= \mathbf{E}\nabla^{\otimes k} f(x+\xi), \end{aligned} \qquad (5.1.5)$$

so average translation operators do not deteriorate the differential properties of a function:

$$\left|\nabla^{\otimes k}\mathbf{E}e^{\langle\xi,\nabla\rangle} f\right|_\infty \leq \left|\nabla^{\otimes k} f\right|_\infty.$$

The moment $\mathbf{E}\langle\xi,\nabla\rangle^k f(x)$ in (5.1.5) can be written as $\langle\mathbf{E}\xi^{\otimes k}, \nabla^{\otimes k} f(x)\rangle$ using expectation of tensor power of the RV and higher-order Fréchet derivative. All operators polynomial in ∇ and translations (by constant and random vectors, or averages of the latter) are commutative.

If $\xi_j \in X$ are independent, then

$$\begin{aligned} \mathbf{E}\prod_{j=1}^k \langle\xi_j,\nabla\rangle^{\kappa_j} &= \prod_{j=1}^k \mathbf{E}\langle\xi_j,\nabla\rangle^{\kappa_j}, \\ \mathbf{E}\exp\left\{\sum_{j=1}^k \langle u_j\xi_j,\nabla\rangle\right\} &= \prod_{j=1}^k \mathbf{E}\exp\{\langle u_j\xi_j,\nabla\rangle\}. \end{aligned} \qquad (5.1.6)$$

(The operators are well defined on sufficiently regular functions.) Thus, sums of independent RV's are associated with products of commuting averages of differential exponentials.

Consider a RV ξ satisfying the Cramér condition (3.6.2). Under this restriction, the Taylor series for $\mathbf{E}e^{a\langle\xi,\nabla\rangle} f$ is uniformly convergent for a function from $\mathcal{M}$ if $|a|\langle f\rangle < A_1$, and the differential exponential $\mathbf{E}e^{z\langle\xi,\nabla\rangle} f$ can be extended to complex values of z belonging to the strip $\mathbf{C}\bigcap\{z : |\mathbf{Im}\, z| < A_1\langle f\rangle^{-1}\}$ by convergent Taylor series (see the formula after (5.1.3)). For a RV satisfying the stronger restriction

$$\forall t > 0 \;\; \mathbf{E}\exp\{t|\xi|\} < \infty, \qquad (5.1.7)$$

the operator

$$\mathbf{E}\exp\{z\langle\xi,\nabla\rangle\} = \sum_{k=0}^{\infty} \frac{z^k}{k!}\mathbf{E}\langle\xi,\nabla\rangle^k \qquad (5.1.8)$$

is, for each value of its complex argument z, a bounded operator on all $\mathcal{M}$. All the customary expressions using Taylor formula remain valid. Condition (5.1.7) holds, e.g., for bounded or Gaussian RV's.

Gaussian Translations Consider a centered Gaussian RV η. The characteristic property of centered Gaussian distributions (cf. Theorem 1.1.3) is expressed by the identity

$$\forall\, u \in [0,\, 1]\ \ \mathbf{E}e^{\langle \eta, \nabla\rangle} = \mathbf{E}e^{\langle u\eta_1 + \sqrt{1-u^2}\eta_2, \nabla\rangle} = \mathbf{E}e^{\langle u\eta, \nabla\rangle}\mathbf{E}e^{\langle \sqrt{1-u^2}\eta, \nabla\rangle}, \quad (5.1.9)$$

where η_1 and η_2 are independent copies of η. The differential operators $\mathbf{E}\langle \eta, \nabla\rangle^k$ are expressed through covariances by familiar formulae for the moments of Gaussian rv's.

Lemma 5.1.1 *For all* $k \in \mathbf{N}$

$$\mathbf{E}\langle \eta, \nabla\rangle^k = \begin{cases} (2l-1)!!\left[\mathbf{E}\langle \eta, \nabla\rangle^2\right]^l & , \textit{ if } k = 2l \textit{ is even,} \\ 0 & , \textit{ if } k \textit{ is odd.} \end{cases}$$

Proof of Lemma 5.1.1. The fact that for odd powers $\mathbf{E}\langle \eta, \nabla\rangle^l = 0$ is an obvious consequence of the symmetry of distribution. Consequently, the Taylor expansion for the average $\mathbf{E}\exp\{\langle u\eta, \nabla\rangle\}$ does not contain odd powers of u.

Put $m_k = \mathbf{E}\langle \eta, \nabla\rangle^{2k}$. To derive a recurrence equation for the operators m_k, recall the characteristic property of the centered Gaussian law in the form

$$\mathbf{E}\exp\left\{\sqrt{u(1+v)}\,\langle \eta, \nabla\rangle\right\} = \mathbf{E}\exp\left\{\sqrt{u}\,\langle \eta, \nabla\rangle\right\}\, \mathbf{E}\exp\left\{\sqrt{uv}\,\langle \eta, \nabla\rangle\right\}.$$

Comparing the expansions in powers of $\sqrt{u}$, $u > 0$, for the left- and right-hand sides, one arrives at the identity for two differential polynomials:

$$\frac{(1+v)^k}{(2k)!} m_k = \sum_{l=0}^{k} \frac{v^{k-l}}{(2l)!(2k-2l)!} m_l m_{k-l}.$$

Differentiation in v at $v = 0$ yields the formula

$$m_k = \frac{1}{k}\frac{(2k)!}{(2k-2)!2!} m_{k-1} m_1 = (2k-1) m_{k-1} m_1,$$

which implies that of the lemma. ○

It follows from the expression for the moments in the lemma that on $\mathcal{M}$ for all $z \in \mathbf{C}$

$$\mathbf{E}\exp\{z\langle \eta, \nabla\rangle\} = \exp\left\{\tfrac{1}{2} z^2 \mathbf{E}\langle \eta, \nabla\rangle^2\right\}. \quad (5.1.10)$$

Thus, in particular, for a centered Gaussian RV the operator $\mathbf{E}e^{z\langle \eta, \nabla\rangle}$ (acting on $\mathcal{M}$) has (also on $\mathcal{M}$) the inverse operator $\mathbf{E}e^{iz\langle \eta, \nabla\rangle}$ with $i = \sqrt{-1}$. The identity

$$\mathbf{E}\exp\{iu\langle \eta, \nabla\rangle\}\, \mathbf{E}\exp\{u\langle \eta, \nabla\rangle\} = 1, \quad u \in \mathbf{R}, \quad (5.1.11)$$

will be quite important for the calculations to follow (although it will rarely be applied directly, as the class $\mathcal{M}$ where it holds is too small).

Note that if the scaling factor u is real, there is equality

$$\mathbf{E}\langle \xi, \nabla\rangle^k e^{u\langle \xi, \nabla\rangle} f(x) = \frac{d^k}{du^k}\mathbf{E}f(x + u\xi) \quad (5.1.12)$$

which applies to all functions satisfying obvious differentiability conditions. The complex exponential and its derivatives $\mathbf{E}\langle\xi,\nabla\rangle^k e^{iu\langle\xi,\nabla\rangle}$ do not admit this interpretation unless the function f is defined on some domain in $X+iX$.

5.1.3 Operator Interpretation of Classical Limit Theorems

By (5.1.6) a sum independent RV's is associated with a product of commuting operator exponentials. In a situation where the summands are small, these exponentials are close to unit operator. This circumstance permits one to interpret certain forms of classical limit theorems of probability theory as asymptotic formulae for products of average translation operators.

Law of large numbers. Consider a sequence of i.i.d. RV's $\xi,\xi_1,\ldots$ Assume that $\mathbf{E}|\xi|^2<\infty$ and $\mathbf{E}\xi=0$. Then

$$|(\mathbf{E}\exp\{u\langle\xi,\nabla\rangle\}-1)f(x)| \le \tfrac{1}{2}u^2\mathbf{E}|\xi|^2\left|\nabla^{\otimes 2}f\right|_\infty,$$

so by (5.1.5)

$$\left|\left[\left(\mathbf{E}\exp\left\{\frac{1}{n}\langle\xi,\nabla\rangle\right\}\right)^n-1\right]f(x)\right| \le \frac{1}{2n}\mathbf{E}|\xi|^2\left|\nabla^{\otimes 2}f\right|_\infty.$$

This is a form of the law of large numbers, though a fairly weak one because the functions with the necessary differentiability properties do not abound in infinite dimensions.

Central limit theorem for i.i.d. RV's. Assume that

$$\mathbf{E}|\xi|^3 < \infty,\quad \mathbf{E}\xi = 0,$$

and that there is a sequence of centered Gaussian X-valued RV's $\eta,\eta_1,\ldots$ with the covariances of ξ:

$$\mathbf{E}\exp\{i\langle\eta,t\rangle\} = \exp\left\{-\tfrac{1}{2}\mathbf{E}\langle\xi,t\rangle^2\right\},\quad t\in X.$$

(This last assumption is automatically fulfilled in Hilbert space).

Consider the normalized sum

$$Z_n = \varepsilon(\xi_1+\cdots+\xi_n),\quad \varepsilon=\tfrac{1}{\sqrt{n}}. \tag{5.1.13}$$

The average translation operator which corresponds to this RV is

$$\mathbf{E}\exp\{\langle Z_n,\nabla\rangle\} = [\mathbf{E}\exp\{\varepsilon\langle\xi,\nabla\rangle\}]^n. \tag{5.1.14}$$

In this representation each factor is close to the average translation operator for a RV with the accompanying Gaussian law: indeed, for each function possessing three Fréchet derivatives, the Taylor formula can be written in the form

$$\mathbf{E}e^{\langle\xi,\nabla\rangle}f(x)-\mathbf{E}e^{\langle\eta,\nabla\rangle}f(x) = \tag{5.1.15}$$

$$= \left[(\mathbf{E}\langle\xi,\nabla\rangle - \mathbf{E}\langle\eta,\nabla\rangle) + \tfrac{1}{2}\left(\mathbf{E}\langle\xi,\nabla\rangle^2 - \mathbf{E}\langle\eta,\nabla\rangle^2\right)\right] f(x)$$
$$+ \int_0^1 \tfrac{1}{2}(1-u)^2 \left(\mathbf{E}\langle\xi,\nabla\rangle^3 f(x+u\xi) - \mathbf{E}\langle\eta,\nabla\rangle^3 f(x+u\eta)\right) du.$$

Consequently,

$$\left|\mathbf{E}\, e^{\varepsilon\langle\xi,\nabla\rangle} f(x) - \mathbf{E}\, e^{\varepsilon\langle\eta,\nabla\rangle} f(x)\right| \le \frac{\varepsilon^3}{3!}\left[\mathbf{E}\,|\xi|^3 + \mathbf{E}\,|\eta|^3\right] \left|\nabla^{\otimes 3} f\right|_\infty \tag{5.1.16}$$

(it follows from known properties of Gaussian laws that $\mathbf{E}\,|\eta|^3 < \infty$).

The "real" average translations do not increase the $|\cdot|_\infty$-norm of derivatives. Hence one can make use of the stability property of centered Gaussian laws in the form

$$\left(\mathbf{E}\exp\left\{n^{-1/2}\langle\eta,\nabla\rangle\right\}\right)^n = \mathbf{E}\exp\{\langle\eta,\nabla\rangle\}$$

to establish the relation

$$\left|\mathbf{E}\exp\{\langle Z_n,\nabla\rangle\} f - \mathbf{E}\exp\{\langle\eta,\nabla\rangle\} f\right|_\infty$$
$$\le \sum_{k=1}^{n}\left|\left[\mathbf{E}e^{\varepsilon\langle\xi,\nabla\rangle}\right]^{n-k}\left(\mathbf{E}e^{\varepsilon\langle\xi,\nabla\rangle} - \mathbf{E}e^{\varepsilon\langle\eta,\nabla\rangle}\right)\left[\mathbf{E}e^{\varepsilon\langle\eta,\nabla\rangle}\right]^{k-1} f\right|_\infty$$
$$\le \frac{\varepsilon}{3!}\left[\mathbf{E}\,|\xi|^3 + \mathbf{E}\,|\eta|^3\right]\left|\nabla^{\otimes 3} f\right|_\infty, \quad \varepsilon = \tfrac{1}{\sqrt{n}}. \tag{5.1.17}$$

The result is a form of the CLT — average values of bounded smooth functions with respect to distributions of normalized sums can be approximated by averages with respect to accompanying Gaussian laws.

5.2 Asymptotic Expansions for Expectations of Smooth Functions

5.2.1 Heuristic Derivation of Asymptotic Expansions

The problem which serves as the starting point for arguments of this subsection is calculation of the expectation $\mathbf{E}f(x+Z_n)$, where Z_n is the normalized sum of i.i.d. RV's (5.1.13). The summands ξ_j are independent copies of the RV ξ which is centered and satisfies the Cramér condition (3.6.2). The RV's η, η_j, etc. are Gaussian with zero mean and same covariances as ξ. The function f is very smooth. It is convenient to assume first that it belongs to $\boldsymbol{\mathcal{M}}$, and later relax this restriction.

The calculation are first done with little attention to justification. Later on the necessary degree of rigor will be restored.

One can employ the "factorization equality" (5.1.11) to represent the average translation of an individual summand in (5.1.13) in the form

$$\mathbf{E}e^{u\langle\xi,\nabla\rangle} = \mathbf{E}e^{u\langle\zeta,\nabla\rangle}\mathbf{E}e^{u\langle\eta,\nabla\rangle}, \quad u \ge 0, \ \zeta = \xi + i\eta, \tag{5.2.1}$$

with ξ and its Gaussian counterpart η independent. Thus, stability of the centered Gaussian law taken into consideration, the operator corresponding to the normalized sum becomes

$$\mathbf{E}\exp\{\langle Z_n,\nabla\rangle\} = [\mathbf{E}\exp\{\varepsilon\langle\zeta,\nabla\rangle\}]^n\,\mathbf{E}\exp\{\langle\eta,\nabla\rangle\}, \quad \varepsilon=\tfrac{1}{\sqrt{n}}. \tag{5.2.2}$$

The Gaussian expectation $\mathbf{E}\exp\{\langle\eta,\nabla\rangle\}f(x)$ is precisely the approximation to $\mathbf{E}f(x+Z_n)$ provided by the central limit theorem. For a smooth function f it is itself a smooth function. It seems probable that one can improve approximation expanding the first operator-valued factor in the right-hand side of (5.2.2) in powers of ε (or, in other words, of $\frac{1}{n}$) and retaining a certain number of terms. Each of these is a differential polynomial, and acting on the CLT approximation it would provide a correction term of appropriate order.

In notation of (5.2.1) and (5.2.2)

$$\begin{aligned} \mathbf{E}e^{\varepsilon\langle\zeta,\nabla\rangle} &= 1+\varepsilon^3\varphi(\varepsilon), \\ \varphi(\varepsilon) &\stackrel{def}{=} \varepsilon^{-3}\left[\mathbf{E}e^{\varepsilon\langle\zeta,\nabla\rangle}-1\right] = \tfrac{1}{3!}\mathbf{E}\langle\zeta,\nabla\rangle^3+\mathcal{O}(\varepsilon), \end{aligned} \tag{5.2.3}$$

since by the choice of its Gaussian component the complex-valued linear combination $\zeta=\xi+i\eta$ satisfies the condition

$$\mathbf{E}\langle\zeta,\nabla\rangle=0,\quad \mathbf{E}\langle\zeta,\nabla\rangle^2=0. \tag{5.2.4}$$

It will be clarified later in what sense the remainder in the right-hand side of (5.2.3) is small.

To obtain an expansion in powers of ε for the function $\left[1+\varepsilon^3\varphi(\varepsilon)\right]^{1/\varepsilon^2}$ (so far it will only be considered as a formal one), first put $\delta=\varepsilon^3\varphi$ and expand in powers of δ (keeping in mind that this parameter has the order ε^3) the function $(1+\delta)^{1/\varepsilon^2}$: after some elementary calculations this yields

$$(1+\varepsilon^3\varphi(\varepsilon))^{1/\varepsilon^2} = \sum_{k=0}^{m}\frac{\varepsilon^k}{k!}\varphi^k(\varepsilon)\prod_{j=1}^{k}(1-[j-1]\varepsilon^2)+\cdots. \tag{5.2.5}$$

In each summand, all functions, both number- and operator-valued, are analytical in ε. It remains to expand the terms of (5.2.5) in powers of ε and collect monomials of same order to obtain the desired representation

$$\mathbf{E}\exp\{\langle Z_n,\nabla\rangle\} = \left[1+\sum_{k=1}^{m}\varepsilon^k\mathcal{P}_k(\nabla)+\cdots\right]\mathbf{E}\exp\{\langle\eta,\nabla\rangle\}, \tag{5.2.6}$$

where $\mathcal{P}_k(\nabla)$ are differential polynomials.

Note that after operator (5.2.6) is applied to a smooth function f, the result is the expansion

$$\mathbf{E}f(Z_n) = \mathbf{E}f(\eta)+\sum_{k=1}^{m}\varepsilon^k\mathbf{E}\mathcal{P}_k(\nabla)f(\eta)+\cdots, \quad \varepsilon=\tfrac{1}{\sqrt{n}}.$$

If one could evaluate the remainder — say, replace "..." by $\mathcal{O}(\varepsilon^{m+\alpha})$ — this error estimate would immediately guarantee uniqueness of the resulting expansion.

Indeed, its successive terms can be identified using the relations

$$\begin{aligned}
\mathbf{E}\mathcal{P}_1(\nabla)f(\eta) &= \lim_{n\to\infty} \frac{1}{\varepsilon}\left[\mathbf{E}f(Z_n) - \mathbf{E}f(\eta)\right], \\
\mathbf{E}\mathcal{P}_2(\nabla)f(\eta) &= \lim_{n\to\infty} \frac{1}{\varepsilon^2}\left[\mathbf{E}f(Z_n) - \mathbf{E}f(\eta) - \varepsilon\mathbf{E}\mathcal{P}_1(\nabla)f(\eta)\right], \ldots
\end{aligned}$$

In finite dimensions, where the Gaussian law $\mathcal{L}(\eta)$ has a smooth density g with respect to the Lebesgue measure in $\mathbf{R}^d$, expansion (5.2.6) can be interpreted as a source of corrections to this latter which improve the precision of approximation. Namely, in this case

$$\mathbf{E}e^{\langle \eta, \nabla\rangle} f(x) = \int_{\mathbf{R}^d} f(x+y)g(y)dy,$$

and integration by parts yields for a differential polynomial of this function the representation

$$\begin{aligned}
\mathcal{P}_k(\nabla)\mathbf{E}e^{\langle \eta, \nabla\rangle} f(x) &= \int_{\mathbf{R}^d} \mathcal{P}_k(\nabla_x)f(x+y)g(y)dy \\
&= \int_{\mathbf{R}^d} f(x+y)\mathcal{P}_k(-\nabla_y)g(y)dy,
\end{aligned}$$

where the "weight" is once again a summable function. Thus, to reproduce with more precision the expectations $\mathbf{E}f(Z_n)$ one can replace the density of Gaussian approximation to $\mathcal{L}(Z_n)$ by the function $\sum_{k=0}^{m} \mathcal{P}_k(-\nabla)g(x)$. The result is an approximation of $\mathcal{L}(Z_n)$ by a measure which is not necessarily itself a distribution.

The last part of the above construction, interpreting improved approximations as integrals with respect to measures that are higher-order approximations to the distribution of the normalized sum, offers considerable difficulties (see Section 5.3 for more details). On the contrary, construction of correction terms for Gaussian approximations to expected values of smooth functions seems to be easy. This part of the job is done below.

5.2.2 Asymptotic Expansion for Individual Summand

Expansion for "imaginary translations" like (5.2.5) can be rigorously justified for functions that have smooth extensions to the complexification of the original space X. The more elementary approach used below avoids using the "imaginary translations". These operator exponentials are replaced by the more tractable differential polynomials — segments of their Taylor expansions. Arguments are less straightforward, but they circumvent excessive restrictions on smoothness of the functions considered.

In this and the next subsections condition (5.2.4) may be partially violated: it is always assumed that $\mathbf{E}\xi = \mathbf{E}\eta = 0$, but the covariance operators of these independent RV's are not necessarily identical, so the operator $\mathbf{E}\langle \xi + i\eta, \nabla\rangle^2$ may be "small" without vanishing altogether.

Lemma 5.2.1 *If $\varepsilon \in [0,\ 1]$, then*

$$\mathbf{E}e^{\varepsilon\langle\xi,\nabla\rangle} = \sum_{k=0}^{m} \frac{\varepsilon^k}{k!}\mathbf{E}\left\langle\xi + i\eta, \nabla\right\rangle^k \mathbf{E}e^{\varepsilon\langle\eta,\nabla\rangle} + \varepsilon^{m+1}\rho_m^{\varphi}(\varepsilon),$$

where ξ and η are independent, as well as ξ and the independent copies η', η'' of η used in the representation of the remainder term

$$\begin{aligned}\rho_m^{\varphi}(\varepsilon) &= \int_0^1 \frac{(1-v)^m}{m!}\mathbf{E}\left\langle\xi, \nabla\right\rangle^{m+1} e^{v\varepsilon\langle\xi,\nabla\rangle}dv \\ &- \int_0^1 \sum_{k=0}^{m} \frac{(1-u)^{m-k}}{k!(m-k)!}\mathbf{E}\left\langle\xi + i\eta', \nabla\right\rangle^k \mathbf{E}\left\langle\eta'', \nabla\right\rangle^{m-k+1} e^{u\varepsilon\left\langle\eta'',\nabla\right\rangle}du.\end{aligned}$$

Remark 5.2.1 All the average translations in the statement of Lemma 5.2.1 are "real". Hence its formula is applicable to any function with the necessary set of Fréchet derivatives satisfying evident summability conditions (cf. (5.1.12)). For $m = 2$ Lemma 5.2.1 yields the following formula:

$$\left[\mathbf{E}e^{\varepsilon\langle\xi,\nabla\rangle} - \mathbf{E}e^{\varepsilon\langle\eta,\nabla\rangle}\right] f(x) = \frac{\varepsilon^2}{2}\mathbf{E}\left\langle\xi + i\eta, \nabla\right\rangle^2 \mathbf{E}e^{\langle\eta,\nabla\rangle} f(x) + \varepsilon^3 \rho_2^{\varphi}(\varepsilon) f(x),$$

where

$$\begin{aligned}\rho_2^{\varphi}(\varepsilon) &= \int_0^1 \frac{(1-u)^2}{2!}\left[\mathbf{E}\left\langle\xi, \nabla\right\rangle^3 e^{u\varepsilon\langle\xi,\nabla\rangle} - \mathbf{E}\left\langle\eta, \nabla\right\rangle^3 e^{u\varepsilon\langle\eta,\nabla\rangle}\right] du \\ &- \frac{1}{2!}\int_0^1 \mathbf{E}\left\langle\xi + i\eta', \nabla\right\rangle^2 \mathbf{E}\left\langle\eta'', \nabla\right\rangle e^{u\varepsilon\left\langle\eta'',\nabla\right\rangle}du,\end{aligned}$$

This is a version of the identity

$$\begin{aligned}\left[\mathbf{E}e^{\varepsilon\langle\xi,\nabla\rangle} - \mathbf{E}e^{\varepsilon\langle\eta,\nabla\rangle}\right] f(x) &= \frac{\varepsilon^2}{2}\mathbf{E}\left\langle\xi + i\eta, \nabla\right\rangle^2 f(x) \qquad (5.2.7) \\ + \ \varepsilon^3 \int_0^1 \frac{(1-u)^2}{2!} &\left[\mathbf{E}\left\langle\xi, \nabla\right\rangle^3 e^{u\varepsilon\langle\xi,\nabla\rangle} - \mathbf{E}\left\langle\eta, \nabla\right\rangle^3 e^{u\varepsilon\langle\eta,\nabla\rangle}\right] f(x)du,\end{aligned}$$

which is very often applied in a slightly different form

$$\begin{aligned}\left[\mathbf{E}e^{\varepsilon\langle\xi,\nabla\rangle} - \mathbf{E}e^{\varepsilon\langle\eta,\nabla\rangle}\right] f(x) &= \frac{\varepsilon^2}{2}\left(\mathbf{E}\left\langle\xi, \nabla\right\rangle^2 - \mathbf{E}\left\langle\eta, \nabla\right\rangle^2\right) f(x) \\ + \ \varepsilon^3 \int_0^1 \frac{(1-u)^2}{2!} &\left[\mathbf{E}\left\langle\xi, \nabla\right\rangle^3 f(x + u\varepsilon\xi) - \mathbf{E}\left\langle\eta, \nabla\right\rangle^3 f(x + u\varepsilon\eta)\right] du.\end{aligned}$$

Lemma 5.2.1 is proved below by an argument that uses identity (5.1.11) and power expansions of operator exponentials, both "real" and "imaginary" average translations. At the price of making calculations slightly more boring, one can get the same result rearranging sums of differential monomials with the help of (5.1.6) and Lemma 5.1.1.

Proof of Lemma 5.2.1. Define a new operator

$$L_m = \sum_{k=0}^{m} \frac{\varepsilon^k}{k!} \mathbf{E} \langle \xi + i\eta, \nabla \rangle^k \sum_{l=0}^{m-k} \frac{\varepsilon^l}{l!} \mathbf{E} \langle \eta, \nabla \rangle^l ,$$

where the inner sum is evidently a segment of Taylor expansion for the exponential $\mathbf{E}e^{\varepsilon\langle \eta, \nabla \rangle}$. After collecting terms with same order in ε, this operator assumes the form

$$\begin{aligned} L_m &= \sum_{p=0}^{m} \frac{\varepsilon^p}{p!} \sum_{k=0}^{p} \frac{p!}{k!(p-k)!} \mathbf{E} \langle \xi + i\eta', \nabla \rangle^k \langle \eta'', \nabla \rangle^{p-k} \\ &= \sum_{p=0}^{m} \frac{\varepsilon^p}{p!} \mathbf{E} \langle \xi + i\eta' + \eta'', \nabla \rangle^p \\ &= \sum_{p=0}^{m} \frac{\varepsilon^p}{p!} \frac{d^p}{du^p} \mathbf{E} \exp \{ u \langle \xi + i\eta' + \eta'', \nabla \rangle \} \Big|_{u=0} . \end{aligned}$$

Recall that all the RV in the argument of the exponential are independent, and that multiplication of the exponent by the imaginary unit inverts the average translation operator for a centered Gaussian RV. Consequently, the last formula implies that

$$L_m = \sum_{p=0}^{m} \frac{\varepsilon^p}{p!} \frac{d^p}{du^p} \mathbf{E} \exp \{ u \langle \xi, \nabla \rangle \} \Big|_{u=0} ,$$

i.e., that this operator is a segment of the Taylor expansion of $\mathbf{E}e^{\varepsilon\langle \xi, \nabla \rangle}$ in the neighborhood of the origin. Hence it follows from (5.1.3) that

$$\mathbf{E} \exp \{ \varepsilon \langle \xi, \nabla \rangle \} - L_m = \varepsilon^{m+1} \int_0^1 \frac{(1-u)^m}{m!} \mathbf{E} \langle \xi, \nabla \rangle^{m+1} \exp \{ \varepsilon u \langle \xi, \nabla \rangle \} \, du.$$

Consider now the difference

$$\begin{aligned} & \sum_{k=0}^{m} \frac{\varepsilon^k}{k!} \mathbf{E} \langle \xi + i\eta', \nabla \rangle^k \mathbf{E} \exp \{ \varepsilon \langle \eta'', \nabla \rangle \} - L_m \\ &= \sum_{k=0}^{m} \frac{\varepsilon^k}{k!} \mathbf{E} \langle \xi + i\eta', \nabla \rangle^k \left[\mathbf{E} \exp \{ \varepsilon \langle \eta'', \nabla \rangle \} - \sum_{p=0}^{m-k} \frac{\varepsilon^p}{p!} \mathbf{E} \langle \eta'', \nabla \rangle^p \right] . \end{aligned}$$

To write down the remaining residuals in the statement of the lemma, it remains to use the Taylor expansion (5.1.3) for the average translation operator $\mathbf{E}e^{\varepsilon\langle \eta'', \nabla \rangle}$. ○

5.2.3 Bergström- and Edgeworth-Type Expansions

The Bergström expansion is the usual intermediate step in construction of asymptotic formulae similar to (5.2.6).

Lemma 5.2.2 *For $\varepsilon = \frac{1}{\sqrt{n}}$ and a natural number $m \le n$*

$$\left[\mathbf{E}e^{\varepsilon\langle\xi,\nabla\rangle}\right]^n - \mathbf{E}e^{\langle\eta,\nabla\rangle} = \sum_{k=1}^{m} \psi_k(\varepsilon)\Delta^k \mathbf{E}e^{\sqrt{1-k/n}\langle\eta,\nabla\rangle} + r_m^B(\varepsilon), \tag{5.2.8}$$

where

$$\psi_k(\varepsilon) = (1/k!)\prod_{j=1}^{k}\left(1-[j-1]\varepsilon^2\right), \quad \Delta(\varepsilon) = \varepsilon^{-2}\left[\mathbf{E}e^{\varepsilon\langle\xi,\nabla\rangle} - \mathbf{E}e^{\varepsilon\langle\eta,\nabla\rangle}\right],$$

and

$$r_m^B(\varepsilon) = \psi_{m+1}(\varepsilon)\Delta^{m+1}\int_0^1 \frac{(1-u)^m}{m!}\left[u\mathbf{E}e^{\varepsilon\langle\xi,\nabla\rangle} + (1-u)\mathbf{E}e^{\varepsilon\langle\eta,\nabla\rangle}\right]^{n-m-1} du.$$

The operator under the integration sign in the remainder term of the lemma is the average translation corresponding to a sum of $(n-m-1)$ i.i.d. RV's whose distribution is the mixture $u\mathcal{L}(\xi) + (1-u)\mathcal{L}(\eta)$. Hence it follows from Lemma 5.2.1 and (5.1.16) that, if condition (5.2.4) is satisfied,

$$\left|r_m^B(\varepsilon)f\right|_\infty \le c\varepsilon^{m+1}\left[\mathbf{E}\left|\xi\right|^3\right]^{m+1}\left|\nabla^{\otimes[3m+3]}f\right|_\infty. \tag{5.2.9}$$

In notation of (5.2.3)

$$\Delta(\varepsilon) = \frac{1}{\varepsilon^2}\left(\mathbf{E}e^{\varepsilon\langle\zeta,\nabla\rangle} - 1\right)\mathbf{E}e^{\varepsilon\langle\eta,\nabla\rangle} = \varepsilon\varphi\mathbf{E}e^{\varepsilon\langle\eta,\nabla\rangle}$$

and $\psi_k(\varepsilon) = 1 + \mathcal{O}(\varepsilon)$, so the order of the first terms in expansion (5.2.8) is, in a loose sense,

$$\psi_k\,\Delta^k\mathbf{E}e^{\sqrt{1-k/n}\langle\eta,\nabla\rangle} = \varepsilon^k\psi_k\varphi^k\mathbf{E}e^{\langle\eta,\nabla\rangle} = \mathcal{O}(\varepsilon^k).$$

Proof of Lemma 5.2.2. Put

$$A(u) = \mathbf{E}\exp\{\varepsilon\langle\xi,\nabla\rangle\}, \quad B(u) = \mathbf{E}\exp\{\varepsilon\langle\eta,\nabla\rangle\},$$

so that $A = B + \varepsilon^2\Delta$. Applying the Taylor formula one obtains the equality

$$\left(B+\varepsilon^2\Delta\right)^n = B^n + \sum_{k=1}^{m}\binom{n}{k}\varepsilon^{2k}\Delta^k B^{n-k}$$
$$+ \binom{n}{m+1}\varepsilon^{2m+2}\int_0^1\frac{(1-u)^m}{m!}\left(B+u\varepsilon^2\Delta\right)^{n-m-1}du.$$

To derive the formula of the lemma, it remains to note that for its choice of ε

$$\binom{n}{k}\varepsilon^{2k} = \psi_k(\varepsilon).$$

The characteristic property of Gaussian laws yields the equality $B^{n-k}(\varepsilon) = \mathbf{E}\exp\left\{\sqrt{1-k/n}\,\langle\eta,\nabla\rangle\right\}$. ○

The Edgeworth expansion. While deriving an expansion like (5.2.6) from Lemma 5.2.2, one should replace the functions $\psi_k(\varepsilon)$ and operators Δ^k by appropriate segments of their Taylor expansions. The calculations are rather cumbersome, and it is convenient to introduce some of the necessary notation beforehand.

The Taylor expansion of ψ_k is

$$\psi_k(\varepsilon) = \sum_{p=0}^{m} \frac{\varepsilon^p}{p!}\psi_k^{(p)}(0) + \rho_{k,m}^{\psi}(\varepsilon), \quad \rho_{k,m}^{\psi}(\varepsilon) = O(\varepsilon^{m+1}).$$

The main terms in the formulae resulting from calculations below include the differential polynomials, defined for $m \in \mathbf{N}$ and $k \le m$,

$$\mathcal{E}_{k,m}(\varepsilon) = \sum_{p=0}^{m-k} \frac{\varepsilon^p}{p!}\psi_k^{(p)}(0) \sum_{l_1=0}^{L_0} \cdots \sum_{l_k=0}^{L_{k-1}} \prod_{j=1}^{k} \left[\varepsilon^{l_j} \frac{\mathbf{E}\,\langle \zeta, \nabla\rangle^{l_j+3}}{(l_j+3)!} \right], \tag{5.2.10}$$

where $\zeta = \xi + i\eta$ and

$$L_0 = m - k - p, \quad L_j = m - k - p - \textstyle\sum_{q=1}^{j} l_q \quad j \ge 1.$$

The order of $\mathcal{E}_{k,m}$ in ε is $m-k$, and the maximal order of derivatives it contains is $m + 2k$.

Notation of Lemma 5.2.1 and subsection 5.2.1 is used without comment, so, e.g.,

$$\frac{1}{\varepsilon}\Delta = \varphi\,\mathbf{E}e^{\langle \varepsilon\eta, \nabla\rangle} = \sum_{l=0}^{m} \varepsilon^l \frac{\mathbf{E}\,\langle \zeta, \nabla\rangle^{l+3}}{(l+3)!}\mathbf{E}e^{\langle \varepsilon\eta, \nabla\rangle} + \varepsilon^{m+1}\rho_{m+3}^{\varphi}(\varepsilon). \tag{5.2.11}$$

Numerical and operator-valued sums and products over empty index sets are ascribed the usual values $\sum_{j\in\emptyset}\alpha_j = 0$ and $\prod_{j\in\emptyset}\alpha_j = 1$.

Lemma 5.2.3 *Assume that* (5.2.4) *is satisfied. Then for each* $m \in \mathbf{N}$ *and* $k = \overline{1,m}$

$$\begin{aligned} \frac{1}{\varepsilon^k}\psi_k(\varepsilon)\Delta^k(\varepsilon) &= \psi_k(\varepsilon)\left[\varphi(\varepsilon)\mathbf{E}e^{\langle \varepsilon\eta, \nabla\rangle}\right]^k \\ &= \mathcal{E}_{k,m}(\varepsilon)\mathbf{E}e^{\sqrt{k/n}\langle \eta, \nabla\rangle} + \varepsilon^{m-k+1}\rho_{k,m}^{E}(\varepsilon), \end{aligned}$$

where the remainder term $\rho_{k,m}^{E}(\varepsilon) = \rho_{k,m}^{E,1}(\varepsilon) + \rho_{k,m}^{E,2}(\varepsilon)$ *is the sum of*

$$\begin{aligned} \rho_{k,m}^{E,1}(\varepsilon) &= \rho_{k,m}^{\psi}(\varepsilon)\left[\varphi(\varepsilon)\mathbf{E}e^{\langle \varepsilon\eta, \nabla\rangle}\right]^k \\ &+ \sum_{p=0}^{m-k} \frac{1}{p!}\psi_k^{(p)}(0)\left[\varphi(\varepsilon)\mathbf{E}e^{\langle \varepsilon\eta, \nabla\rangle}\right]^{k-1}\rho_{m-k-p+3}^{\varphi}(\varepsilon) \end{aligned}$$

and

$$\rho^{E,2}_{k,m}(\varepsilon) = \sum_{p=0}^{m-k} \frac{1}{p!}\psi_k^{(p)}(0)\sum_{j=2}^{k}\left[\varphi(\varepsilon)\mathbf{E}e^{\langle\varepsilon\eta,\nabla\rangle}\right]^{k-j}$$
$$\times \sum_{l_1=0}^{L_0}\cdots\sum_{l_{j-1}=0}^{L_{j-2}}\left[\prod_{q=1}^{j-1}\frac{\mathbf{E}\langle\zeta,\nabla\rangle^{l_q+3}}{(l_q+3)!}\mathbf{E}e^{\langle\varepsilon\eta,\nabla\rangle}\right]\rho^{\varphi}_{L_{j-1}+3}(\varepsilon)$$

with L_j, $j \geq 0$, *as in* (5.2.10).

Proof of Lemma 5.2.3. To obtain the expression of the lemma, in the product considered first ψ_k should be replaced by a segment of its Taylor series, and then each factor $\varphi\mathbf{E}e^{\langle\varepsilon\eta,\nabla\rangle}$ by an appropriate expansion of Lemma 5.2.1.

The first part of the remainder includes parts that appear when only the numerical factor or only one factor $\varphi\mathbf{E}e^{\langle\varepsilon\eta,\nabla\rangle}$ is expanded.

To get the rest of the expansion, it is convenient to diminish the order in φ step by step. Consider the sum

$$S^*_{j-1} = \left[\varphi(\varepsilon)\mathbf{E}e^{\langle\varepsilon\eta,\nabla\rangle}\right]^{k-(j-1)}\sum_{p=0}^{m-k}\frac{\varepsilon^p}{p!}\psi_k^{(p)}(0)$$
$$\times \sum_{l_1=0}^{L_0}\cdots\sum_{l_{j-1}=0}^{L_{j-2}}\prod_{q=1}^{j-1}\left[\varepsilon^{l_q}\frac{\mathbf{E}\langle\zeta,\nabla\rangle^{l_q+3}}{(l_q+3)!}\mathbf{E}e^{\langle\varepsilon\eta,\nabla\rangle}\right].$$

After one more factor $\left[\varphi(\varepsilon)\mathbf{E}e^{\langle\varepsilon\eta,\nabla\rangle}\right]$ is replaced by its expansion (5.2.11), the sum becomes

$$S^*_{j-1} = \left[\varphi(\varepsilon)\mathbf{E}e^{\langle\varepsilon\eta,\nabla\rangle}\right]^{(k-j)}\sum_{p=0}^{m-k}\frac{\varepsilon^p}{p!}\psi_k^{(p)}(0)$$
$$\times \sum_{l_1=0}^{L_0}\cdots\sum_{l_{j-1}=0}^{L_{j-2}}\prod_{q=1}^{j-1}\left[\varepsilon^{l_q}\frac{\mathbf{E}\langle\zeta,\nabla\rangle^{l_q+3}}{(l_q+3)!}\mathbf{E}e^{\langle\varepsilon\eta,\nabla\rangle}\right]$$
$$\times \left(\sum_{l_j=0}^{L_{j-1}}\left[\varepsilon^{l_j}\frac{\mathbf{E}\langle\zeta,\nabla\rangle^{l_j+3}}{(l_j+3)!}\mathbf{E}e^{\langle\varepsilon\eta,\nabla\rangle}\right]+\varepsilon^{L_{j-1}+1}\rho^{\varphi}_{L_{j-1}+3}\right) = S^*_j + R_j.$$

The term R_j, which is a sum of products of $\left[\varphi(\varepsilon)\mathbf{E}e^{\langle\varepsilon\eta,\nabla\rangle}\right]^{k-j}$ with differential monomials and $\rho^{\varphi}_{L_{j-1}+3}$, produces the corresponding terms in the remainder. ○

Lemma 5.2.4 *For each* $m \leq n$ *and* $\varepsilon = \frac{1}{\sqrt{n}}$, *there is equality*

$$\left[\mathbf{E}e^{\varepsilon\langle\xi,\nabla\rangle}\right]^n = \left(1+\sum_{k=1}^{m}\varepsilon^k\mathcal{E}_{k,m}(\varepsilon)\right)\mathbf{E}e^{\langle\eta,\nabla\rangle}+r^B_m(\varepsilon)$$
$$+ \varepsilon^{m+1}\sum_{k=1}^{m}\rho^E_{k,m}(\varepsilon)\mathbf{E}e^{\sqrt{1-k/n}\langle\eta,\nabla\rangle},$$

where notation is that of Lemma 5.2.2 *and Lemma* 5.2.3.

It is easily seen that the maximal order of the moments $\mathbf{E}\xi^{\otimes k}$ used to construct the expansion of the lemma up to terms $\mathcal{O}(\varepsilon^m)$ is $m+2$. To show that the remainder is negligible if the operator expansion is applied to a very smooth function, one needs bounds on $\mathbf{E}|\xi|^{m+3}$.

Proof of Lemma 5.2.4. To prove the lemma, it suffices to change from Δ to φ of (5.2.3) in the Bergström expansion of Lemma 5.2.2 and replace the resulting expressions $\psi_k\varphi^k$ by asymptotic expansions of Lemma 5.2.3. ○

5.3 Smooth Mappings to Finite Dimensions

5.3.1 Approximation via Asymptotic Expansion for CF

It is often difficult to describe or compare infinite-dimensional distributions because many probabilistic tools elaborated for Euclidean spaces fail. (For instance, the method of characteristic functions loses its convenient inversion formula.) However, most standard probabilistic techniques are applicable if the problem is simplified to considering images of infinite-dimensional laws under a regular mapping with finite-dimensional range

$$X \ni x \mapsto \phi(x) = \left(\phi^{(j)}(x), j=\overline{1,d}\right) \in \mathbf{R}^d,\ d \geq 1. \tag{5.3.1}$$

(The inner product and norm in $\mathbf{R}^d$ are denoted $(\cdot,\cdot)$ and $\|\cdot\| = \sqrt{(\cdot,\cdot)}$.)

For example, one can evaluate proximity between infinite-dimensional distributions $\boldsymbol{\Phi}$ and $\boldsymbol{\Gamma}$ in terms of some distance between the distributions $F_\phi = \boldsymbol{\Phi}\circ\phi^{-1}$ and $G_\phi = \boldsymbol{\Gamma}\circ\phi^{-1}$. These finite-dimensional laws are the distributions of the $\mathbf{R}^d$-valued RV's $\phi(\zeta)$ and $\phi(\eta)$ obtained from the RV's $\zeta,\eta \in X$ with distributions $\mathcal{L}(\zeta) = \boldsymbol{\Phi}$, $\mathcal{L}(\eta) = \boldsymbol{\Gamma}$. The above approach is natural, for instance, from the statistical viewpoint: when results of observations are processed, samples of whatever nature are usually compressed to a set of real-valued statistics, necessarily finite, which is easier to treat and store.

In the same situation, it may prove easier to construct approximations to the finite-dimensional distribution F_ϕ using an image of a standard measure G_ϕ than try to reproduce the infinite-dimensional law $\boldsymbol{\Phi}$ itself in some simpler way.

Approximations to the distribution of the normalized sum (5.1.13) Z_n are treated below within the framework of the above approach. A natural first approximation to $\mathcal{L}(Z_n)$ is the accompanying Gaussian law G, and η is a RV with distribution $\mathcal{L}(\eta) = G$.

It is natural to consider the problem in slightly greater generality, dealing with weighted measures defined for $B \in \mathcal{B}(\mathbf{R}^d)$ by the equalities

$$F_n(B) = \mathbf{E}\Psi(Z_n)\mathbf{1}_B(\phi(Z_n)), \tag{5.3.2}$$

$$G(B) = \mathbf{E}\Psi(\eta)\mathbf{1}_B(\phi(\eta)). \tag{5.3.3}$$

Both the mapping ϕ and the function Ψ of (5.3.2)–(5.3.3) are smooth, and Ψ

does not grow too fast at infinity. (Restrictions will be stipulated more precisely later on.)

The Fourier images of both above measures are, respectively,

$$f_n(t) = \mathbf{E} \exp\{i(t, \phi(x + Z_n))\} \Psi(x + Z_n)\big|_{x=0}, \tag{5.3.4}$$

$$\widehat{g}(t) = \mathbf{E} \exp\{i(t, \phi(x + \eta))\} \Psi(x + \eta)\big|_{x=0}. \tag{5.3.5}$$

The function $f_n(t)$ is the expectation of a smooth function that can be treated by techniques developed in Section 5.1, and there is a regular way of constructing, e.g., Edgeworth-type approximations for it. These are typically expansions of the form

$$f_n(t) = \sum_{l,m \le L} \varepsilon^m g_{l,m}(t) + \cdots, \quad \varepsilon = \tfrac{1}{\sqrt{n}},$$

$$g_{l,m}(t) = \left\langle t^{\otimes l}, \mathbf{E} e^{i(t,\phi(\eta))} \Psi_{l,m}(\eta) \right\rangle,$$

where the "monomials" $g_{l,m}$ result from differentiation in $x \in X$ of the imaginary exponential of (5.3.4). (The remainder terms are not considered for the time being.) Assembling terms of same order in ε, one can write f_n as

$$f_n(t) = \sum_{m \le M} \varepsilon^m \widetilde{Q}_m(t) + \cdots$$

where the terms $\widetilde{Q}_m(t)$ are expectations of products of $e^{i(t,\phi(\eta))} \Psi_{l,m}(\eta)$ with polynomials in t whose coefficients depend on η.

In many situations of interest both the limit Fourier transform $\widehat{g}(t)$ and the correction terms above are summable functions of t, rapidly decaying at infinity. The measure G is thus absolutely continuous with bounded continuous density $g = (dG/d\text{mes})$, and the faster the decay of $\widehat{g}$, the better are its analytical properties. The corrections to G are once again signed measures with regular densities.

It may not be possible to identify corrections to G in any specific way in the general case. However, one can go further if the mapping ϕ is linear and $\Psi \equiv 1$. The measure G is then a Gaussian distribution, and one encounters once again the interpretation of corrections described in subsection 5.2.1. Namely, if $\widehat{f} = \mathcal{F} f$ is the Fourier transform of a function f, then $Q(it)\widehat{f}(t)$ corresponds to $Q(-\nabla)f$ for a polynomial Q. Thus, in the classical Edgeworth expansion corrections to Gaussian density g can be calculated by differentiating it, and are consequently products of $g(x)$ with some polynomials.

The summability properties of the Gaussian approximation $\widehat{g}$ to the Fourier transform (5.3.4) of (5.3.2) and its Edgeworth corrections are good under fairly natural assumptions. However, this may not be the case with the Fourier image (5.3.4) itself, which is usually less tractable than $\widehat{g}$. For this reason, evaluation of accuracy for approximations to measure (5.3.2) has to be restricted to classes of sufficiently regular sets B. Calculations of this sort have to employ the so-called smoothing, a standard tool if estimates for the error term are envisaged in the setting of multidimensional central limit theorem. This apparatus is exposed in the next subsection.

5.3.2 Smoothing Inequality and Fourier Transforms

Consider a function $h : \mathbf{R}^d \to \mathbf{R}$ which is summable with respect to the Lebesgue measure dx. Denote its translations and oscillation on a ball of radius $\varepsilon > 0$ by

$$h \circ S_y(x) \equiv h(x+y), \quad \operatorname{osc}_\varepsilon h(x) \stackrel{def}{=} \sup_{|x_i - x| \le \varepsilon} |h(x_1) - h(x_2)|. \tag{5.3.6}$$

It is evident that $\operatorname{osc}_\varepsilon h$ is nonnegative and $\operatorname{osc}_\varepsilon[\pm h \circ S_y](x) = [\operatorname{osc}_\varepsilon h] \circ S_y(x)$. Moreover, if $\|z\| \le \varepsilon$, then

$$\begin{aligned} |h(x+z) - h(x)| &\le \operatorname{osc}_\varepsilon h(x), \\ h(x+z) + \operatorname{osc}_\varepsilon h(x+z) &\ge h(x). \end{aligned} \tag{5.3.7}$$

Consider a measure F and a measure (or signed measure) G on $\mathcal{B}(\mathbf{R}^d)$. The purpose of this subsection is to develop certain technical means for estimating the difference $\int_{\mathbf{R}^d} h(x)\,[F(dx) - G(dx)]$ in terms of the Fourier transforms $\widehat{f} = \mathcal{F}F$, $\widehat{g} = \mathcal{F}G$, and $\widehat{h} = \mathcal{F}h$. The situations of interest is one where it is difficult to evaluate $\widehat{f} = \mathcal{F}F$ and $\widehat{g} = \mathcal{F}G$ for very large values of argument $\|t\|$.

The lemma that follows is a version of the so-called "smoothing inequality". It reduces the task to comparing the integrals $\int \tilde{h} dF$ and $\int \tilde{h} dG$ for a smooth function $\tilde{h}$ approximating h.

Below, the approximation for h is the convolution

$$h * k_\varepsilon(x) = \int_{\mathbf{R}^d} h(x+z)\, k_\varepsilon(z)\, dz, \quad k_\varepsilon(z) = \varepsilon^{-d} k\left(\tfrac{1}{\varepsilon} z\right). \tag{5.3.8}$$

The choice of the kernel k is subject to the conditions

$$\begin{aligned} k(z) &\ge 0, \quad \textstyle\int_{\mathbf{R}^d} k(z)dz = 1, \\ \widehat{k}(t) &\stackrel{def}{=} \textstyle\int_{\mathbf{R}^d} e^{i(t,z)} k(z)dz = 0 \text{ if } \|t\| \ge K_0. \end{aligned} \tag{5.3.9}$$

Lemma 5.3.1 *Suppose that F is a nonnegative measure and G a signed measure in $\mathbf{R}^d$ If $\alpha = \alpha(\varepsilon) = \int_{|x| \le \varepsilon} k_\varepsilon(x)\, dx > \frac{1}{2}$, then*

$$\begin{aligned} &\sup_{y \in \mathbf{R}^d} \left| \int_{\mathbf{R}^d} h \circ S_y(x) F(dx) - \int_{\mathbf{R}^d} h \circ S_y(x) G(dx) \right| \\ \le\; &\frac{1}{2\alpha - 1} \sup_{y \in \mathbf{R}^d} \left| \int_{\mathbf{R}^d} k_\varepsilon * h \circ S_y(x)\, [F(dx) - G(dx)] \right| \\ +\; &\frac{1}{2\alpha - 1} \sup_{y \in \mathbf{R}^d} \left| \int_{\mathbf{R}^d} k_\varepsilon * \operatorname{osc}_\varepsilon h \circ S_y(x)\, [F(dx) - G(dx)] \right| \\ +\; &\frac{1+\alpha}{2\alpha - 1} \sup_{y \in \mathbf{R}^d} \int_{\mathbf{R}^d} \operatorname{osc}_{2\varepsilon} h \circ S_y(x)\, |G(dx)|. \end{aligned}$$

As usual, $|dG|$ is the total variation of the signed measure G.

Proof of Lemma 5.3.1. Fix an arbitrary value of $y_0 \in \mathbf{R}^d$ and abbreviate for the time being $h \circ S_{y_0}$ to $\tilde{h}$, $\operatorname{osc}_\varepsilon h \circ S_{y_0}$ to $\operatorname{osc}_\varepsilon$, and $[F(dx) - G(dx)]$ to $\Delta(dx)$.

It follows from the definition of convolution that

$$\int_{\mathbf{R}^d} k_\varepsilon * \left(\widetilde{h} + \operatorname{osc}_\varepsilon\right)(x)\Delta(dx)$$

$$= \left(\int_{\|z\|\le\varepsilon} + \int_{\|z\|>\varepsilon}\right) k_\varepsilon(z)dz \int_{\mathbf{R}^d} \left[\widetilde{h}(x+z) + \operatorname{osc}_\varepsilon(x+z)\right] \Delta(dx).$$

Since F is a positive measure, (5.3.7) implies that for $\|z\| \le \varepsilon$

$$\int_{\mathbf{R}^d} \left[\widetilde{h} + \operatorname{osc}_\varepsilon\right](x+z)\Delta(dx) \ge \int_{\mathbf{R}^d} \widetilde{h}(x)\Delta(dx) - \int_{\mathbf{R}^d} \operatorname{osc}_\varepsilon(x+z)G(dx)$$

$$- \int_{\mathbf{R}^d} [\widetilde{h}(x+z) - \widetilde{h}(x)]G(dx) \ge \int_{\mathbf{R}^d} \widetilde{h}(x)\Delta(dx) - 2\int_{\mathbf{R}^d} \operatorname{osc}_{2\varepsilon}(x)\,|G(dx)|\,.$$

For $\|z\| > \varepsilon$ there is inequality (note that $S_{y'} \circ S_{y''} = S_{y'+y''}$)

$$\int_{\mathbf{R}^d} \left[\widetilde{h} + \operatorname{osc}_\varepsilon\right](x+z)\Delta(dx)$$

$$\ge -\sup_{y\in\mathbf{R}^d} \left|\int_{\mathbf{R}^d} h \circ S_y(x)\Delta(dx)\right| - \sup_{y\in\mathbf{R}^d} \int_{\mathbf{R}^d} \operatorname{osc}_{2\varepsilon} h \circ S_y h(x)\,|G(dx)|\,.$$

(Since $F(dx) \ge 0$, it is excluded from the second summand on the right-hand side.)

The bounds obtained are uniform in the argument z in their respective domains, so integration in z yields the following estimate, valid for all translations by y_0 (until the end of the proof, the integration domain, which is $\mathbf{R}^d$, is omitted in the notation):

$$\sup_{y_0\in\mathbf{R}^d} \left|\int k_\varepsilon * h \circ S_{y_0}(x)\Delta(dx)\right| + \sup_{y_0\in\mathbf{R}^d} \left|\int k_\varepsilon * \operatorname{osc}_\varepsilon h \circ S_{y_0}(x)\Delta(dx)\right|$$

$$\ge \alpha \int h \circ S_{y_0}(x)\,[F(dx) - G(dx)] - (1+\alpha) \sup_{y\in\mathbf{R}^d} \int \operatorname{osc}_{2\varepsilon} h \circ S_y(x)\,|G(dx)|$$

$$- (1-\alpha) \sup_{y\in\mathbf{R}^d} \left|\int h \circ S_y(x)\Delta(dx)\right|.$$

The functions $-h \circ S_{y_0}$ have same oscillation as h (appropriately translated). Thus, the result of the above calculations applies also to $-h \circ S_{y_0}$ and yields the following lower bound: for all $y_0 \in \mathbf{R}^d$

$$\alpha \int h \circ S_{y_0}\,[F(dx) - G(dx)] \ge -(1-\alpha) \sup_{y\in\mathbf{R}^d} \left|\int h \circ S_y(x)\Delta(dx)\right|$$

$$- (1+\alpha) \sup_{y\in\mathbf{R}^d} \int \operatorname{osc}_{2\varepsilon} h \circ S_y(x)\,|G(dx)|$$

$$- \sup_{y\in\mathbf{R}^d} \left|\int k_\varepsilon * h \circ S_y(x)\Delta(dx)\right| - \sup_{y\in\mathbf{R}^d} \left|\int k_\varepsilon * \operatorname{osc}_\varepsilon h \circ S_y(x)\Delta(dx)\right|.$$

The final result is the bound for the absolute value of the integral considered:

$$\begin{aligned}
\forall\, y_0 \in \mathbf{R}^d \qquad & \alpha \left| \int h \circ S_{y_0}(x)\,[F(dx) - G(dx)] \right| \\
\leq\ & (1-\alpha) \sup_{y\in\mathbf{R}^d} \left| \int h \circ S_y(x)\,[F(dx) - G(dx)] \right| \\
+\ & (1+\alpha) \sup_{y\in\mathbf{R}^d} \int \operatorname{osc}_{2\varepsilon} h \circ S_y(x)\,|G(dx)| \\
+\ & \sup_{y\in\mathbf{R}^d} \left| \int k_\varepsilon * h \circ S_y(x) \Delta(dx) \right| + \sup_{y\in\mathbf{R}^d} \left| \int k_\varepsilon * \operatorname{osc}_\varepsilon h(x) \Delta(dx) \right| .
\end{aligned}$$

Since the parameter $y_0 \in \mathbf{R}^d$ is free, this last inequality yields the estimate of the lemma. ○

Lemma 5.3.1 leads to the following bound in terms of Fourier transforms $\widehat{f}$ and $\widehat{g}$.

Lemma 5.3.2 *For a function $h : \mathbf{R}^d \to \mathbf{R}$,*

$$\begin{aligned}
\left| \int_{\mathbf{R}^d} h(x)\,[F(dx) - G(dx)] \right| \ \leq\ & c_1 \sup_{y\in\mathbf{R}^d} \int_{\mathbf{R}^d} \operatorname{osc}_{2\varepsilon} h \circ S_y(x)\,|G(dx)| \\
+\ & c_2 \int_{\|t\| \leq \frac{1}{\varepsilon}} \left(\left|\widehat{h}(t)\right| + \left|\widehat{\operatorname{osc}_\varepsilon h}(t)\right| \right) \left|\widehat{f}(t) - \widehat{g}(t)\right| dt.
\end{aligned}$$

In the statement of the lemma, c_i are, as usual, positive constants.

Proof of Lemma 5.3.2 is a straightforward application of Lemma 5.3.1, the Parseval equality, and the fact that the absolute value of $\widehat{k}_\varepsilon(t) = \widehat{k}(\varepsilon t)$ never exceeds one in the points of space where it does not vanish altogether. ○

Note that the Fourier transforms of h and its oscillation can further be evaluated in terms of L_1-norms of h and its oscillation

$$|h|_1 = \int_{\mathbf{R}^d} |h(x)|\,dx, \quad |\operatorname{osc}_\varepsilon h|_1 = \int_{\mathbf{R}^d} \operatorname{osc}_\varepsilon h(x)\,dx, \quad \varepsilon > 0. \tag{5.3.10}$$

Namely,

$$\forall\, t \in \mathbf{R}^d \quad \left|\widehat{h}(t)\right| \leq |h|_1, \quad \left|\widehat{\operatorname{osc}_\varepsilon h}(t)\right| \leq |\operatorname{osc}_\varepsilon h|_1 . \tag{5.3.11}$$

5.4 Fourier Transforms Near the Origin

Fourier transforms (5.3.4)–(5.3.5) are usually chosen as the means of constructing asymptotic expansions for measure (5.3.2). Lemma 5.3.2 reduces the evaluation of approximation error to the comparison of the complex-valued functions (5.3.4) with asymptotic expansions based on (5.3.5).

Lemma 5.3.2 shows that one should be able to get nontrivial estimates of proximity between CF's for values of argument $\|t\|$ varying in a very wide range to make approximation error small.

Known estimates for very large values of $||t||$ are based on conditions which guarantee some degree of smoothness of the distribution approximated. So far, these conditions seem to be far from their ultimate form. Some of the problems arising in their derivation are discussed in later sections.

On the contrary, moment restrictions (5.4.2) and formulae of Section 5.2 suffice for relatively small values of argument. The range treated below is $||t|| = \mathcal{O}\left(\tau^{-\delta}\right)$, $0 < \delta \ll 1$.

5.4.1 Regularity Assumptions

Below $\xi_1, \xi_2, \ldots$ are independent copies of a RV ξ with values in a real separable Hilbert space X. The norm in X is denoted $|\cdot|$ and the inner product $\langle\cdot,\cdot\rangle$ to avoid confusion with the inner product $(\cdot,\cdot)$ and norm $||\cdot||$ in $\mathbf{R}^d$. Notation $|\cdot|$ is employed also for norms of polylinear forms over X, $||\cdot||$ for norm in tensor powers of $\mathbf{R}^d$, and either of these notations can be used for the norm in a product of tensor powers of these spaces.

It is the purpose of the calculations to follow to derive some bounds for the Fourier transforms (5.3.4)–(5.3.5). This problem is considered in a slightly more general setting.

The functions Ψ and ϕ depend on $\varepsilon = \frac{1}{\sqrt{n}}$ and a new small parameter τ. The distribution of $\xi = \xi_{[\tau]}$ may also depend on τ. The small parameter τ will always agree with the "principal" one:

$$\varepsilon = \tfrac{1}{\sqrt{n}}, \quad \tau = r\varepsilon, \quad r(n) \geq 1, \quad \lim_{n\to\infty} \tau(n) = o(1). \tag{5.4.1}$$

This additional complication of the scheme is introduced with a view to increase the range of possible applications for asymptotic expansions below. The above kind of dependence appears naturally, e.g., in problems about large deviations. To stress that the asymptotic expansions below are more complicated, notation (5.1.13) of the normalized sum Z_n is changed to

$$\zeta_n = \varepsilon\left(\xi_1 + \cdots + \xi_n\right), \quad \xi_i = \xi_{[\tau],i}.$$

No attempt is made to minimize the moment restrictions on $|\xi|$, and the summands are centered: it is assumed that uniformly in $\tau \in [0,\,1]$

$$\forall\, m \in \mathbf{N} \;\; \mathbf{E}\left|\xi_{[\tau]}\right|^m \leq \mu_m < \infty, \;\; \mathbf{E}\xi_{[\tau]} = 0. \tag{5.4.2}$$

Dependence of the distribution of $\xi_{[\tau]}$ on τ is supposed to be regular. All tensor-valued moments of $\xi_{[\tau]}$ can be expanded in powers of τ in a neighborhood of zero: for all $k, \kappa \in \mathbf{N}$

$$\begin{aligned} \boldsymbol{\mu}_k(\tau) \stackrel{def}{=} \mathbf{E}\xi_{[\tau]}^{\otimes k} &= \textstyle\sum_{\iota=0}^{\kappa} \tau^{\iota}\boldsymbol{\mu}_{k,\iota} + \rho_{k,\kappa}(\tau), \\ |\rho_{k,\kappa}(\tau)| &\leq c_{k,\kappa}\tau^{\kappa+1}. \end{aligned} \tag{5.4.3}$$

A special part will be played by the "unperturbed" covariance operator

$$\mathcal{V} = \boldsymbol{\mu}_{2,0}(0), \quad \langle \mathcal{V}t, t\rangle = \mathbf{E}\langle \xi_{[0]}, t\rangle^2. \tag{5.4.4}$$

In what follows, $\eta_1, \eta_2, \dots$ are independent copies of the centered Gaussian RV's η with CF

$$\mathbf{E} \exp\{i\langle\eta, t\rangle\} = \exp\left\{-\tfrac{1}{2}\langle \mathcal{V}t, t\rangle\right\}. \tag{5.4.5}$$

In all calculations all the RV's ξ_j and η_k are independent.

The measure F_n is defined by (5.3.2) with Z_n changed to ζ_n, and G by (5.3.3). (Later G and similar notations will refer to a measure or signed measure obtained by differentiation of translations of G as in Lemma 5.4.3 rather than to this measure proper). It is convenient to rewrite Fourier transforms (5.3.4)–(5.3.5) as

$$f_n(t;\tau,\varepsilon) = \mathbf{E}\exp\left\{i\left(t, \phi\left(\zeta_{[\tau],n};\tau,\varepsilon\right)\right)\right\}\Psi\left(\zeta_{[\tau],n};\tau,\varepsilon\right) = \mathbf{E}H\left(\zeta_n\right) \tag{5.4.6}$$

and

$$\widehat{g}(t;\tau,\varepsilon) = \mathbf{E}\exp\{i(t,\phi(\eta;\tau,\varepsilon))\}\Psi(\eta;\tau,\varepsilon) = \mathbf{E}H(\eta), \tag{5.4.7}$$

where

$$H(x) \stackrel{def}{=} \exp\{i(t,\phi(x))\}\Psi(x). \tag{5.4.8}$$

(Small parameters are dropped in the last formula to make it easier to read).

Some notation is necessary to formulate restrictions on the functions $\phi^{(j)}$ and Ψ.

For arbitrary values of natural numbers k, l, p, q satisfying the condition $k + l + p + q \le n$, and real-valued vectors

$$u = (u_j) \in [0,\ 1]^k, \quad v = (v_j) \in [0,\ 1]^l,$$

define the mixed sums by the equality

$$\begin{aligned} W_{k,l,p,q;n}[u,v] = \varepsilon\Big(\textstyle\sum_{j=1}^k u_j\xi_j &+ \textstyle\sum_{j=1}^l v_j\eta_{k+j} \\ + \textstyle\sum_{j=1}^p \xi_{k+l+j} &+ \textstyle\sum_{j=1}^q \eta_{k+l+p+j}\Big). \end{aligned} \tag{5.4.9}$$

It follows from known estimates for moments of the norms of sums of independent RV's that for each n and $\tau \le 1$

$$\mathbf{E}\left|W_{k,l,p,q}[u,v]\right|^B \le M(B), \quad B \ge 2, \tag{5.4.10}$$

where the positive constant M depends only on B and can be expressed in terms of bounds on the moments $\mathbf{E}\left|\xi_{[\tau]}\right|^m$, $m \le B$.

Here and below no attempt is made at keeping track of the numerical values of constants. Thus, constants which do not depend on the distribution of ξ or n are denoted c, c', etc., and same notations may refer to constants with different values in different parts of an argument. Constants are enumerated only locally and only when this is necessary to avoid misunderstanding.

Derivatives in X-valued arguments and real parameters are

$$\begin{aligned} \langle a, \nabla\rangle^k f(x) &= \left\langle a^{\otimes k}, \nabla^{\otimes k} f(x)\right\rangle = \frac{d^k}{dt^k} f(x+ta)\Big|_{t=0}, \\ \partial f(x,\tau,\varepsilon) &= \frac{\partial}{\partial\tau} f(x,\tau,\varepsilon), \quad \bar{\partial} f(x,\tau,\varepsilon) = \frac{\partial}{\partial\varepsilon} f(x,\tau,\varepsilon). \end{aligned} \tag{5.4.11}$$

Notation of small parameter ε is sometimes changed to w to stress that one deals with a function defined not only for $\varepsilon = \frac{1}{\sqrt{n}}$.

Estimate (5.4.3) implies, in particular, that for any function

$$\left|\mathbf{E}\left\langle \xi + i\eta, \nabla\right\rangle^2 f(x)\right| \leq M\tau \left|\nabla^{\otimes 2} f(x)\right|. \tag{5.4.12}$$

All coordinates of the mapping ϕ in (5.4.6)-(5.4.7) are supposed to have derivatives of all orders in all their arguments at all points where they are defined. In "spatial" variables the domain is all the space X. In real variables which play the part of small parameters this is a neighborhood of the segment [0, 1]. The smoothness condition used below is the following: for each collection of nonnegative integers κ, λ, ν there is a number $B(\kappa, \lambda, \nu) \geq 0$ such that

$$\sum_{j=1}^{d} \left|\partial^\kappa \bar{\partial}^\lambda \nabla^{\otimes\nu} \phi^{(j)}(x; \tau, w)\right| \leq \bar{\phi}_{\kappa,\lambda,\nu} \left[1 + |x|\right]^{B(\kappa,\lambda,\nu)}. \tag{5.4.13}$$

(Here and elsewhere zero order derivative is the function itself.)

A similar condition for the factor Ψ in (5.4.6) employs a function $\Gamma : X \to \mathbf{R}_+$ such that for some $\alpha > 0$

$$\mathbf{E}\left[\Gamma\left(W_{k,l,p,q;n}\left[u, v\right]\right)\right]^{1+\alpha} \leq \boldsymbol{\Gamma} < \infty, \quad k + l + p + q \leq n, \tag{5.4.14}$$

uniformly in all the parameters of the sum. The restriction on Ψ is as follows: for each collection of nonnegative integer numbers κ, λ, ν satisfying the inequality $\kappa + \lambda + \nu \leq m$ there exist numbers $\overline{B} = \overline{B}(m) \geq 0$, $\alpha = \alpha(m)$, and a function $\Gamma = \Gamma_m$ obeying the inequality in (5.4.14) such that

$$\left|\partial^\kappa \bar{\partial}^\lambda \nabla^{\otimes\nu} \Psi(x)\right| \leq \overline{\Psi}_{\kappa,\lambda,\nu} \left[1 + |x|\right]^{\overline{B}} \Gamma(x). \tag{5.4.15}$$

The very specific form of regularity condition (5.4.15) is quite natural, e.g., in some problem about probabilities of moderately large deviations, where Γ appears naturally in the form of an exponential factor.

Both averages (5.4.6) and (5.4.7) are fairly complicated expressions. In calculations they are often differentiated. It is convenient to describe the structure of derivatives in a separate proposition.

Lemma 5.4.1 *Derivatives of the product $H(x) = e^{i(t,\phi)}\Psi$ in (5.4.8) are polynomials in t multiplied by $e^{i(t,\phi)}$: whatever the choice of $k, l, m \in \mathbf{N}$,*

$$\bar{\partial}^k \partial^l \left\langle a, \nabla\right\rangle^m e^{i(t,\phi)}\Psi = \sum_{\kappa=0}^{k+l} \left\langle g^{(\kappa)}_{k,l,m}, t^{\otimes\kappa} \otimes a^{\otimes m}\right\rangle e^{i(t,\phi)},$$

where the coefficients of the polynomial are smooth functions with values in appropriate spaces of polylinear forms over X and $\mathbf{R}^d$. They satisfy the following condition: for all $k', l', m' \in \mathbf{N}$ and $x \in X$

$$\left\|\bar{\partial}^{k'} \partial^{l'} \left\langle b, \nabla\right\rangle^{m'} g^{(\kappa)}_{k,l,m}\right\| \leq c^{k',l',m'}_{k,l,m}(\kappa)|b|^{m'} \left[1 + |x|\right]^A \Gamma(x),$$

where the exponents $A = A(k, l, m, k', l', m', \kappa)$ *are nonnegative, and notation is that of* (5.4.14)-(5.4.15).

The exponents A and other constants can be evaluated explicitly through those in (5.4.13) and (5.4.15).

Proof of Lemma 5.4.1 is an evident application of known formulae for the derivatives of products and exponentials in combination with conditions (5.4.13) and (5.4.15).

Note that all summands are linear in Ψ or its derivatives. ○

5.4.2 Bounds on Remainder Terms

Bergström Expansion. The following lemma evaluates the remainder term in the expansion of $f_n(t) - g(t)$ resulting from (5.2.7), the simplest "individual" expansion of Lemma 5.2.1. In its statement H is the function of (5.4.8), the normalized sum $\zeta_n = \zeta_{[\tau],n}$ consists of copies of $\xi = \xi_{[\tau]}$ (cf. (5.4.2)), and

$$\Delta = \varepsilon^{-2}\left[\mathbf{E}e^{\varepsilon\langle\xi_{[\tau]},\nabla\rangle} - \mathbf{E}e^{\varepsilon\langle\eta,\nabla\rangle}\right].$$

Note that condition (5.2.4) may be violated, but $\mathbf{E}(\xi + i\eta)^{\otimes 2} = \mathcal{O}(\tau)$ by (5.4.2) and (5.4.3).

Lemma 5.4.2 *If conditions* (5.4.2), (5.4.3), (5.4.13)-(5.4.15) *are satisfied and* $\varepsilon/\tau \le 1$, *then for the function* H *of* (5.4.8) *the remainder term in expansion* (5.2.8),

$$r_m^B H(0) = \mathbf{E}H\left(\zeta_{[\tau],n}\right) - \sum_{k=1}^{m} \psi_k(\varepsilon)\,\Delta^k H\left(x + \varepsilon \textstyle\sum_{j=1}^{n-k} \eta_j\right)\Big|_{x=0},$$

admits the estimate

$$\left|r_m^B H(0)\right| \le M\tau^{m+1}\left[1 + \|t\|\right]^{3m+3}.$$

The constant $M = M(m)$ *depends only on* m, *the exponents in smoothness conditions* (5.4.13)-(5.4.15), *and the values of constants in moment conditions* (5.4.3) *for orders* $m \le m^* = m^*(m)$.

Proof of Lemma 5.4.2. By Lemma 5.2.2 the remainder term of the Bergström expansion (in operator form) admits the representation

$$r_m^B = \frac{\psi_{m+1}(\varepsilon)}{(m+1)!}\Delta^{m+1}\sum_{l=0}^{n-m-1} c_{l,n-m}\left[\mathbf{E}e^{\varepsilon\langle\xi,\nabla\rangle}\right]^l\left[\mathbf{E}e^{\varepsilon\langle\eta,\nabla\rangle}\right]^{n-m-l-1}, \quad (5.4.16)$$

where the constants are nonnegative and have bounded sum:

$$c_{l,n-m} = \frac{(m+1)!(n-m-1)!}{l!(n-m-l-1)!}\int_0^1 \frac{u^l(1-u)^{n-l-1}}{m!}du,$$

with

$$\sum_{l=0}^{n-m-1} c_{l,n-m} = (m+1)\int_0^1 (1-u)^m du = 1.$$

Since condition (5.2.4) may be violated, equality (5.2.7) yields for Δ the decomposition

$$\Delta = \Delta_0 + \varepsilon\left[\Delta_{1,\xi} - \Delta_{1,\eta}\right], \tag{5.4.17}$$

where $\Delta_0 = \mathbf{E}\langle \xi + i\eta, \nabla\rangle^2$ is a second-order differential operator, and the terms containing third-order derivatives are

$$\Delta_{1,\omega} = \int_0^1 \frac{(1-u)^2}{2!}\mathbf{E}\langle\omega,\nabla\rangle^3 e^{u\varepsilon\langle\omega,\nabla\rangle} du, \quad \omega = \xi, \eta.$$

An elementary calculation shows that

$$\Delta^{m+1} = \sum_{\iota+\kappa+\lambda=m+1} \frac{(m+1)!}{\iota!\kappa!\lambda!}(-1)^\lambda \varepsilon^{\kappa+\lambda}\Delta_0^\iota \Delta_{1,\xi}^\kappa \Delta_{1,\eta}^\lambda,$$

so for an arbitrary smooth function f and $x \in X$

$$\Delta^{m+1} f(x) = \sum_{\iota+\kappa+\lambda=m+1} (-1)^\lambda \left(\frac{\varepsilon}{2!}\right)^{\kappa+\lambda}\left[\mathbf{E}\langle\xi+i\eta,\nabla\rangle^2\right]^\iota$$

$$\times \int_{[0,1]^\kappa} du \prod_{j=1}^{\kappa}(1-u_j)^2 \int_{[0,1]^\lambda} dv \prod_{j=1}^{\lambda}(1-v_j)^2 \tag{5.4.18}$$

$$\times \left[\mathbf{E}\prod_{j=1}^{\kappa}\langle\xi_j,\nabla\rangle^3 \prod_{j=1}^{\lambda}\langle\eta_{\kappa+j},\nabla\rangle^3 f\left(x+\varepsilon\textstyle\sum_{j=1}^{\kappa} u_j\xi_j + \varepsilon\sum_{j=1}^{\lambda} v_j\eta_{\kappa+j}\right)\right].$$

Each of the operators in the sum of the right-hand side of (5.4.16) can be interpreted as the average translation by a mixed sum (5.4.9):

$$\left[\mathbf{E}\exp\{\varepsilon\langle\xi,\nabla\rangle\}\right]^l \left[\mathbf{E}\exp\{\varepsilon\langle\eta,\nabla\rangle\}\right]^{n-l-m-1} = \mathbf{E}\exp\{\langle W_{0,0,p,q;n}[0,0],\nabla\rangle\}$$

(here and below $p = l$, $q = n-m-l-1$). Combined with (5.4.18), this formula yields the following expression for the operator remainder r_m^B applied to an arbitrary smooth function f:

$$r_m^B f(x) = \frac{\psi_{m+1}(\varepsilon)}{(m+1)!}\sum_{l=0}^{n-m-1} c_{l,n-m}\Delta^{m+1}\mathbf{E}f\left(x + W_{0,0,p,q;n}[0,0]\right), \tag{5.4.19}$$

where $W_\bullet$ are mixed sums (5.4.9) and

$$\Delta^{m+1}\mathbf{E}f\left(x+W_{0,0,p,q;n}[0,0]\right) = \sum_{\iota+\kappa+\lambda=m+1}(-1)^\lambda\left(\frac{\varepsilon}{2!}\right)^{\kappa+\lambda}$$

$$\times \int_{[0,1]^\kappa} du\prod_{j=1}^{\kappa}(1-u_j)^2\int_{[0,1]^\lambda} dv\prod_{j=1}^{\lambda}(1-v_j)^2\left[\mathbf{E}\langle\xi+i\eta,\nabla\rangle^2\right]^\iota$$

$$\times \quad \left[\mathbf{E} \prod_{j=1}^{\kappa} \langle \xi_j , \nabla \rangle^3 \prod_{j=1}^{\lambda} \langle \eta_{j+\kappa}, \nabla \rangle^3 f\left(x + W_{\kappa,\lambda,p,q;n}[u,v]\right) \right].$$

Note that in the last factor of the right-hand side the mixed sum contains summands $u_j \xi_j$ and $v_j \eta_{\kappa+j}$, so it is not independent from the random coefficients of differential operators.

Inequality (5.4.12) taken into consideration, the resulting estimate for the remainder term in the Bergström expansion is

$$\begin{aligned} \left| r_m^B f(x) \right| &\leq M' \max_* \tau^\iota \varepsilon^{\kappa+\lambda} \\ &\times \mathbf{E} \prod_{j=1}^{\kappa} |\xi_j|^3 \prod_{j=1}^{\lambda} |\eta_{\kappa+j}|^3 \left| \nabla^{\otimes(2\iota+3\kappa+3\lambda)} f(x+W) \right|. \end{aligned} \tag{5.4.20}$$

The maximum is over all nonnegative integers ι, κ, λ satisfying the restriction $\iota + \kappa + \lambda = m + 1$, and all vectors $u \in [0,\, 1]^\kappa$, $v \in [0,\, 1]^\lambda$. Once again, the mixed sum $W = W_{\kappa,\lambda,p,q;n}[u,v]$ is not independent of the RV's in the coefficient that precedes it.

Consider now the case of function $f = H = e^{i(t,\phi)}\Psi$ (see (5.4.8)) and $x = 0$. By the rule for differentiating products, each differentiation of the exponential adds a summand containing the factor $i\,(t, \langle \omega, \nabla \rangle\, \phi)$ to the expression of the derivative considered. After all differentiations are completed, the resulting expression is the product of $\exp\{i\,(t,\phi)\}$ with a sum of "monomials" $\langle t^{\otimes k(\bullet)}, G_\bullet \rangle$ where $G_\bullet$ is some combination of derivatives of ϕ and Ψ, linear in Ψ or its derivatives. Upper bounds for the monomials follow from conditions (5.4.13) and (5.4.15).

The final result amounts to the inequality

$$\begin{aligned} \left| r_m^B \left[e^{i(t,\phi)} \Psi \right](x) \Big|_{x=0} \right| &\leq M\,[1 + \|t\|]^{3m+3} \,[\tau + \varepsilon]^{m+1} \\ &\times \max_{\iota+\kappa+\lambda=m+1} \Big\{ \mathbf{E} \textstyle\prod_{j=1}^{\kappa} |\xi_j|^3 \prod_{j=1}^{\lambda} |\eta_{\kappa+j}|^3 \\ &\times [1 + |W_{\kappa,\lambda,p,q;n}[u,v]|]^{A(\iota,\kappa,\lambda)}\, \Gamma\left(W_{\kappa,\lambda,p,q;n}[u,v]\right) \Big\}. \end{aligned} \tag{5.4.21}$$

To arrive at the estimate of the lemma, it remains to use the Hölder inequality, inequality (5.4.10), and condition (5.4.14). ○

Edgeworth-Type Expansion. Below, as in Lemma 5.4.2 condition (5.2.4) may be violated, and distributions depend on one more small parameter. Hence Edgeworth-type expansions of Section 5.2 have to be replaced by much more complicated combinations of differential polynomials depending on ϵ and the new parameter τ. It seems probable that convenient asymptotic formulae could be constructed under special assumptions about the relative magnitudes of parameters. For instance, one can seek double asymptotic expansions in powers of the "principal" parameter τ with corrections which depend on another parameter $\frac{1}{r} = \varepsilon/\tau$ in a situation where $\tau \gg \varepsilon$. However, elaboration of universal approximations seems a hopeless task. For this reason no attempt is made here to write out detailed asymptotic expansions. Some error estimates are exposed below to give an idea of the difficulties encountered.

The "individual" expansions of Lemma 5.2.1 remain valid if the remainder terms are changed to

$$\bar{\rho}^{\varphi}_{m}(\varepsilon) = \begin{cases} (3/\varepsilon)\mathbf{E}\langle \xi + i\eta, \nabla\rangle^2 + \rho^{\varphi}_{2}(\varepsilon), & m = 2, \\ \rho^{\varphi}_{m}(\varepsilon), & m \geq 3. \end{cases} \tag{5.4.22}$$

To make valid the expansions of Lemmas 5.2.3–5.2.4, one should replace the monomials $\mathcal{D}_l = \mathbf{E}\langle \xi + i\eta, \nabla\rangle^l$ by

$$\overline{\mathcal{D}_l} = \begin{cases} (3/\varepsilon)\mathbf{E}\langle \xi + i\eta, \nabla\rangle^2 + \mathcal{D}_l, & m = 3, \\ \mathcal{D}_l, & m \geq 4. \end{cases} \tag{5.4.23}$$

It should be kept in mind that the coefficients in the new expansions can be further decomposed in powers of the new small parameter τ, and remainder terms both in the Bergström expansion (5.2.8) and Edgeworth expansions of Lemmas 5.2.3–5.2.4 have order τ^{m+1} and not ε^{m+1} because (5.4.22) contains the "large" additional summand $\mathcal{O}(\tau/\varepsilon)$.

Define the "perturbed versions" $\overline{\mathcal{E}}_{k,m}$ of operators in Lemmas 5.2.3-5.2.4 by same expressions with $\mathbf{E}\langle \xi + i\eta, \nabla\rangle^k$ replaced by operators $\overline{\mathcal{D}_k}$ of (5.4.23). The following lemma seems to be a natural intermediate step towards a full asymptotic expansion of $\left[\mathbf{E}e^{\varepsilon\langle\zeta,\nabla\rangle}\right]^n$ in powers of τ and $\frac{1}{r}$.

Lemma 5.4.3 *Put*

$$\widehat{g}_m(t) = \left(\sum_{k=0}^{m} \varepsilon^k \overline{\mathcal{E}}_{k,m}(\varepsilon)\right) \mathbf{E}e^{i\langle t, \phi(x+\eta)\rangle}\Psi(x+\eta)\Big|_{x=0},$$

where $\overline{\mathcal{E}}_{k,m}$ *are the differential polynomials of Lemma* 5.2.3. *For each* $m \in \mathbf{Z}_+$

$$\left|\mathbf{E}e^{i\langle t, \phi(\zeta_{[r],n})+\rangle}\Psi\left(\zeta_{[r],n}\right) - \widehat{g}_m(t)\right| \leq M\tau^{m+1}\left[1 + \|t\|\right]^{3m+3}.$$

Proof of Lemma 5.4.3 is practically a repetition of the calculations used to prove Lemma 5.4.2. Its estimate follows from representations of Lemmas 5.2.3–5.2.4 for the remainder terms (with changes described in (5.4.22)) and the bound of Lemma 5.4.2 on the error of the Bergström approximation. ○

5.5 Reduction to Pure CF

5.5.1 Notation and Restrictions

Rapid decay of CF (1.1.6) at infinity is a well-known property of Gaussian law. A similar behavior is typical of the CF's of certain functions of Gaussian argument — the quadratic forms of Theorem 1.1.7 decay very rapidly in very high dimensions. In the study of a measure, the rapid decay of its Fourier transform is most important as it guarantees the existence of a regular density.

Under certain assumptions of regularity, the complicated Fourier transforms of (5.4.6)–(5.4.7) and their derivatives go to zero quite fast. These functions, which have the form

$$\mathbf{E}H(x+\bullet), \quad H(x) = \exp\{i\,(t, \phi(x))\}\,\Psi(x), \quad \bullet = \eta, \zeta_n,$$

inherit this type of behavior from the pure CF's

$$\mathbf{E}\exp\{i\,(t, \phi(x+\bullet))\}.$$

Reduction to this special case is through calculations very similar to expansions of Section 5.4. It seems reasonable to separate this argument from the study of CF's behavior for $\|t\| \gg 1$, which uses other technical means. The reduction procedure is exposed in the remaining part of this section.

The object considered below are the products of monomials in t with imaginary exponentials which result from application of Lemma 5.4.1. They can be written in the form

$$\|t\|^{\overline{m}} \left\langle G, \vartheta^{\otimes\overline{m}} \right\rangle \exp\{i\,(t, \phi\,)\}, \quad \vartheta = \|t\|^{-1} t,$$

where the coefficients G (of evident algebraic nature) are functions that satisfy, as well as their derivatives of all orders, growth conditions like

$$\|G(x)\| \le \overline{G}\,[1+|x|]^B\, \Gamma(x), \quad B \ge 0.$$

Thus, it suffices to show that $\mathbf{E}H(x+\bullet)$ decays faster than any power of $\|t\|^{-1}$ under restrictions (5.4.13)–(5.4.15) to establish the same kind of behavior also for all the derivatives of this function (see Lemma 5.4.1) — if no attempt is made to keep track of constants in various bounds.

In the calculations of expectations below, the random argument of H will usually be the sum of a "large" RV Z with a "small" independent perturbation $\delta\zeta$, $0 < \delta \ll 1$:

$$\bullet \;=\; \mathrm{Z} + \delta\zeta, \quad \mathbf{E}\zeta \;=\; 0, \quad \mathbf{E}\,|\zeta|^m + \mathbf{E}\,|\mathrm{Z}|^m \;\le\; A_m, \quad m = 1, 2, \ldots \tag{5.5.1}$$

To make the presentation clearer, no attempt is made at minimizing the restrictions upon the RV's considered.

The functions ϕ and Ψ in (5.3.2)–(5.3.3) obey conditions (5.4.13) and (5.4.15) as well as an analog of (5.4.14):

$$\exists\,\alpha > 0\;\; \forall\,\lambda \in [0,\,1]\;\; \mathbf{E}\,[\Gamma\,(\mathrm{Z} + \lambda\delta\zeta)]^{1+\alpha} \;\le\; \overline{\Gamma} \;<\; \infty. \tag{5.5.2}$$

The requirements of the present section are met, e.g., in problems concerning large deviations.

5.5.2 Reduction to Polynomial Imaginary Exponent

The calculations that follow reduce the task of evaluating $\mathbf{E}H\,(\mathrm{Z}+\delta\zeta)$, $H = \Psi e^{i(t,\phi)}$, to a similar problem where both the imaginary exponent and the factor preceding the exponential are polynomial in $\delta\zeta$ of (5.5.1).

For the exponent in (5.3.4)–(5.3.5), Taylor's expansion in powers of δ is

$$(t, \phi(Z + \delta\zeta)) = \left(t, F_m(Z; \delta\zeta) + \delta^{m+1} \rho_m^{\phi}(Z, \zeta, \delta)\right), \qquad (5.5.3)$$

where

$$F_m(Z; \delta\zeta) = \sum_{k=0}^{m} \frac{\delta^k}{k!} \langle \zeta, \nabla \rangle^k \phi(Z),$$

$$\rho_m^{\phi}(Z, \zeta, \delta) = \int_0^1 \frac{(1-\lambda)^m}{m!} \langle \zeta, \nabla \rangle^{m+1} \phi(Z + \lambda\delta\zeta)\, d\lambda.$$

(Recall that $\phi \in \mathbf{R}^d$ is a vector, as also the expressions above.)
Similarly

$$\Psi(Z + \delta\zeta) = \Psi_m(Z; \zeta, \delta) + \delta^{m+1} \rho_m^{\Psi}(Z; \zeta, \delta), \qquad (5.5.4)$$

where

$$\Psi_m(Z; \zeta, \delta) = \sum_{k=0}^{m} \frac{\delta^k}{k!} \langle \zeta, \nabla \rangle^k \Psi(Z),$$

$$\rho_m^{\Psi}(Z; \zeta, \delta) = \int_0^1 \frac{(1-\lambda)^m}{m!} \langle \zeta, \nabla \rangle^{m+1} \Psi(Z + \lambda\delta\zeta)\, d\lambda.$$

In the above expansions F_m and Ψ_m are polynomials in $\zeta \in X$. Below this circumstance is, if necessary, stressed by the notation

$$F_m = \sum_{k=0}^{m} \frac{\delta^k}{k!} \langle \zeta^{\otimes k}, \nabla^{\otimes k} \phi \rangle, \quad \Psi_m = \sum_{k=0}^{m} \frac{\delta^k}{k!} \langle \zeta^{\otimes k}, \nabla^{\otimes k} \Psi \rangle. \qquad (5.5.5)$$

Lemma 5.5.1 *Assume that condition* (5.5.1) *is satisfied, and the functions* ϕ, Ψ *of* (5.3.4)–(5.3.5) *obey restrictions* (5.4.13), (5.4.15) *and* (5.5.2). *There exist positive constants* c_i *such that if* $\delta^{m+1} \|t\| \le c_1$, *then*

$$\left|\mathbf{E} H(Z + \delta\zeta) - \mathbf{E} \exp\{i(t, F_m(Z; \delta\zeta))\} \Psi_m(Z; \zeta, \delta)\right| \le c_3 \delta^{m+1} [1 + \|t\|].$$

Proof of Lemma 5.5.1 . It follows from the obvious inequality

$$\left|e^{i(A+a)}(B + b) - e^{iA} B\right| \le |b| + |a|\,|B|, \quad a, A, b, B \in \mathbf{R},$$

that

$$\begin{aligned} & \left|H(Z + \delta\zeta) - e^{i(t, F_m)} \Psi_m(Z, \zeta, \delta)\right| \\ \le\; & \delta^{m+1} \|t\| \left\|\rho_m^{\phi}(Z, \zeta, \delta)\right\| + \delta^{m+1} \left|\rho_m^{\Psi}(Z, \zeta, \delta)\right| \\ \le\; & c' \delta^{m+1} |\zeta|^{m+1} [1 + \|t\|] [1 + |Z| + \delta |\zeta|]^{A(m)} \int_0^1 \Gamma(Z + \lambda\delta\zeta)\, d\lambda, \end{aligned}$$

where the constants c' and $A(m)$ are determined by the choice of m and restrictions upon the growth of ϕ, Ψ, and their derivatives.

The estimate of the lemma now follows from Hölder's inequality. Select some $\alpha > 0$ to satisfy (5.5.2) and $p, q > 0$ for which $1/(1+\alpha) + 1/p + 1/q = 1$. Then

$$\begin{aligned}\left|\mathbf{E}H(\mathrm{Z}+\delta\zeta) - \mathbf{E}e^{(t,F_m)}\Psi_m\right| &\le c\delta^{m+1}\left[1+\|t\|\right] \\ &\times \max_{\lambda\in[0,\,1]}\left(\mathbf{E}\Gamma^{1+\alpha}(\mathrm{Z}+\lambda\delta\zeta)\right)^{1/(1+\alpha)} \\ &\times \left(\mathbf{E}\,|\zeta|^{(m+1)p}\right)^{1/p}\left(\mathbf{E}\left[1+|\mathrm{Z}|+|\delta\zeta|\right]^{Aq}\right)^{1/q}.\end{aligned}$$

This proves the lemma. ○

5.5.3 Products of Polynomials with Imaginary Exponentials

It is the purpose of this subsection to show that one can derive bounds on expectations $\mathbf{E}e^{i(t,F_m)}\Psi_m$ of Lemma 5.5.1 from estimates for "pure" CF's $\mathbf{E}e^{i(t,F_m')}$ and inequalities for moments of norms for sums of independent RV's. (Derivation of estimates for the CF's proper will be discussed later.) The functions below may depend on additional variables, e.g., small parameters τ, $\frac{1}{r}$, etc. All conditions are supposed to hold uniformly in these parameters (invariably omitted in notation), as well as the bounds obtained using them.

The next lemma reduces the degree m of the polynomial exponent in approximation of Lemma 5.5.1 to a smaller value v (usually $v = 1, 2$). This is achieved by one more Taylor expansion of the imaginary exponential. Put

$$\mathcal{P}^{(v)}_{m,q}(\mathrm{Z};\zeta,\delta) \stackrel{def}{=} \sum_{p=0}^{q}\frac{1}{p!}\left(t, F_m(\mathrm{Z};\delta\zeta) - F_v(\mathrm{Z};\delta\zeta)\right)^p\Psi_m(\mathrm{Z};\zeta,\delta).$$

Lemma 5.5.2 *Under restrictions* (5.5.1)–(5.5.2) *and* (5.4.13), (5.4.15) *there exists a positive constant* c *such that for* $\delta \in [0,\,1]$

$$\left|\mathbf{E}\exp\{i(t,F_m)\}\Psi_m - \mathbf{E}\exp\{i(t,F_v)\}\mathcal{P}^{(v)}_{m,q}\right| \le c\left[\delta^{v+1}\|t\|\right]^{q+1}.$$

The arguments of F_m, Ψ_m, etc. in the statement of the lemma are as in the preceding definition of $\mathcal{P}^{(v)}_{m,q}(\mathrm{Z};\zeta,\delta)$. The factors accompanying the imaginary exponential in Lemma 5.5.2 are polynomial in the variables $\zeta \in X$ and $t \in \mathbf{R}^d$. The coefficients of these polynomials depend on $\mathrm{Z} \in X$. When the necessity arises to stress the structure of these expressions, this is done using notation (cf. (5.5.5))

$$\mathcal{P}^{(v)}_{m,q}(\mathrm{Z};\zeta,\delta) = \sum_{p=0}^{q}\sum_{k=0}^{m(p)}\delta^k\left\langle\zeta^{\otimes k}\otimes t^{\otimes p}, \mathcal{P}^{(v)}_{p,q,k,m}(\mathrm{Z})\right\rangle. \tag{5.5.6}$$

As functions of Z, the coefficients in (5.5.6) admit the bounds

$$\left\|\mathcal{P}^{(v)}_{p,q,k,m}(\mathrm{Z})\right\| \le \bar{\mathbf{P}}^{(v)}_{p,q,k,m}\left[1+|\mathrm{Z}|\right]^{A_{p,q,k,m}}\Gamma(\mathrm{Z}), \tag{5.5.7}$$

inherited from assumptions on growth of ϕ, Ψ, and their derivatives. Notation $\|\cdot\|$ is used for norms in all spaces of polylinear functions over X and $\mathbf{R}^d$. The constants $\bar{\mathbf{p}}^{(v)}_{p,q,k,m}$ and exponents A are determined by constants in restrictions (5.4.13), (5.4.15), and (5.5.2).

Proof of Lemma 5.5.2. It follows from Taylor's expansion for an exponential that

$$\begin{aligned} \left|e^{i(t,F_m)}\Psi_m\right. &- \left. e^{i(t,F_v)}\left[\sum_{p=0}^{q}\tfrac{1}{p!}\left(t,F_m-F_v\right)^p\right]\Psi_m\right| \\ &\le \frac{1}{(q+1)!}\left|\left(t,F_m-F_v\right)^{q+1}\Psi_m\right|, \end{aligned}$$

and obviously $|(t, F_m - F_v)| \le \|t\|\,\|F_m - F_v\|$. With respect to ζ, the expressions $F_m - F_v$ and Ψ_m are polynomials which satisfy the inequalities

$$\begin{aligned} \|F_m - F_v\| &\le \bar{c}\left[\delta^{v+1}|\zeta|^{v+1} + \delta^m|\zeta|^m\right](1+|\mathrm{Z}|)^{A(m)} \\ &\le c'\delta^{v+1}|\zeta|^{v+1}(1+|\mathrm{Z}|+\delta|\zeta|)^{A'(m)}, \\ |\Psi_m| &\le c''(1+|\mathrm{Z}|+\delta|\zeta|)^{A''(m)}\,\Gamma(\mathrm{Z}). \end{aligned}$$

The inequality of the lemma follows from (5.5.1) and the Hölder inequality (cf. the proof of Lemma 5.5.1). ○

Lemma 5.5.2 reduces the problem of evaluating the expectations $\mathbf{E}H$ to the corresponding problem for the averages $\mathbf{E}\,\varphi(t)$,

$$\varphi(t) = \exp\{i(t, F(\zeta;\mathrm{Z}))\}\left\langle \zeta^{\otimes p}, \mathcal{P}(\mathrm{Z})\right\rangle, \tag{5.5.8}$$

where the random arguments ζ and Z are independent, and the coefficients $\mathcal{P}(\mathrm{Z})$ have bounded moment of some order $1+\alpha > 1$. The fact that the factor accompanying the imaginary exponential is polynomial in its argument ζ permits to reduce the task to estimating separately the "pure" CF $\mathbf{E}\exp\{i(t, F(\zeta; y))\}$ and the moments of $|\zeta|$ and $|\mathrm{Z}|$. This opportunity arises if the argument ζ enjoys a property akin to infinite divisibility, viz., if for large natural m there are decompositions

$$\mathcal{L}(\zeta) = \mathcal{L}\left(\zeta^{(1)} + \cdots + \zeta^{(m)}\right) = \mathcal{L}^{*m}\left(\zeta^{(1)}\right), \tag{5.5.9}$$

where $\zeta^{(j)}$ are i.i.d. RV's (the asterisk denotes convolution).

Lemma 5.5.3 *Consider the expectation* $\mathbf{E}\,\varphi$ *with* φ *the product of* (5.5.8) *whose degree of homogeneity in* ζ *equals* p. *Suppose that decomposition* (5.5.9) *is possible for some* m *with summands satisfying the condition*

$$\forall\, k \in \mathbf{N}\;\; \mathbf{E}\left|\zeta^{(j)}\right|^k \le A'_k. \tag{5.5.10}$$

If $m > p$, *then*

$$|\varphi(t)| \le c\mathbf{E}\,|\mathcal{P}(\mathrm{Z})|\,\phi^+(\mathrm{Z}),\;\; \phi^+(y) = \sup_{a\in X}\left|\mathbf{E}\exp\left\{i\left(t, F(a+\zeta^{(1)}; y)\right)\right\}\right|.$$

Proof of Lemma 5.5.3. Consider the expansion

$$\left(\zeta^{(1)} + \cdots + \zeta^{(m)}\right)^{\otimes p} = \sum_{l_1=1}^{m} \cdots \sum_{l_p=1}^{m} \otimes_{\kappa=1}^{p} \zeta^{(l_\kappa)}. \tag{5.5.11}$$

The monomials above are homogeneous in each of the variables $\zeta^{(j)}$, $j = \overline{1,m}$, and the degrees of homogeneity are

$$p[j] = \sum_{k=1}^{p} \mathbf{1}_{\{j[k]=j\}} = \text{card}\left\{\kappa \in \{1,\ldots,p\} : l_\kappa = j\right\}. \tag{5.5.12}$$

It is obvious that if $m > p$, no term in (5.5.11) can contain all subsummands of (5.5.9) with nonzero powers (5.5.12). The number of summands in (5.5.11) is bounded. Thus, it suffices to evaluate a generic summand to derive the estimate of the lemma.

As no summand of (5.5.9) enjoys preference, no generality is lost in evaluating the expression

$$\varphi_1(t; a, y) = \exp\left\{i(t, F(\zeta^{(1)} + a; y))\right\} \left\langle \otimes_{j=2}^{m} [\zeta^{(j)}]^{\otimes p[j]}, \mathcal{P}(y)\right\rangle,$$

where $a = \sum_{j=2}^{m} \zeta^{(j)}$ and the RV $\zeta^{(1)}$ is present only in the imaginary exponent.

To obtain the inequality of the lemma, first fix the arguments Z, $\zeta^{(k)}$, $k = \overline{2,m}$, and calculate the conditional expectation averaging only with respect to $\zeta^{(1)}$. Next, replace the conditional CF by its modulus and the product of $\zeta^{(k)}$ ($k \geq 2$) by the product of norms, and compute conditional expectation for fixed value of Z. Since the rv's $|\zeta^{(k)}|^{p[k]}$ enter only as independent factors, the conditional expectations can be evaluated using (5.5.10). This yields the final estimate. ○

5.6 Almost Linear Mappings

The CF's of linear functions of Gaussian RV's decay very rapidly as their arguments go to infinity. This section considers some conditions which ensure similar behavior of CF's for RV's that are smooth functions (5.3.1) of Gaussian or "nearly Gaussian" ones. The situation considered below is one where the mapping is nearly linear, so nontrivial bounds on CF's can be obtained using the linear approximation F_1 of Lemmas 5.5.1–5.5.3.

5.6.1 Gaussian Case

As in (5.5.1), it is assumed that the argument of the smooth mapping considered is the sum of a "large" RV $\mathbf{Z}$ with independent "perturbation" $\delta\zeta$, $\delta \ll 1$. In this subsection, the latter is centered Gaussian: $\zeta = \eta$, $\mathbf{E}\eta = 0$, with the CF of η given by (5.4.5) with covariance operator $\mathcal{V} = \mathbf{E}\eta^{\otimes 2}$. Under this restriction, fast decay of CF's is typical of smooth mappings to finite dimensions which satisfy certain nondegeneracy conditions.

Consider the RV $\langle \eta, \acute{\nabla} \rangle \phi(x) \in \mathbf{R}^d$ and the real-valued rv

$$Y[\eta; x, t] = \langle \eta, \nabla \rangle (t, \phi(x)) = \left(t, \frac{1}{\delta} [F_1(x; \delta\eta) - F_1(x; 0)] \right),$$

For fixed $x \in X$ and $t = \left(t^{(j)}, j = \overline{1,d}\right) \in \mathbf{R}^d$, the latter is a centered Gaussian rv whose variance is a quadratic form in t:

$$\mathbf{D}Y = \left| \mathcal{V}^{1/2} \nabla (t, \phi(x)) \right|^2 = \sum_{j,k=1}^{d} t^{(j)} t^{(k)} \left\langle \mathcal{V} \nabla \phi^{(j)}(x), \nabla \phi^{(k)}(x) \right\rangle .$$

The nondegeneracy condition used below is as follows: whatever the choice of $B > 0$, there exists a constant $C = C(B)$ such that for all $t \in \mathbf{R}^d$

$$\mathbf{E} \exp \left\{ - \left| \mathcal{V}^{1/2} \nabla (t, \phi(\mathrm{Z})) \right|^2 \right\} \leq C / [1 + \|t\|]^B . \tag{5.6.1}$$

Note that it imposes restrictions both on the derivatives of ϕ and on the distribution of Z.

Lemma 5.6.1 *If η is a centered Gaussian RV with covariance operator $\mathcal{V}$ and condition* (5.6.1) *is satisfied, then for each $a \in X$ and $A > 0$ there exists a constant $\bar{c} = \bar{c}(A)$ such that in all* $\mathbf{R}^d$

$$|\mathbf{E} \exp \{ i (t, F_1 (\mathrm{Z}; \delta[a + \eta])) \}| \leq \bar{c} / [1 + \delta \|t\|]^A .$$

The constant in the lemma depends on the choice of A, but not on a.

Proof of Lemma 5.6.1. In the right-hand side of the identity

$$\begin{aligned} \exp \left\{ i(t, F_1(\mathrm{Z}; \delta[a + \eta])) \right\} &= \exp \left\{ i(t, \phi(\mathrm{Z})) \right\} \exp \left\{ i\delta \langle a, \nabla \rangle (t, \phi(\mathrm{Z})) \right\} \\ &\times \exp \left\{ i\delta \langle \eta, \nabla \rangle (t, \phi(\mathrm{Z})) \right\} \end{aligned}$$

all factors have absolute value 1 and are independent (conditionally, given Z). With respect to the distribution of η, the exponent in the last of them is a centered normal rv, so

$$\mathbf{E} \left\{ \exp \left\{ i\delta \langle \eta, \nabla \rangle (t, \phi(\mathrm{Z})) \right\} | \mathrm{Z} \right\} = \exp \left\{ -\tfrac{1}{2} \delta^2 | \mathcal{V}^{1/2} \nabla (t, \phi(\mathrm{Z})) |^2 \right\} .$$

To derive the inequality of the lemma, it remains to calculate first the conditional expectation of the product above given Z. The bound of the lemma follows from (5.6.1) after the absolute values of the factors not containing η are evaluated by 1, and the resulting estimate is averaged with respect to distribution of Z. ○

It may not be easy to verify condition (5.6.1), but it is natural in some cases of interest. The situation treated below occurs, e.g., in the study of large deviations for Gaussian RV's.

Consider a function whose Fréchet derivatives are uniformly bounded from the second on:

$$\Phi : X \to \mathbf{R}, \quad \left| \nabla^{\otimes k} \Phi(x) \right| \leq \bar{\Phi}_k, \quad x \in X, \quad k = 2, 3, \ldots \tag{5.6.2}$$

Take a point $\hat{x} \in X$ where the gradient of Φ does not vanish. Put $e_0 =$

$|\nabla\Phi(x)|^{-1}\nabla\Phi(x)$, and select a finite set of orthogonal unit vectors e_j, $j=\overline{1,d}$, orthogonal to $\nabla\Phi(\hat{x})$, so that $\langle e_j, e_k\rangle = \delta_{jk}$, $j,k=\overline{0,d}$.

Define, for large values of r, the mapping of X to $\mathbf{R}^{d+1}$ by the formula

$$\phi(x) \stackrel{def}{=} \frac{r\left[\Phi\left(\hat{x}+\frac{1}{r}x\right)-\Phi(\hat{x})\right]}{|\nabla\Phi(\hat{x})|}e_0+\sum_{j=1}^{d}\langle x, e_j\rangle e_j. \tag{5.6.3}$$

This mapping is almost the orthogonal projection onto the $(d+1)$-dimensional subspace generated by the vectors e_j, which is identified with $\mathbf{R}^{d+1}$ in the calculations below. Only one coordinates of (5.6.3) contains a nonlinear perturbation.

In the sequel, the point $\hat{x}$ is a smooth function of small parameters τ and $\frac{1}{r}$. All conditions imposed on Φ with argument dependent on $\hat{x}$ are supposed to hold uniformly in values of these small parameters. The mention of the small parameters is usually suppressed (both in text and in notation) unless this may cause ambiguity.

For a vector $t=\left(t^{(j)}, j=\overline{0,d}\right)\in\mathbf{R}^{d+1}$,

$$\langle\zeta,\nabla\rangle\left(t,\phi(\mathrm{Z})\right) = \frac{\langle\zeta,\nabla\rangle\,\Phi\left(\hat{x}+\frac{1}{r}\mathrm{Z}\right)}{|\nabla\Phi(\hat{x})|}t^{(0)}+\sum_{j=1}^{d}\langle\zeta, e_j\rangle t^{(j)}, \tag{5.6.4}$$

so in condition (5.6.1) the exponent is

$$(Q[\mathrm{Z}]t,t) = \left|\mathcal{V}^{1/2}\left[t^{(0)}\widetilde{e}_0(\mathrm{Z})+\textstyle\sum_{j=1}^{d}t^{(j)}e_j\right]\right|^2, \tag{5.6.5}$$

with $\widetilde{e}_0(\mathrm{Z})=|\nabla\Phi(\hat{x})|^{-1}\nabla\Phi\left(\hat{x}+\frac{1}{r}\mathrm{Z}\right)$.

Lemma 5.6.2 *Suppose that in* (5.6.1) *the argument* Z *is a centered Gaussian RV with covariance operator* $\theta^2\mathcal{V}$, $\theta\in[\frac{1}{2},1]$.

Mapping (5.6.3) *obeys condition* (5.6.1) *if*

$$\forall\, t\in\mathbf{R}^{d+1}\quad (Q[0]t,t)\ \geq\ \bar{\phi}_0\,\|t\|^2, \tag{5.6.6}$$

and for arbitrary (large) values of $B>0$ *there exist constants* $\bar{\phi}_1$ *and* C_ϕ *such that for all* $u\in[0,1]$, $s\in\mathbf{R}$, *and values of small parameters satisfying* $\tau,\frac{1}{r}\leq\bar{\phi}_1$

$$\begin{aligned}&\mathbf{E}\exp\left\{-\tfrac{1}{2}s^2\left|\mathcal{V}^{1/2}\nabla^{\otimes 2}\Phi\left(\hat{x}+u\tfrac{1}{r}\eta_2\right)\eta_1\right|^2\right\}\\ &=\textstyle\prod_{j=1}^{\infty}1/\sqrt{1+s^2\widehat{\sigma}_j^2\left(\frac{1}{r}u\eta\right)}\ \leq\ C_\Phi/\left[1+s^2\right]^B,\end{aligned} \tag{5.6.7}$$

where η_1 *and* η_2 *are independent centered Gaussian RV's with covariance operator* $\mathcal{V}$, *and* $\widehat{\sigma}_j^2(y)$, $j\in\mathbf{N}$, *are eigenvalues of the operator*

$$\widehat{\mathcal{V}}(y)=\left(\mathcal{V}^{1/2}\nabla^{\otimes 2}\Phi(\hat{x}+y)\,\mathcal{V}^{1/2}\right)^2.$$

In the lemma, the second Fréchet derivative is considered as a bounded operator from X to X.

Its first assumption (5.6.6) is that mapping (5.6.3) would be nondegenerate without the nonlinear perturbation (recall that, in accordance with an earlier agreement, nondegeneracy is uniform in small parameters).

The presence of a nonlinear perturbation makes necessary the second assumption (5.6.7). This is easily seen to be a restriction on the covariance operator $\widehat{\mathcal{V}}(y)$ of another Gaussian RV, $\mathcal{V}^{1/2}\nabla^{\otimes 2}\Phi(\hat{x}+y)\,\eta_1$. It stipulates that for small values of y this vector rarely hits small neighborhoods of zero. Small probabilities of landing in small spheres are typical of Gaussian distributions in very high dimensions. In fact, the condition requires that neither the subspace $\nabla^{\otimes 2}\Phi(\hat{x})X$ nor the support of $\mathcal{L}(\eta)$ should be finite-dimensional.

Proof of Lemma 5.6.2. Quadratic form (5.6.5) of condition (5.6.6) can be written as

$$(Q[y]t,t) = \left|\mathcal{V}^{1/2}\left[\hat{t}+s\psi(y)\right]\right|^2, \tag{5.6.8}$$

where

$$\begin{aligned} t &= \left(t^{(j)}, j=\overline{0,d}\right) \in \mathbf{R}^{d+1}, \\ s &= s(t) \stackrel{def}{=} (r\,|\nabla\Phi(\hat{x})|)^{-1}t^{(0)} \in \mathbf{R}, \\ \hat{t} &= t^{(0)}e_0 + \textstyle\sum_{j=1}^{d} t^{(j)}e_j \in X, \\ \psi(y) &\stackrel{def}{=} r\left[\nabla\Phi\left(\hat{x}+\tfrac{1}{r}y\right) - \nabla\Phi(\hat{x})\right], \end{aligned}$$

and by (5.6.2)

$$|\psi(y)| \le c\,|y|\,. \tag{5.6.9}$$

It is necessary to discern between the "large" and "small" values of s.

a) If $|s| \le \sqrt{\|t\|}$, the sum of $\mathcal{V}^{1/2}\hat{t}$ and $s\mathcal{V}^{1/2}\psi$ can be small only if $|y| \ge c\,\|t\|^{1/2}$ (see (5.6.9)). Since $\mathcal{L}(\mathrm{Z}) = \mathcal{L}(\theta\eta)$, by the exponential Chebyshev inequality and, e.g., Theorem 2.1.1

$$\mathbf{E}\exp\left\{\, -\tfrac{1}{2}\,(Q[\mathrm{Z}]t,t)\,\right\} \le e^{-c'\|t\|} + \mathbf{P}\left\{|\eta| \ge c''\,\|t\|^{1/2}\right\} \le 2e^{-c'''\|t\|},$$

so (5.6.1) holds for each B if the constant $C(B)$ is appropriately chosen.

b) The converse case, $|s| > \sqrt{\|t\|}$, needs a different argument.

While calculating the expectation in (5.6.1), one can assume that $\mathrm{Z} = \sqrt{q}\eta' + \sqrt{\theta^2-q}\eta''$, where η', η'' are independent centered Gaussian RV's with covariance operator $\mathcal{V}$ and $q = |s|^{-4/3}$, so that $|s|\sqrt{q} = q^{-1/4}$ is large and $|s|\,q = q^{1/4}$ small.

Write out the first terms of Taylor's expansion for $\psi(\sqrt{q}\eta' + \sqrt{\theta^2-q}\eta'')$ in $\sqrt{q}\eta'$ and rewrite the vector in (5.6.8) using this expansion:

$$\begin{aligned} \psi(\sqrt{q}\eta' + \sqrt{\theta^2-q}\eta'') &= A + \sqrt{q}\,\Phi''\eta' + \rho_\psi(\eta',\eta'',q), \\ \hat{t} + s\psi(\mathrm{Z}) &= s\sqrt{q}\,[A^+ + \Phi''\eta'] + s\rho_\psi(\eta',\eta'',q), \end{aligned} \tag{5.6.10}$$

where

$$A = r\left[\nabla\Phi\left(\hat{x} + \tfrac{1}{r}\sqrt{\theta^2-q}\eta''\right) - \nabla\Phi(\hat{x})\right],$$

$$\Phi'' = \nabla^{\otimes 2}\Phi\left(\hat{x} + \tfrac{1}{r}\sqrt{\theta^2-q}\eta''\right), \quad A^+ = (s\sqrt{q})^{-1}[\hat{t} + sA],$$

both A^+ and Φ'' are independent from η' by the construction, and the remainder term admits the estimate

$$|\rho_\psi(\eta', \eta'', q)| \le cr^{-1} q \, |\eta'|^2 . \tag{5.6.11}$$

Abbreviate the conditional probability given the value of η'' to $\mathbf{P}''(\cdot) \stackrel{def}{=} \mathbf{P}\{\cdot|\eta''\}$. It follows from the above that for a small $\gamma > 0$

$$\begin{aligned} & \mathbf{P}''\left\{(Q\,[Z]\,t,t) \le q^{2\gamma}\right\} \qquad (5.6.12) \\ = \;& \mathbf{P}''\left\{\left|\mathcal{V}^{1/2}\left(s\sqrt{q}[A^+ + \Phi''\eta'] + s\rho_\psi\right)\right| \le q^\gamma\right\} \\ \le \;& \mathbf{P}''\left\{|s|\sqrt{q}\left|\mathcal{V}^{1/2}[A^+ + \Phi''\eta']\right| \le \tfrac{3}{2}q^\gamma\right\} + \mathbf{P}''\left\{|s|\left|\mathcal{V}^{1/2}\rho_\psi\right| \ge \tfrac{1}{2}q^\gamma\right\} \\ \le \;& \mathbf{P}''\left\{\left|\mathcal{V}^{1/2}\left[A^+ + \Phi''(\eta'')\eta'\right]\right| \le \tfrac{3}{2}q^{\gamma+1/4}\right\} + \mathbf{P}\left\{|\eta'|^2 \ge c'rq^{\gamma-1/4}\right\}. \end{aligned}$$

(Recall that $|s| = q^{-3/4}$ and $r \ge 1$).

The second probability in the right-hand side is small by known bounds on large deviation probabilities for Gaussian RV's:

$$\mathbf{P}\left\{|\eta'|^2 \ge c'rq^{\gamma-1/4}\right\} \le c''\exp\left\{-c'''q^{-1/4+\gamma}\right\}. \tag{5.6.13}$$

While evaluating the first term in the right-hand side of (5.6.12), it is convenient to "duplicate" η' into two independent copies ξ_i , $i = 1, 2$, which are independent also of η'', and set $\varepsilon = \frac{3}{2}q^{\gamma+1/4}$:

$$\begin{aligned} & \mathbf{P}''\left\{\left|\mathcal{V}^{1/2}\left[A^+ + \Phi''\eta'\right]\right| \le \varepsilon\right\} \\ = \;& \sqrt{\mathbf{P}''\left(\left\{\left|\mathcal{V}^{1/2}[A^+ + \Phi''\xi_1]\right| \le \varepsilon\right\} \bigcap \left\{\left|\mathcal{V}^{1/2}[A^+ + \Phi''\xi_2]\right| \le \varepsilon\right\}\right)} \\ \le \;& \sqrt{\mathbf{P}''\left\{\left|\mathcal{V}^{1/2}\Phi''\tilde{\xi}\right| \le \sqrt{2}\varepsilon\right\}}, \quad \tilde{\xi} = \tfrac{1}{\sqrt{2}}(\xi_1 - \xi_2), \end{aligned}$$

where the symmetrization in the right-hand side has the original Gaussian distribution with the covariance operator $\mathcal{V}$: by the characteristic property of Gaussian laws $\mathcal{L}(\tilde{\xi}) = \mathcal{L}(\eta)$.

The condition of the lemma stipulates that the RV in the right-hand side of the last inequality rarely lands in a small neighborhood of zero. By a version of Chebyshev's inequality, $e\,\mathbf{P}\{|Y| \le \kappa\} \le \mathbf{E}\exp\left\{-(Y/\kappa)^2\right\}$ for any rv $Y \in \mathbf{R}$ and $\kappa > 0$, so the probability above admits the estimate

$$\mathbf{P}''\left\{\left|\mathcal{V}^{1/2}\left[A^+ + \Phi''\eta'\right]\right| \le \varepsilon\right\} \le \hat{c}\sqrt{\mathbf{E}''\exp\left\{-\frac{1}{2\varepsilon^2}\left|\mathcal{V}^{1/2}\Phi''\eta'\right|^2\right\}}. \tag{5.6.14}$$

Since $\varepsilon = \frac{3}{2}|s|^{-(1/3+4\gamma/3)} \le \tilde{c}\,\|t\|^{-(1/6+\gamma 2/3)}$ by the choice of q, the bound of the lemma for $|s| \ge \sqrt{\|t\|}$ follows from inequalities (5.6.12), (5.6.14), and (5.6.7) after averaging with respect to the distribution of η''. ○

A bound on Fourier transforms. It is convenient to place here an estimate for the more complicated Fourier transforms of measures (5.3.3) $\mathbf{E}\Psi \exp\{i\,(t, \phi)\}$ which results from calculations of this and preceding sections.

Consider a mapping $\phi : X \to \mathbf{R}^d$, a function Ψ obeying restrictions (5.4.13) and (5.4.15), and a centered Gaussian RV η with covariance operator $\mathcal{V}$. As before, the mapping and the function can smoothly depend on parameters τ, $\frac{1}{\tau}$, etc., and ∂, $\bar{\partial}$ are derivatives in these variables. To avoid encumbering formulae, parameters are omitted in notation. All conditions and resulting bounds are uniform in values of these parameters from some standard neighborhood of zero, say, $[-\frac{1}{2}, \frac{1}{2}]$.

Lemma 5.6.3 *Assume that condition* (5.6.1) *holds for* $Z = \theta\eta$ *for each* $B > 0$ *with a common constant for all values of* $\theta \in [\frac{1}{2}, 1]$, *as well as conditions* (5.5.2), (5.4.13), *and* (5.4.15).

For arbitrary numbers $k, l, m \in \mathbf{Z}_+$ *and* $B > 0$ *there exists a constant* $c = c_{k,l,m,B} > 0$ *such that the inequality*

$$\left|\mathbf{E}\partial^k \bar{\partial}^l \nabla^{\otimes m}\, e^{i(t,\phi(\eta))}\, \Psi(\eta)\right| \le c/\left[1 + \|t\|\right]^B$$

holds for all $t \in \mathbf{R}^d$.

Proof of Lemma 5.6.3. It suffices to consider only the case of $\|t\| > 1$ since otherwise the estimates of the lemma follow immediately from its conditions.

By Lemma 5.4.1 all derivatives considered in the lemma are, up to an imaginary exponential factor, polynomials in t whose powers do not exceed m, and the coefficients satisfy the same conditions (5.4.13), (5.4.15), and (5.5.2). Consequently, bounds have to be derived only for the expectations $\mathbf{E}e^{i(t,\phi)}\Psi$, with Ψ and ϕ treated as the generic notation for functions with the above-mentioned properties.

Fix the desired power $B > 0$ in the final bound on the decay rate at infinity. By the characteristic property of Gaussian laws (see Theorem 1.1.3), for each $\delta \in [0, 1]$ there is equality of distributions

$$\mathcal{L}(\eta) = \mathcal{L}\left(\delta\eta' + \sqrt{1-\delta^2}\,\eta''\right),$$

where η', η'' are independent copies of η.

Centered Gaussian RV's satisfy the inequalities of (5.5.1), so one can apply Lemma 5.5.1 and then Lemma 5.5.2 to reduce the problem of evaluating $\mathbf{E}e^{i(t,\phi)}\Psi$ to its analog for $\mathbf{E}e^{i(t,F_1)}\mathcal{P}^{(1)}_{M,q}$, with exponent $i(t, F_1)$ linear in "perturbation" $\delta\zeta = \delta\eta'$ and the pre-exponential factor $\mathcal{P}^{(1)}_{M,q}$ polynomial in it. Below, the choice of the small parameter used in expansions of these lemmas is $\delta = \|t\|^{-2/3}$, so as $\|t\| \nearrow \infty$

$$\delta\|t\| = \|t\|^{1/3} \nearrow \infty, \quad \delta^2\|t\| = \|t\|^{-1/3} \searrow 0. \tag{5.6.15}$$

For this choice of the parameter, Lemma 5.5.1 and Lemma 5.5.2 yield the following bound for the error of approximation:

$$\left|\mathbf{E}e^{i(t,\phi)}\Psi - \mathbf{E}e^{i(t,F_1)}\mathcal{P}^{(1)}_{M,q}\right| \le c\left(\delta^{M+1}\|t\| + \left[\delta^2\|t\|\right]^{q+1}\right) \le c/\left(1+\|t\|\right)^B,$$

provided that M and q in the approximation are chosen large enough. Note that degrees of expansions of Lemma 5.5.1 and Lemma 5.5.2 are determined by the desired decay rate $B > 0$ only.

It remains to deduce an estimate for expectation (5.5.8) $\mathbf{E}e^{i(t,F_1)}\mathcal{P}$, with $\mathcal{P}$ one of the monomials of (5.5.6). This can be obtained through the agency of Lemma 5.5.3.

Indeed, expansion (5.5.9) for Z satisfying condition (5.5.10) exists by the stability of centered Gaussian laws:

$$\forall D \in \mathbf{N} \quad \mathcal{L}(\eta) = \mathcal{L}\left(\tfrac{1}{\sqrt{D}}\left[\eta^{(1)} + \cdots + \eta^{(D)}\right]\right),$$

where $\eta^{(j)}$ are independent copies of the Gaussian RV η.

Hence by Lemma 5.5.3, Hölder's inequality, and Lemma 5.6.1

$$\begin{aligned}\left|\mathbf{E}e^{i(t,F_1)}\mathcal{P}\right| &\le \left(\mathbf{E}\,|\mathcal{P}|\right)^{1/\vartheta}\left(\mathbf{E}\left[\sup_a\left|\mathbf{E}\exp\left\{i\left(t, F_1\left(y; a+\delta\eta\right)\right)\right\}\right|\right]^{\bar\vartheta}\right)^{1/\bar\vartheta} \\ &\le c\left(1+\|t\|\right)^{A(q,M)}\left(1+\tfrac{1}{\sqrt{D}}\delta\|t\|\right)^{-B'} \le c'\left[1+\|t\|\right]^{A(q,M)-B'/3},\end{aligned}$$

where B' can be chosen arbitrarily large, and the choice of $\vartheta, \bar\vartheta > 1$, such that $1/\vartheta + 1/\bar\vartheta = 1$, is determined by the exponent in condition (5.5.2). This proves the lemma. ○

5.6.2 Sums in the Quasi-Gaussian Range.

The properties of Gaussian laws used in subsection 5.6.1 are very rapid decay of CF at infinity and low concentration in small spheres (if dimension number is high). Both of these are, to an extent, shared by distributions of sums of independent RV's in a situation where Gaussian laws can serve as approximations.

Analogs of Lemma 5.6.1 and Lemma 5.6.2 with sums of RV's replacing the Gaussian arguments η and Z can be derived by calculations similar to those of subsection 5.6.1, although only for moderately large values of $\|t\|$.

Quasi-Gaussian bounds on CF's of sums. By analogy with the estimates of subsection 5.6.1, it is convenient to make random the argument of the CF's considered.

Lemma 5.6.4 *Assume that the RV's $\xi_1, \xi_2, \ldots$ are independent, and each ξ_j is either centered Gaussian with covariance operator $\mathcal{B}_j$ or satisfies the conditions*

$$\begin{aligned} &\mathbf{E}\xi_j = 0, \quad \mathbf{E}\left\langle \xi_j, f\right\rangle^2 \ge \left\langle \mathcal{B}_j f, f\right\rangle, \\ \forall f \in X \quad &\mathbf{E}\left|\left\langle \xi_j, f\right\rangle^3\right| \le \Lambda\,|f|\left\langle \mathcal{B}_j f, f\right\rangle, \end{aligned} \tag{5.6.16}$$

where $\mathcal{B}_j$ are nonnegative definite operators in X.

There exists a constant $c > 0$ *such that, if the RV* Y *is independent of* $\xi_j, j = \overline{1,m}$, *then for all values of* $t \in \mathbf{R}$ *and* $y > 0$ *obeying the restriction* $\Lambda\,|t|\,y \le c$

$$\left|\mathbf{E}\exp\left\{it\left\langle Y, \sum_{j=1}^m \xi_j\right\rangle\right\}\right| \le \mathbf{P}\{|Y| \ge y\} + \mathbf{E}\exp\left\{-\tfrac{1}{4}t^2 \sum_{j=1}^m \langle \mathcal{B}_j Y, Y\rangle\right\}.$$

Proof of Lemma 5.6.4. Fix the value of Y such that $|Y| < y$ and retain notation $\mathbf{E}$ for conditional expectations given the value of Y. By independence, these are integrals with respect to the joint distribution of ξ_j for functions of $Y, \xi_1, \ldots$

Put $\varphi_j(h) = \mathbf{E}e^{\langle \xi_j, h\rangle}$. By independence

$$f(Y) = \mathbf{E}\exp\left\{it \sum_{j=1}^m \langle \xi_j, Y\rangle\right\} = \prod_{j=1}^m \varphi_j(tY).$$

Consider the CF of a non-Gaussian RV. It follows from Taylor's expansion for an exponential that

$$\varphi_j(h) = 1 - \tfrac{1}{2}\mathbf{E}\langle \xi_j, h\rangle^2 + r(h), \quad |r(h)| \le \tfrac{1}{6}\mathbf{E}\left|\langle \xi_j, h\rangle^3\right| \le \tfrac{1}{6}\Lambda\,|h|\,\langle \mathcal{B}_j h, h\rangle.$$

Obviously, if $\frac{1}{6}\Lambda\,|Y|\,|t| \le \frac{1}{4}$, then

$$|\varphi_j(tY)| \le 1 - \tfrac{1}{4}\langle \mathcal{B}_j Y, Y\rangle \le \exp\left\{-\tfrac{1}{4}t^2\langle \mathcal{B}_j Y, Y\rangle\right\}.$$

This bound evidently holds if ξ_j is Gaussian.

The inequality of the lemma now follows from the decomposition

$$\mathbf{E}\exp\left\{\,it\sum_{j=1}^m \langle \xi_j, Y\rangle\,\right\} = \mathbf{E}\left[\mathbf{1}_{\{|Y|<y\}} + \mathbf{1}_{\{|Y|\ge y\}}\right]\exp\left\{\,it\sum_{j=1}^m \langle \xi_j, Y\rangle\,\right\}. \ \bigcirc$$

Quasi-Gaussian bounds on concentration. In high dimensions, a RV with Gaussian or similar distribution rarely lands in a small sphere. The effect is the more marked the greater the number of dimensions. Sums of independent RV's enjoy a similar property.

Lemma 5.6.5 *Let* $s \in \mathbf{R}, U > 0$ *be real numbers,* $\mathcal{A} : X \to \mathbf{R}^d$ *a bounded operator, and* $\xi_j \in X$ *some independent RV's that satisfy conditions of Lemma* 5.6.4.

There exist constants $c_i > 0$ *such that if* $\|\mathcal{A}\|U\Lambda\,|s|\,/\sqrt{n} \le c_1$, *then the normalized sum* $\zeta_n = n^{-1/2}(\xi_1 + \cdots + \xi_n)$ *admits the estimate*

$$\begin{aligned}\sup_{a\in X}\mathbf{E}\exp\left\{-\tfrac{1}{2}s^2\left\|\mathcal{A}(a+\zeta_n)\right\|^2\right\} &\le \mathbf{E}\exp\left\{-\tfrac{1}{4}s^2\left\|\mathcal{A}\mathrm{H}\right\|^2\right\} \\ &+ c_2(d)U^{d-2}e^{-U^2/2},\end{aligned}$$

where H *is a centered Gaussian RV with the covariance operator*

$$\mathcal{V}_{\mathrm{H}} = \tfrac{1}{n}(\mathcal{B}_1 + \cdots + \mathcal{B}_n).$$

Proof of Lemma 5.6.5. For an arbitrary $a \in X$, the exponential in the right-hand side of lemma's estimate can be represented in the form

$$\mathbf{E}\exp\left\{-\tfrac{1}{2}s^2\left\|\mathcal{A}(a+\zeta_n)\right\|^2\right\} = J_{<U} + J_{>U}, \tag{5.6.17}$$

where the notation

$$J(C) = (2\pi)^{-d/2} \int_C \mathbf{E} \exp\{i(s\mathcal{A}(a+\zeta_n), u)\}\, e^{-\|u\|^2/2} du, \quad C \subset \mathbf{R}^d,$$

is abbreviated to $J_{>U} = J(\{\|u\| \geq U\})$ and $J_{<U} = J(\{\|u\| \leq U\})$ for integrals over a solid sphere and its complement.

To evaluate the integral over distant values of the argument, note that the absolute value of any CF is bounded by 1 and apply standard estimates for the "tails" of standard Gaussian distribution:

$$|J_{>U}| \leq cU^{d-2}e^{-U^2/2}, \quad c = c(d),\ U \geq 1. \tag{5.6.18}$$

By Lemma 5.6.4, the CF in the integrand of $J_{<U}$, where $\|u\| \leq U$, satisfies the inequality

$$\begin{aligned} |\mathbf{E} \exp\{i(s\mathcal{A}(a+\zeta_n), u)\}| &= |\mathbf{E} \exp\{i(s\mathcal{A}\zeta_n, u)\}| \\ &\leq \exp\{-\tfrac{1}{4}s^2(\mathcal{V}_{\mathrm{H}}\mathcal{A}^* u, \mathcal{A}^* u)\} \end{aligned}$$

provided that $\|\mathcal{A}\|\, U\Lambda\, |s|/\sqrt{n} \leq c_1$. Consequently, completing the integration domain to all space, one arrives at the estimate

$$\begin{aligned} |J_{<U}| &\leq (2\pi)^{-d/2} \int_{\mathbf{R}^d} \mathbf{E} \exp\left\{\tfrac{i}{\sqrt{2}} s(\mathcal{A}\mathrm{H}, u) - \tfrac{1}{2}\|u\|^2\right\} du \qquad (5.6.19) \\ &= \exp\left\{-\tfrac{1}{4}s^2\|\mathcal{A}\mathrm{H}\|^2\right\}. \end{aligned}$$

This completes the proof. ○

Corollary 5.6.1 *Assume that the conditions of Lemma* 5.6.4 *are satisfied and for some* $\alpha > 0$

$$\mathcal{A}\mathcal{V}_{\mathrm{H}}\mathcal{A}^* \geq \alpha^2 I_d,$$

where I_d *is the identity operator in* $\mathbf{R}^d$. *Then for* $r \geq c'\Lambda(\ln n)/\sqrt{n}$ *there is inequality*

$$\sup_{a \in X} \mathbf{P}\{\|\mathcal{A}(a+\zeta_n)\| \leq r\} \leq c''\left[(r/\alpha)^d + \exp\left\{-c'''(\ln n)^2\right\}\right].$$

Proof of Corollary 5.6.1. By a generalization of Chebyshev's inequality and Lemma 5.6.4,

$$\begin{aligned} \mathbf{P}\{\|\mathcal{A}(a+\zeta_n)\| \leq r\} &\leq \sqrt{e}\, \mathbf{E} \exp\left\{-\frac{1}{2r^2}\|\mathcal{A}(a+\zeta_n)\|^2\right\} \\ &\leq \sqrt{e}\, \mathbf{E} \exp\left\{-\tfrac{1}{4}r^{-2}\|\mathcal{A}\mathrm{H}\|^2\right\} + cU^{d-2}e^{-U^2/2} \end{aligned}$$

if $\|\mathcal{A}\|\, \Lambda U/(r\sqrt{n}) \leq c'$. The expectation in the right-hand side can be calculated explicitly:

$$\mathbf{E} \exp\left\{-\tfrac{1}{4}r^{-2}\|\mathcal{A}\mathrm{H}\|^2\right\} = \det\left(I_d + \tfrac{1}{2}r^{-2}\mathcal{A}\mathcal{V}_{\mathrm{H}}\mathcal{A}^*\right)^{-1/2}$$

$$\leq \left(1+\tfrac{1}{2}\alpha^2/r^2\right)^{-d/2} \leq c\left(r/\alpha\right)^d .$$

To complete the proof, it remains to choose $U = \ln n$ and fix appropriate values of the constants. ○

Inequalities for CF's of smooth mappings. In calculations of CF's that follow the argument of the smooth mapping ϕ is a sum of independent RV's $\delta\zeta + Z$, partitioned into the "bulk" summand and perturbation as in (5.5.1) to employ the technique of subsection 5.6.1. The partition is based on two integer-valued sequences $K = K(n) \gg 1$ and $M = M(n) \gg 1$, for which the parameters

$$\delta = \sqrt{K/n}, \quad \gamma = \sqrt{M/n}, \tag{5.6.20}$$

are small for large n, and the summands in $\delta\zeta + Z$ are

$$\zeta = \zeta_K, \quad Z = \gamma\zeta_M + Z^0, \tag{5.6.21}$$

where

$$\zeta_K = \tfrac{1}{\sqrt{K}}\left(\xi_1 + \cdots + \xi_K\right), \quad \zeta_M = \tfrac{1}{\sqrt{M}}\left(\xi_{K+1} + \cdots + \xi_{K+M}\right),$$

and all the RV's Z^0, ξ_j are independent. Restrictions on K and M will be specified later on.

All RV's ξ_j have finite expectations $\mathbf{E}\left|\xi_j\right|^2$ and are centered: $\mathbf{E}\xi_j = 0$. They are either Gaussian with covariance operator $\mathcal{V}$ or satisfy condition (5.6.16) with $\mathcal{B}_j = \mathcal{V}$. These requirements are met if, e.g., the independent RV's ξ_j, not necessarily identically distributed, all have covariance operator $\mathcal{V}$ and are a.s. bounded:

$$\mathbf{P}\left\{|\xi_j| \leq \Lambda\right\} = 1. \tag{5.6.22}$$

It will be shown later on how to relax this restriction.

Lemma 5.6.6 *Assume that mapping* (5.3.1) *satisfies condition* (5.4.13) *and*

$$\left|\nabla\phi^{(j)}(y)\right| \leq \bar{\phi}_1, \quad y \in X, \; j = \overline{1,d}.$$

Then there exist constants $c_i > 0$ *such that for all* $y \in X$

$$\left|\mathbf{E}\exp\left\{i\left(\, t, \langle\delta\zeta, \nabla\rangle\,\phi(y)\,\right)\right\}\right| \leq \exp\left\{\; -c_1\delta^2\left|\sum_{j=1}^d t^{(j)}\mathcal{V}^{1/2}\nabla\phi^{(j)}(y)\right|^2 \;\right\}$$

if $\delta\zeta$ *is the RV of* (5.6.21) *and* $\Lambda\,\|t\| \leq c_2\sqrt{n}$.

Proof of Lemma 5.6.6. Put $Y = n^{-1/2}\nabla_y\left(t, \phi(y)\right)$. Under the restrictions of the lemma, the inequality $|Y| \leq \bar{\phi}_1\|t\|/\sqrt{n}$ holds for all y, so Lemma 5.6.4 can be used to evaluate the value of the CF for the sum of K RV's included in $\delta\zeta$. To complete the proof, it suffices to replace K/n by δ^2 in the resulting inequality. ○

To generalize Lemma 5.6.1 to sums of independent RV's, it is necessary to have some analog of condition (5.6.1). It seems that a condition of this type can hardly be both compact and easy to verify. For this reason only the special mapping (5.6.3) is treated in some detail below.

The following condition will be used when the RV's of (5.6.21) are not necessarily Gaussian: for each $B > 0$ there exists a constant c_B such that

$$\mathbf{P}\left\{|\mathrm{Z}^0| \geq R\right\} \leq c_B/R^B, \quad \mathbf{P}\left\{|\zeta_M| \geq R\right\} \leq c_B/R^B, \quad R \geq 1. \tag{5.6.23}$$

The latter restriction is satisfied if Z^0 is a normalized sum of centered RV's with uniformly bounded moments of arbitrary order. For instance, for bounded RV's satisfying (5.6.22) Theorem 3.3.2 yields the estimate

$$\mathbf{P}\left\{|\zeta_n| \geq R\right\} \leq 2\exp\left\{-\tfrac{1}{2}(R/\Lambda)^2\left(1+\tfrac{c}{\sqrt{n}}(R/\Lambda)\right)^{-1}\right\}, \quad R > 0. \tag{5.6.24}$$

Condition (5.6.6) is replaced by a more complicated one. It is assumed that for an arbitrary number $B' > 0$ there exists an orthogonal projection onto a subspace $\mathcal{P} = \mathcal{P}(B') : X \to X$ with finite number of dimensions $D = D(B') = \dim(\mathcal{P}X) < \infty$, such that for $\frac{1}{r}, r \leq \bar{\phi} = \bar{\phi}(B')$, $u \in [0,\, 1]$, $s \in \mathbf{R}$,

$$\mathbf{E}\exp\left\{-\tfrac{1}{2}s^2\left|\mathcal{P}\mathcal{V}^{1/2}\nabla^{\otimes 2}\Phi\left(\hat{x}+\tfrac{1}{r}u\mathrm{Z}^0\right)\eta\right|^2\right\} \leq c_B/\left[1+s^2\right]^{B'}, \tag{5.6.25}$$

where η, a centered Gaussian RV with covariance operator $\mathcal{V}$, is independent of Z^0.

The left-hand side in condition (5.6.25) can be written more explicitly in terms of the covariance operator of the Gaussian RV $\mathcal{P}\mathcal{V}^{1/2}\nabla^{\otimes 2}\Phi(\hat{x}+y)\eta$,

$$\mathcal{W}_{\mathcal{P},\Phi}(y) = \mathcal{P}\mathcal{U}^2\mathcal{P}, \quad \mathcal{U} = \mathcal{V}^{1/2}\nabla^{\otimes 2}\Phi\left(\hat{x}+y\right)\mathcal{V}^{1/2}.$$

This operator maps the D-dimensional subspace $\mathcal{P}X$ onto itself, and is evidently symmetrical and nonnegative-definite. Denote its eigenvalues by $\widetilde{\sigma}_j^2(y)$, $j = \overline{1, D}$. After the conditional expectation given the value of Z^0 is calculated, the condition acquires the form

$$\begin{aligned}&\mathbf{E}\exp\left\{-\tfrac{1}{2}s^2\left|\mathcal{P}\mathcal{V}^{1/2}\nabla^{\otimes 2}\Phi\left(\hat{x}+\tfrac{1}{r}u\mathrm{Z}^0\right)\eta\right|^2\right\}\\ &= \mathbf{E}\prod_{j=1}^{D} 1/\sqrt{1+s^2\widetilde{\sigma}_j^2\left(\tfrac{1}{r}u\mathrm{Z}^0\right)} \leq c_{B'}/\left[1+s^2\right]^{B'}.\end{aligned}$$

Condition (5.6.25) is easiest to verify when the function Φ is quadratic, because then the second Fréchet derivative is a constant linear operator. Note that this is the case if one considers squared norms of sums of independent RV's, as, e.g., in the study of the ω^2-statistics.

Lemma 5.6.7 *Assume that mapping* (5.3.1) ϕ *and the independent RV's* ξ_j, Z^0 *satisfy conditions* (5.6.6), (5.6.23), *and* (5.6.25), *and, in addition, the gradients of coordinates of* ϕ *are bounded (cf. Lemma* 5.6.6*).*

For an arbitrary $B > 0$ *there exist constants* $c_i = c_i(B)$ *such that if* $t \in \mathbf{R}^d$ *and* γ *of* (5.6.20) *satisfy the restrictions* $\|t\|\Lambda/\sqrt{n} \leq c_1$ *and* $\Lambda \ln n \leq c_2\gamma^{3/2}\sqrt{n}$, *then there is inequality*

$$\left|\mathbf{E}\exp\left\{i\left(\, t, \langle\delta\zeta_K, \nabla\rangle\, \phi(\mathrm{Z})\, \right)\right\}\right| \le c_3\left(\delta^B + \gamma^B + 1/\left[1+\delta\|t\|\right]^B\right).$$

Notation here is that of (5.6.20)–(5.6.21).

In the proof below no attempt is made at keeping track of the numerical values of constants, so same notation may refer to different numbers in different parts of the argument.

Proof of Lemma 5.6.7. Since the gradient of ϕ is bounded and ζ_K, $\mathrm{Z} = \gamma\zeta_M + \mathrm{Z}^0$ are independent, Lemma 5.6.4 yields the inequality

$$\left|\mathbf{E}\exp\left\{i\left(t, \langle\delta\zeta_K, \nabla\rangle\, \phi\left(\gamma\zeta_M + \mathrm{Z}^0\right)\right)\right\}\right| \le \exp\left\{-\tfrac{1}{4}\left(Q\left[\mathrm{Z}\right]t, t\right)\right\},$$

where the quadratic form in the exponent is described in (5.6.8) and Z in (5.6.21).

Fix the desired exponent $B > 0$ in the final estimate of the lemma. Further calculations follow the proof of Lemma 5.6.2. Abbreviations $\mathbf{P}^0\{\cdot\} = \mathbf{P}\left\{\cdot \mid \mathrm{Z}^0\right\}$, $\mathbf{E}^0\{\cdot\} = \mathbf{E}\left\{\cdot \mid \mathrm{Z}^0\right\}$ are used for conditional probabilities and expectations given the value of Z^0.

(a) Select "small" values of s from (5.6.8) by the inequality $|s| \le \delta\|t\|$. In this range, condition (5.6.6) implies the inequality

$$\left|\mathcal{V}^{1/2}\left[\hat{t} + s\psi(y)\right]\right| \ge c'\|t\|\left(1 - c''\,|y|\,|s| \,/\, \|t\|\right) \ge c'\|t\|(1 - c''\delta|y|).$$

If $y = \mathrm{Z}$, the probability that this RV assumes large values, say, $|\mathrm{Z}| \ge 1/(2c''\delta)$, is negligible by condition (5.6.23), so

$$\begin{aligned}\left|\mathbf{E}^0\exp\left\{-\tfrac{1}{4}\delta^2\left(Q\left[\mathrm{Z}\right]t, t\right)\right\}\right| &\le e^{-c(\delta\|t\|)^2} + \mathbf{P}^0\left\{|\mathrm{Z}| \ge \bar{c}/\delta\right\} \qquad (5.6.26)\\ &\le \exp\left\{-c\left(\delta\,\|t\|\right)^2\right\} + \hat{c}\delta^{B'},\end{aligned}$$

where the exponent $B' > 0$ can be chosen arbitrarily at the price of changing the constants in the right-hand side. This proves the inequality of the lemma for "small" values of $|s|$.

(b) If $|s| > \delta\,\|t\|$, write the two-term Taylor expansion (5.6.10) as

$$\hat{t} + s\psi\left(\mathrm{Z}\right) = s\gamma\left[A^+ + \Phi''\zeta_M\right] + s\rho_\psi, \qquad (5.6.27)$$

where both $\Phi'' = \nabla^{\otimes 2}\Phi\left(\hat{x} + \frac{1}{r}\mathrm{Z}^0\right)$ and A^+ are independent from ζ_M, and $|\rho_\psi| \le c\gamma^2\left|\zeta_M\right|^2/r$.

Fix some $\varepsilon > 0$ (its value will be selected later). It follows from (5.6.8) and (5.6.27) that

$$\begin{aligned}&\mathbf{P}^0\left\{\left(Q\left[\mathrm{Z}\right]t, t\right) \le \varepsilon^2\|t\|^2\right\} \qquad (5.6.28)\\ \le\ &\mathbf{P}^0\left\{\left|\mathcal{V}^{1/2}\left(s\gamma\left[A^+ + \Phi''\zeta_M\right]\right)\right| \le \frac{3}{2}\varepsilon\,\|t\|\right\} + \mathbf{P}^0\left\{\left|\mathcal{V}^{1/2}\rho_\psi\right| \ge \frac{1}{2}\frac{\varepsilon\|t\|}{|s|}\right\}\\ \le\ &\mathbf{P}^0\left\{\left|\mathcal{V}^{1/2}\left[A^+ + \Phi''\zeta_M\right]\right| \le \frac{3}{2}\frac{\varepsilon\,\|t\|}{\gamma|s|}\right\} + \mathbf{P}^0\left\{\left|\zeta_M\right|^2 \ge c\frac{\varepsilon\,\|t\|}{\gamma^2\,|s|}\right\}.\end{aligned}$$

The second summand in the right-hand side of (5.6.28) is a probability of large deviation, and an estimate for it follows from (5.6.23): if $\gamma^2 |s| / (\varepsilon \|t\|) < 1$,

$$\mathbf{P}^0 \left\{ |\zeta_M|^2 \geq c\varepsilon \|t\| / (\gamma^2 |s|) \right\} \leq c' \left(\varepsilon \|t\| / (\gamma^2 |s|) \right)^{-B'} < 1 \tag{5.6.29}$$

where the power B' can be chosen arbitrarily large.

The first summand in the right-hand side of (5.6.29) is the probability that ζ_M lands in a small sphere. It can be evaluated using Corollary 5.6.1. Select the finite-dimensional projection $\mathcal{P}$ of condition (5.6.25) corresponding to a sufficiently large power $B' > 0$. Orthogonal projections are contracting, so

$$\begin{aligned} & \mathbf{P}^0 \left\{ \left| \mathcal{V}^{1/2} \left[A^+ + \Phi'' \zeta_M \right] \right| \leq \tfrac{3}{2} \varepsilon \|t\| / (\gamma |s|) \right\} \\ \leq \; & \mathbf{P}^0 \left\{ \left| \mathcal{P} \mathcal{V}^{1/2} \left[A^+ + \Phi'' \zeta_M \right] \right| \leq \tfrac{3}{2} \varepsilon \|t\| / (\gamma |s|) \right\}. \end{aligned} \tag{5.6.30}$$

The projection in the right-hand side is onto the D-dimensional subspace $\mathcal{P}X$, which can be identified with $\mathbf{R}^D$, so by Corollary 5.6.1

$$\begin{aligned} & \mathbf{P}^0 \left\{ \left| \mathcal{V}^{1/2} \left[A^+ + \Phi'' \zeta_M \right] \right| \leq \tfrac{3}{2} \varepsilon \|t\| / (\gamma |s|) \right\} \\ \leq \; & c_1 \left([\varepsilon \|t\| / (\gamma |s|)]^D + \exp \left\{ -c_2 [\ln M]^2 \right\} \right) \end{aligned} \tag{5.6.31}$$

if

$$\frac{\varepsilon \|t\|}{\gamma |s|} \geq c_3 \frac{\Lambda \ln n}{\gamma \sqrt{n}} \geq c_3 \frac{\Lambda \ln M}{\sqrt{M}} \quad \Longleftrightarrow \quad \frac{\varepsilon \|t\|}{|s|} \geq c_3 \frac{\Lambda \ln n}{\sqrt{n}}.$$

The second term in the right-hand side of (5.6.31) is negligible because $\ln M \geq c \ln(1/\gamma)$. It remains to choose the parameter ε. Put $\varepsilon = \gamma^{3/2} |s| / \|t\|$. This selection yields the bound of the lemma. ○

5.6.3 Central Limit Theorem with Error Estimate

The method exposed in subsection 5.3.2 is a version of the method of characteristic functions currently applied to estimation of error terms in limit theorems. This subsection shows how the technical means developed in Section 5.5 and Section 5.6 can be used in this setting. The theorem stated and proved below provides an estimate for the convergence rate in central limit theorem. No attempt is made to reach utmost generality, and the rather special situation treated here has been chosen with a view to later application to large deviations.

Below, X is a separable Hilbert space with norm $|\cdot| = \sqrt{\langle \cdot, \cdot \rangle}$. The problem considered is that of approximating the distribution of the normalized sum $\zeta_n[\xi]$ of independent copies of the RV ξ by a Gaussian distribution $\mathcal{L}(\eta)$. This latter is not precisely the accompanying Gaussian law, but one quite close to it.

The summands are supposed to be centered, $\mathbf{E}\xi = 0$, and satisfy the Cramér condition (3.6.2). The distribution of a summand may depend on a number of parameters, more specifically, on the small parameters τ and $\frac{1}{r}$ (cf. (5.4.3)–(5.4.5)). The constants in (3.6.2) are same for all values of parameters.

Proximity to the Gaussian law is understood as in subsection 5.3.1. More specifically, the problem considered here is evaluation of the quantity (cf. (5.3.2)–

(5.3.3) and (5.3.4)–(5.3.5))

$$\Delta_{\phi,\Psi} \stackrel{def}{=} \mathbf{E}\, g \circ \phi(\zeta_n)\, \Psi(\zeta_n) - \mathbf{E}\, g \circ \phi(\eta)\, \Psi(\eta), \tag{5.6.32}$$

where $\phi : X \to \mathbf{R}^{d+1}$ is a smooth mapping, Ψ a smooth positive function, and g one more function, which is bounded and summable with respect to the Lebesgue measure. The mapping ϕ studied here has special form (5.6.3), where the function Φ satisfies condition (5.6.2). Restrictions on the function $g : \mathbf{R}^{d+1} \to \mathbf{C}$ will be specified later.

In the sequel, both ϕ and Ψ smoothly depend on some parameters (among them the small parameters τ and $\frac{1}{r}$). For instance, the point $\hat{x}$ in definition (5.6.3) is a smooth function of them. All conditions imposed on functions with this kind of dependence are supposed to hold uniformly in values of these small parameters from a neighborhood of zero and all values of the remaining ones. To make reading easier, the mention of parameters is suppressed (both in text and in notation) whenever this is unlikely to cause misunderstanding.

The function Ψ also has derivatives of all orders satisfying condition (5.4.15), and the "larger-than-polynomial" factor in this bound obeys condition (5.4.14) if the random argument is one of the normalized mixed sums (5.4.9). These latter have bounded moments of the norm of all orders (see (5.4.10)).

In order to obtain nontrivial bounds on Fourier transforms (5.3.4), one should impose certain nondegeneracy conditions on the mapping ϕ. These are (5.6.5)–(5.6.6) and a version of (5.6.25).

Namely, consider the operator

$$\mathcal{W}(y; \mathcal{V}_1, \mathcal{V}_2) \stackrel{def}{=} \mathcal{S}(y; \mathcal{V}_1, \mathcal{V}_2)\, \mathcal{S}^*(y; \mathcal{V}_1, \mathcal{V}_2), \tag{5.6.33}$$

where $\mathcal{S}(y; \mathcal{V}_1, \mathcal{V}_2) \stackrel{def}{=} \mathcal{V}_1^{1/2} \nabla^{\otimes 2}\Phi(\hat{x} + y)\, \mathcal{V}_2^{1/2}$ and $\mathcal{V}_i$ are nonnegative operators with finite trace (i.e., covariance operators of some Gaussian RV's). It is assumed below that for Φ of (5.6.3) and the covariance operator $\mathcal{V}$ of (5.4.4) the operator $\mathcal{W}(0; \mathcal{V}, \mathcal{V})$ has an infinite set of strictly positive eigenvalues. In other words, for each number $D \in \mathbf{N}$ there exists an orthogonal projection operator $\mathcal{P} = \mathcal{P}[D]$ such that

$$\mathcal{P}\mathcal{W}(0; \mathcal{V}, \mathcal{V})\mathcal{P} \geq \lambda^2 \mathcal{P}, \quad \dim(\mathcal{P}X) = D, \quad \lambda = \lambda(D) > 0. \tag{5.6.34}$$

Condition (5.6.33) is stable in the following sense: by the regularity assumptions about Φ for each fixed $D \in \mathbf{N}$ there are positive constants $\widetilde{\phi}_i$ such that for the same projection $\mathcal{P}(D)$

$$\mathcal{P}\mathcal{W}(y; \mathcal{V}_1, \mathcal{V}_2)\mathcal{P} \geq \tfrac{1}{2}\lambda^2 \mathcal{P} \tag{5.6.35}$$

if

$$\|\mathcal{V}_1 - \mathcal{V}\| + \|\mathcal{V}_2 - \mathcal{V}\| \leq \widetilde{\phi}_1, \quad |y| \leq \widetilde{\phi}_2.$$

Theorem 5.6.1 *Consider the normalized sums $\zeta_n[\xi]$ of centered i.i.d. RV's with parameter-dependent distribution $\mathcal{L}(\xi)$ which satisfies the Cramér condition* (3.6.2) *and* (5.4.3). *Consider, moreover, a centered Gaussian RV η with* *distribution* (5.4.4)–(5.4.5).

Assume that mapping ϕ *of* (5.6.3) *and function* Ψ *obey restrictions* (5.6.2) and (5.4.13)–(5.4.15), *as well as the nondegeneracy conditions* (5.6.5)–(5.6.6) *and* (5.6.33)–(5.6.34) *(with* Z^0 *equal to normalized mixed sums* (5.4.9)*)*.

For each $B > 0$ *there exist positive constants* c_i *such that for* $\frac{1}{r}, \tau \leq c_0$ *and* $\varepsilon = \frac{1}{\sqrt{n}}$

$$|\mathbf{E}g \circ \phi(\zeta_n) \Psi(\zeta_n) - \mathbf{E}g \circ \phi(\eta) \Psi(\eta)| \leq \frac{c_1}{r^B} \sup |g(x)|$$
$$+ \quad c_2(\tau + \varepsilon)(|g|_1 + |\mathrm{osc}_\varepsilon g|_1) + c_3 \sup_{y \in \mathbf{R}^{d+1}} \mathbf{E} \,\mathrm{osc}_{2\varepsilon}\, g(y + \phi(\eta)) \Psi(\eta).$$

Proof of Theorem 5.6.1. (a) It follows from the Cramér condition that the rv's $|\zeta_n|$ have bounded moments of arbitrary order, so for an arbitrary exponent $A > 0$ there is a constant $\bar{c}(A)$ such that

$$\mathbf{E}\,|g \circ \phi(\zeta_n)| \Psi(\zeta_n) \mathbf{1}_{\{|\zeta_n| \geq 1/\sqrt{r}\}} + \mathbf{E}\,|g \circ \phi(\eta)| \Psi(\eta) \mathbf{1}_{\{|\eta| \geq 1/\sqrt{r}\}} \leq \frac{c'(A)}{r^A} \sup |g|.$$

(The calculation uses Chebyshev's inequality and condition (5.4.14)–(5.4.15).) Thus, as soon as the dimension number D is selected in condition (5.6.34), the function Φ can be replaced in calculations by one which satisfies (5.6.34) for all $y \in X$ as soon as

$$\|\mathcal{V}_1 - \mathcal{V}\| + \|\mathcal{V}_2 - \mathcal{V}\| \leq \varepsilon(D).$$

The purpose is served, e.g., by the function

$$\widetilde{\Phi}(x) = U(x)\Phi(x) + (1 - U(x)) \left[\Phi(\hat{x}) + \langle x - \hat{x}, \nabla\rangle \Phi(\hat{x}) + \tfrac{1}{2} \langle x - \hat{x}, \nabla\rangle^2 \Phi(\hat{x})\right],$$

where $U(x) = \bar{U}(x - \hat{x})$ is a cutoff function for the spherical ρ-neighborhood of zero with $\left|\nabla^{\otimes k} U\right| \leq u_k/\rho^k$ for all $k \in \mathbf{N}$. To get the necessary properties, ρ should be taken small enough, $\rho \leq \rho_0(D)$.

Below Φ is supposed to satisfy (5.6.35) in all space. If the above-described alteration is necessary to achieve this, notation is not changed to avoid encumbering calculations. Inequality (5.6.35) will be used with a fixed value of the dimension number D whose choice will be specified later.

(b) The limit measure (5.3.3) has bounded density. This follows from the estimates of Lemma 5.6.3 for its Fourier transform. Consequently, one can use the smoothing inequality of Lemma 5.3.1 and Lemma 5.3.2. The desired degree of accuracy dictates the choice of "frequency range". Thus the problem is reduced to evaluation of the quantity

$$\Delta_n(t) = \mathbf{E}\exp\{i(t, \phi(\zeta_n))\} \Psi(\zeta_n) - \mathbf{E}\exp\{i(t, \phi(\eta))\} \Psi(\eta) \qquad (5.6.36)$$

for $t \in \mathbf{R}^{d+1}$ such that $\|t\| \leq c\sqrt{n}$.

(c) Put $H(x) = H(x;t) = \exp\{i(\phi(x), t)\} \Psi(x)$. Under restrictions of the theorem this is a smooth summable function, as also each of the functions $\overline{H}_{k,n}(x) = \mathbf{E}H(W_{k,n} + x)$, where

$$W_{k,n} = \varepsilon \textstyle\sum_{1 \leq j < k} \xi_j + \varepsilon \sum_{k < j \leq n} \eta_j, \quad \varepsilon = \frac{1}{\sqrt{n}},$$

(cf. (5.4.9)). The expectation $\overline{H}$ has continuous Fréchet derivatives which grow no faster than polynomials. In this notation the quantity (5.6.36) can be written as

$$\begin{aligned}\Delta_n(t) &= \mathbf{E}H(\zeta_n) - \mathbf{E}H(\eta) \\ &= \sum_{k=1}^n \left[\mathbf{E}H(W_{k,n} + \varepsilon\xi_k) - \mathbf{E}H(W_{k,n} + \varepsilon\eta_k)\right] \qquad (5.6.37) \\ &= \sum_{k=1}^n \left[\mathbf{E}\overline{H}_{k,n}(\varepsilon\xi_k) - \mathbf{E}\overline{H}_{k,n}(\varepsilon\eta_k)\right].\end{aligned}$$

In turn, each summand can be represented using Taylor's formula. After deleting the equal terms in the difference, the resulting formula is

$$\begin{aligned}&\mathbf{E}\overline{H}_{k,n}(\varepsilon\xi_k) - \mathbf{E}\overline{H}(\varepsilon\eta_k) \\ = \;& \tfrac{1}{2}\varepsilon^2\left[\mathbf{E}\langle\xi_k,\nabla\rangle^2\overline{H}_{k,n}(0) - \mathbf{E}\langle\eta_k,\nabla\rangle^2\overline{H}_{k,n}(0)\right] \qquad (5.6.38) \\ + \;& \varepsilon^3\int_0^1 \tfrac{1}{2}(1-u)^2\left[\mathbf{E}\langle\xi_k,\nabla\rangle^3\overline{H}(u\varepsilon\xi_k) - \mathbf{E}\langle\eta_k,\nabla\rangle^3\overline{H}(u\varepsilon\eta_k)\right]du.\end{aligned}$$

The right-hand side in this equality is clearly of order $\mathcal{O}([\tau+\varepsilon]/n)$, which would suffice for the estimate of the theorem if there were no integration in t on an interval of length $\sqrt{n}$. To conserve the order of the estimate in n and τ and add favorable dependence on $\|t\|$, it is necessary to consider with more attention the behavior of $\overline{H}_{k,n}(x;t)$ and its derivatives in x for large values of $\|t\|$.

To this end one can split the sum $W_{k,n}$ according to (5.6.21) with $\delta = \|t\|^{-2/3}$ (the second parameter γ will be chosen later on).

Application of Lemma 5.5.1 and Lemma 5.5.2 permits one to replace the derivatives of $\exp\{i(t,\phi)\}\Psi$ in (5.6.38) by the products of polynomials in $\zeta_K)$ with the exponential $\exp\{i(t,\delta\langle\zeta_K,\nabla\rangle\phi(\mathrm{Z}))\}$ (whose exponent is linear in ζ_K). To make the error small, one should take $v=1$ and large m,q. The resulting error does not exceed $c(\tau+\varepsilon)[1+\|t\|]^{-B'}$, where the exponent $B'>0$ can be made arbitrarily large at the price of including more terms in approximation. This means raising the order of polynomial factor in $\|t\|$ and other arguments; however, it depends only on the desired value of B'.

Next, Lemma 5.5.3 reduces the problem to evaluation of the "pure CF's". These are the expectations of $\exp\{i(t,\delta\langle\zeta_{K'},\nabla\rangle\phi(\mathrm{Z}))\}$ (with ζ_K replaced by a similar sum with the number of summands K' having same order).

At this stage one can have recourse to Lemma 5.6.7 and thus use the nondegeneracy condition. The parameter γ can be chosen of order $\gamma\sim 1/\|t\|^\kappa$ with a small exponent $\kappa>0$ (e.g., $\kappa=0.3$), so that $\gamma^{3/2}\sqrt{n}\geq n^{(1-3\kappa)/2}\gg 1$ for large n.

Note that the dimension number D in condition (5.6.34) can be chosen arbitrarily large. This permits one to suppress the polynomial growth of the factors preceding the imaginary exponential in approximations described above. The final estimate for the difference in (5.6.38) is $c(\tau+\varepsilon)[1+\|t\|]^{-B}$, where the exponent $B>0$ can be made arbitrarily large, at the usual price of spoiling the constant. This function is summable, so the theorem is proved. ○

5.7 Bounds on CF of Squared Norm

5.7.1 A Symmetrization Inequality for CF's

In calculations of Section 5.6, the smooth mappings were almost linear, quadratic and higher-order terms playing mostly the part of small perturbations. When the nonlinear part of the mapping was really important, the effect used to evaluate CF's of "nearly quadratic" mappings was the low concentration of Gaussian distributions in small spheres, very marked in high dimensions. Exposition of Section 5.6 emphasized this fact.

The same effect can be exploited using an alternative technique based on an elegant symmetrization inequality (due to F. Götze). It proves adequate when Gaussian RV's or sums undergo a transformation that can be approximated by a polynomial one with terms of highest order $\upsilon \geq 2$ enjoying special nondegeneracy conditions.

Only quadratic functions are treated in detail below. The calculations that follow could be used, e.g., in the setting of Lemma 5.6.2 and Lemma 5.6.7.

In this section X is a separable real Hilbert space with inner product and norm $|\cdot| = \sqrt{\langle\cdot,\cdot\rangle}$. As usual, symmetrizations of RV's $\xi, \eta, \dots$ are $\widetilde{\xi} = \xi' - \xi'', \widetilde{\eta} = \eta' - \eta'', \dots$ where $(\xi', \xi''), (\eta', \eta''), \dots$ are pairs of independent copies of the corresponding RV's. When the "original" RV's are independent, the pairs of symmetrization components are also supposed independent (even if this is not stipulated explicitly).

Notation $\mathbf{E}_\xi f(\xi, \dots) = \int_X f(x, \dots)\mathbf{P}\{\xi \in dx\}$ is used for the conditional expectation corresponding to fixed values of all the unspecified random variables.

Lemma 5.7.1 *If Ξ, Υ, and Θ are independent RV's with values from the Hilbert space X, then*

$$\forall \tau \in \mathbf{R} \quad \left|\mathbf{E}\exp\{i\tau|\Xi + \Upsilon + \Theta|^2\}\right| \leq \left[\mathbf{E}\exp\{2i\tau\langle\widetilde{\Xi}, \widetilde{\Upsilon}\rangle\}\right]^{1/4}.$$

Proof of Lemma 5.7.1. By the Hölder inequality

$$\begin{aligned}
\phi &\stackrel{def}{=} |\mathbf{E}\exp\{i\tau|\Xi + \Upsilon + \Theta|^2\}| = \left|\mathbf{E}\mathbf{E}_\Xi \exp\{i\tau|\Xi + \Upsilon + \Theta\,|^2\}\right| \\
&\leq \left[\mathbf{E}\left|\mathbf{E}_\Xi \exp\{i\tau|\Xi + \Upsilon + \Theta|^2\}\right|^2\right]^{1/2} \\
&= \left[\mathbf{E}\exp\{i\tau\left(|\Xi' + \Upsilon + \Theta|^2 - |\Xi'' + \Upsilon + \Theta|^2\right)\}\right]^{1/2}.
\end{aligned}$$

Apply the elementary identity $|a|^2 - |b|^2 = \langle a - b, a + b\rangle$ to transform the exponent. The resulting expression is linear in $\widetilde{\Xi}$. The calculation can be repeated to exclude the RV Θ and to arrive at the estimate of the lemma:

$$\begin{aligned}
\phi &\leq \left[\mathbf{E}\exp\{i\tau\langle\widetilde{\Xi}, \Xi' + \Xi'' + 2\Upsilon + 2\Theta\rangle\}\right]^{1/2} \\
&\leq \left[\mathbf{E}\mathbf{E}_{\Upsilon',\Upsilon''}\exp\{2i\tau\langle\widetilde{\Xi}, \Upsilon' - \Upsilon''\rangle\}\right]^{1/4} = \left[\mathbf{E}\exp\{2i\tau\langle\widetilde{\Xi}, \widetilde{\Upsilon}\rangle\}\right]^{1/4}. \bigcirc
\end{aligned}$$

The above lemma does not use any special properties of the RV's Ξ, Υ, or Θ. Note that the last of them plays the part of ballast — one can, e.g., conveniently collect in it the rests of truncation procedures and like. This circumstance can be used, e.g., to stretch the range of argument's values for which the nontrivial bounds on CF can be obtained by reduction to the Gaussian case via Lemma 5.6.4 (see (5.7.3)–(5.7.4) below).

Lemma 5.7.1 is clearly applicable to Gaussian RV's, and the estimate that it yields is only slightly worse than those following from known explicit formulae for CF's. Suppose that Ξ and Υ are centered Gaussian with same covariance operator $\mathcal{V}$. In this case $\mathcal{L}(\widetilde{\Xi}) = \mathcal{L}(\widetilde{\Upsilon}) = \mathcal{L}(\sqrt{2}\Xi)$, so the estimate of Lemma 5.7.1 is

$$\begin{aligned} \left|\mathbf{E}\exp\{i\tau|\Xi+\Upsilon+\Theta|^2\}\right| &\le \left[\mathbf{E}\exp\{-8\tau^2|\mathcal{V}^{1/2}\Xi|^2\}\right]^{1/4} \\ &\le \textstyle\prod_{j=1}^{\infty}\left[1+16\tau^2\sigma_j^4\right]^{-1/8}, \end{aligned}$$

where σ_j^2 are eigenvalues of $\mathcal{V}$ and σ_j^4 those of the covariance operator of the RV $\mathcal{V}^{1/2}\Xi$. In absence of Θ the exact formula of Theorem 1.1.7 is

$$\begin{aligned} \left|\mathbf{E}\exp\{i\tau|\Xi+\Upsilon|^2\}\right| &= \left|\textstyle\prod_{j=1}^{\infty} 1/\sqrt{1-i\tau 2\sqrt{2}\sigma_j^2}\right| \\ &= \textstyle\prod_{j=1}^{\infty}\left[1+8\tau^2\sigma_j^4\right]^{-1/4}. \end{aligned}$$

5.7.2 Bounds on CF of Squared Norm for Sums

In estimates for CF's of sums Lemma 5.7.1 can be used in combination with Lemma 5.6.4 and concentration estimates of Lemma 5.6.5 or Corollary 5.6.1. By contrast with the Gaussian case, the resulting bounds are applicable only in the range of moderately large values of CF's argument. Only the case of RV's with equal covariance operators is treated here to avoid unessential complications.

Estimate for CF: uniformly bounded third moments. Assume that the RV's ξ_j satisfy the conditions

$$\mathbf{E}\xi_j = 0,\ \ \forall f\in X\ \ \mathbf{E}\langle\xi_j,f\rangle^2 \ge \langle\mathcal{V}f,f\rangle,\ \ \forall l\in\mathbf{N}\ \ \mathbf{E}|\xi_j|^l \le \mu_l, \tag{5.7.1}$$

and additional restrictions of (5.6.16) concerning relations between covariance operators and third-order moments:

$$\forall f\in X\ \ \mathbf{E}\left|\langle\xi_j,f\rangle^3\right| \le \Lambda\,|f|\,\langle\mathcal{V}f,f\rangle. \tag{5.7.2}$$

The operator $\mathcal{V}$ is nonnegative definite with finite trace and an infinite set of strictly positive eigenvalues: $\dim\{x:\ \langle\mathcal{V}x,x\rangle>0\}=\infty$.

Condition (5.7.2) is difficult to verify and may prove unrealistic in the infinite-dimensional setting, and an additional argument permits to replace it with the more standard (5.7.5). Yet it is convenient to begin with an intermediate result whose derivation is straightforward.

The object considered below are the normalized sum of n independent RV's $\xi_\bullet$ and its smaller parts: for $\varepsilon = \frac{1}{\sqrt{n}}$

$$\zeta_n = \varepsilon \sum_{j=1}^{n} \xi_j, \quad \zeta_{l,m} = \zeta_{l,m;n}[\xi] \stackrel{def}{=} \varepsilon \sum_{j=l+1}^{m} \xi_{j0}, \quad l < m.$$

The next proposition is a combination of Lemma 5.7.1 with bounds on large deviation probabilities and concentrations, which can serve as a source of more special estimates in the setting of central limit theorem.

Lemma 5.7.2 *Select arbitrary positive numbers A and B. There exist positive constants c_i such that, for any natural numbers k, m satisfying the restrictions $k + m \le n$, and positive y, δ, the CF of $|\zeta_n[\xi]|^2$ admits the estimate*

$$\begin{aligned}\left|\mathbf{E}\exp\{it|\zeta_n[\xi]|^2\}\right| &\le c_1\left[1/y^A + \delta^B + \exp\left\{-c_2(\ln k)^2\right\}\right] \\ &+ \bar{c}_1 \exp\left\{-c_3\delta^2(km/n^2)t^2\right\}\end{aligned}$$

if $y\,|t|\,\sqrt{k/n} \le c_4\sqrt{n}$ and $\delta \ge (c_5 \ln k)/\sqrt{k}$, where the values of the constants depend only on covariances of ξ_j, constants in restrictions (5.7.1) *on higher-order moments of $|\xi_j|$, and Λ.*

Proof of Lemma 5.7.2. Decompose the normalized sum into independent summands:

$$\zeta_n = \Xi + \Upsilon + \Theta, \quad \Xi = \zeta_{1,m;n}, \quad \Upsilon = \zeta_{m+1,m+k;n}, \quad \Theta = \zeta_{m+k+1,n;n}.$$

These RV's obey conditions of Lemma 5.7.1, whose inequality will serve as the base of the calculations to follow.

Known bounds on moments of the norm for sums of independent RV's and Chebyshev's inequality show that condition (5.7.1) provides bounds similar to (5.6.23) for large deviation probabilities of all summands in decomposition above and their symmetrizations. Thus, for an arbitrary $A > 0$

$$\mathbf{E}\left|\widetilde{\Upsilon}\right|^A \le C(A)(k/n)^{A/2}, \quad \mathbf{P}\left\{\left|\widetilde{\Upsilon}\right| \ge y\sqrt{k/n}\right\} \le C'(A)/y^A.$$

If Lemma 5.6.4 is applied to replace the m summands in $\widetilde{\Xi}$ by centered Gaussian RV's with same covariances, Lemma 5.7.1 and this estimate show that for $\Lambda y|t|\sqrt{k/n} \le c_2\sqrt{n}$

$$\begin{aligned}\left|\mathbf{E}\exp\{i\tfrac{t}{n}|\zeta_n|^2\}\right| &\le \left[\mathbf{E}\exp\{i2t(\widetilde{\Xi}, \widetilde{\Upsilon})\}\right]^{1/4} \\ &\le \left[\mathbf{E}\exp\{-c_1\tfrac{m}{n}t^2|\mathcal{V}^{1/2}\widetilde{\Upsilon}|^2\}\right]^{1/4} + C'(A)/y^A.\end{aligned}$$

The RV $(n/k)^{1/2}\,\widetilde{\Upsilon} = k^{-1/2}\sum_{j=1}^{k}\widetilde{\xi}_{m+j}$ is itself a sum of k independent RV's normalized in a standard way. A repetition of the argument from the proof of Lemma 5.6.7 reduces the problem of evaluating probability of landing in a small sphere to its finite-dimensional analog. Thus, Corollary 5.6.1 is applicable and yields the estimate

$$\mathbf{P}\left\{|\mathcal{V}^{1/2}\widetilde{\Upsilon}| \le \delta\sqrt{k/n}\right\} \le c\left[e^{-c''[\ln k]^2} + \delta^B\right]$$

if $\delta \geq (c\Lambda \ln k)/\sqrt{k}$, where the exponent $B > 0$ can be chosen arbitrarily large. This proves the lemma. ○

Lemma 5.7.2 yields for the CF $\mathbf{E}\exp\left\{it|a+\zeta_n|^2\right\}$ some estimates that are uniform in $a \in X$ and nontrivial for the values of the argument satisfying the restriction

$$|t| \leq n^{1-\gamma} \tag{5.7.3}$$

with arbitrary $\gamma \in (0,\, 1)$.

Consider first "small" values of t from the interval $|t| \leq \sqrt{n^{1-\gamma/2}}$. Put

$$k = [n/2], \quad m = n-k, \quad |y| = [1+|t|]^{\alpha}, \quad \delta = 1/[1+|t|]^{\beta}$$

with some positive exponents α, β and apply Lemma 5.7.2. If $\beta < 1$ and $\alpha < \gamma/(2-\gamma)$, the restrictions of Corollary 5.7.2 are satisfied for large n, and the resulting bound is

$$\left|\mathbf{E}\exp\left\{it\,|a+\zeta_n|^2\right\}\right| \leq c/[1+|t|]^{A'}, \quad n \geq n_0. \tag{5.7.4}$$

The remaining values of t satisfy the inequality $n^{\frac{1}{2}(1-\gamma/2)} \leq |t| \leq n^{1-\gamma}$. In this case the choice of partition is different. Take some positive numbers $\overline{\alpha}$ and $\overline{\kappa}$ which obey the conditions (whose compatibility is easily verified)

$$\overline{\kappa} < \overline{\alpha}/2, \quad \overline{\kappa}+\overline{\alpha}/2 < \gamma/(1-\gamma), \quad \overline{\alpha}/2 > \gamma/(2-\gamma)+\overline{\kappa}.$$

Put

$$k = \left[|t|^{\overline{\alpha}}\right], \quad y = |t|^{\overline{\kappa}}, \quad \delta = |t|^{-\overline{\kappa}}, \quad m = n-k$$

and apply Lemma 5.7.2. The resulting estimate is once again (5.7.4). ○

Reduction to bounded RV's. The assumption that the RV's used to construct sums are centered plays no part in the calculations of this section. It is included for the sake of convenience only. On the contrary, some form of condition (5.7.2) or (5.6.16) for the symmetrized RV's is essential. Condition (5.6.16) is valid, e.g., if the RV's considered are a.s. bounded as in (5.6.22). Restrictions on the relation between covariances and third-order moments seem quite severe in high or infinite dimensions if the RV's considered are unbounded. Yet a much more general situation can be reduced to one where condition (5.6.16) is satisfied.

Assume that the distribution of i.i.d. RV's $\xi, \xi_1, \ldots$ satisfies the condition

$$L = \mathbf{E}\,|\xi|^3 < \infty, \quad \mathbf{E}\xi = 0, \quad \mathbf{E}\,\langle\xi,t\rangle^2 \geq \langle\mathcal{V}t,t\rangle \geq 0, \tag{5.7.5}$$

with the spectrum of the finite-trace operator $\mathcal{V}$ subject to the restriction

$$\dim\{x:\ \mathcal{V}x \neq 0\} = +\infty.$$

Bounds on the CF $\mathbf{E}\exp\left\{it\,|\zeta_n\,[\xi]|^2\right\}$ can be obtained under condition (5.7.5) by means of the following reduction procedure.

Fix a positive number r and put

$$G(B) = \mathbf{P}\{ \xi \in B \mid |\xi| \le r \}, \quad B \in \mathcal{B}(X). \tag{5.7.6}$$

The original distribution can be represented in the form

$$\mathcal{L}(\xi) = pG + (1-p)H, \tag{5.7.7}$$

where $p = \mathbf{P}\{|\xi| \le r\}$ and H is some other distribution in X. For large r the event in the condition has probability close to one, so $p \ge \frac{1}{2}$. Decomposition (5.7.7) can be used to construct a special RV with the distribution of ξ.

Let $\xi^{(0)}, \xi^{(1)} \in X$ and the rv $\nu \in \{0,1\}$ be independent with the distributions

$$\mathcal{L}\left(\xi^{(j)}\right) = \begin{cases} G, & j=1, \\ H, & j=0, \end{cases} \qquad \mathbf{P}\{\nu = j\} = \begin{cases} p, & j=1, \\ 1-p, & j=0. \end{cases}$$

The following RV has the distribution of ξ:

$$\xi = \xi^{(\nu)} = \nu\xi^{(1)} + (1-\nu)\xi^{(0)}. \tag{5.7.8}$$

In what follows, $\xi^{(0)}, \xi^{(0)}_1, \ldots \in X$, $\xi^{(1)}, \xi^{(1)}_1, \ldots \in X$, and $\nu, \nu_1, \ldots \in \mathbf{R}$ are independent sequences of i.i.d. random variables with properties stipulated in (5.7.8).

Denote by σ_j^2 the jth largest eigenvalue of the operator $\mathcal{V}$ from (5.7.5) (counting multiple eigenvalues as series of equal ones).

Lemma 5.7.3 *The positive parameter r in (5.7.6) can be chosen in such a way that representation (5.7.8) can be realized with the RV $\xi^{(1)}$ obeying the conditions*

$$\mathbf{E}|\langle \xi^{(1)} - \mathbf{E}\xi^{(1)}, t\rangle|^3 \le \Lambda\,|t|\,\mathbf{E}\langle \xi^{(1)} - \mathbf{E}\xi^{(1)}, t\rangle^2,$$
$$\Lambda \le cL/\sigma_k^2, \qquad \sigma_k^2[\xi^{(1)}] \ge \tfrac{1}{2}\sigma_k^2,$$

where $\sigma_k^2[\xi^{(1)}]$ is the k-th largest eigenvalue of the covariance operator of $\xi^{(1)}$.

Proof of Lemma 5.7.3. To meet the requirements of the lemma, one can put $r = 8L/\sigma_k^2$.

Denote $\xi' = \xi^{(1)} - \mathbf{E}\xi^{(1)}$. By the construction, this RV is bounded, so it fulfills the first condition of the lemma.

By the definition of $\xi^{(1)}$, its covariance operator satisfies the inequality

$$\mathbf{E}\langle \xi', t\rangle^2 \ge \left(1 - \mathbf{P}\{|\xi| \ge r\}\right)^{-1}\left(\mathbf{E}\langle \xi, t\rangle^2 - 2r^{-1}|t|^2\,\mathbf{E}|\xi|^3\right).$$

For the choice of the parameter used here, it follows from the Chebyshev inequality that

$$\mathbf{P}\{|\xi| \le r\} \ge 1 - (\tfrac{1}{8})^3\sigma_k^6/L^2 \ge 1 - (\tfrac{1}{8})^3 \ge \tfrac{1}{2},$$

as by the Hölder inequality $\sigma_k^6 = \left(\mathbf{E}|\xi|^2\right)^3 \le \left(\mathbf{E}|\xi|^3\right)^2 = L^3$. Consequently, for all t

$$\mathbf{E}\langle \xi', t\rangle^2 \ge \left(1 - 8^{-3}\right)^{-1}\left(\mathbf{E}\langle \xi, t\rangle^2 - \tfrac{1}{4}\,|t|^2\sigma_k^2\right).$$

Restricted to the vectors t from the k-dimensional invariant subspace of $\mathcal{V}$ that is spanned by the eigenvectors associated to its k largest eigenvalues, the above estimate implies the relation $\mathbf{E}\langle \mathcal{P}\xi', t\rangle^2 \geq \frac{1}{2}\sigma_k^2 |t|^2$, so the bound of the kth eigenvalue of the covariance operator of $\xi^{(1)}$ follows from the max-min property of eigenvalues. ○

An estimate for the CF of the original sum $\zeta_n[\xi]$ can now be derived from Lemma 5.7.2 and the following proposition.

Lemma 5.7.4 *Let $\xi, \xi_1, \ldots$ be i.i.d. RV's admitting representation* (5.7.7)–(5.7.8). *The following inequality holds for all $t \in \mathbf{R}$:*

$$\begin{aligned}\left|\mathbf{E}\exp\left\{it\,|\zeta_n[\xi]|^2\right\}\right| &\leq \left[\mathbf{E}\exp\left\{2it\langle\zeta_{1,k;n}[\widetilde{\xi}^{(1)}], \zeta_{k+1,k+m;n}[\widetilde{\xi}^{(1)}]\rangle\right\}\right]^{1/4} \\ &+ \exp\{-c_1 n\},\end{aligned}$$

where $k+m \leq [n/4]$ are arbitrary natural numbers and $\widetilde{\xi}^{(1)}$ is the symmetrization of the RV $\xi^{(1)}$ in representation (5.7.8).

Proof of Lemma 5.7.4. Consider a reconstruction of the i.i.d. sequence ξ_i in form (5.7.8) from independent i.i.d. sequences $\xi_j^{(1)}$, $\xi_j^{(0)}$, and ν_j.

Conditional distribution of $S_{1,n}[\xi]$ given the values of all ν_j is that of the sum $S_{1,\nu}[\xi^{(1)}] + S_{1,n-\nu}[\xi^{(0)}]$, where $\bar{\nu} = S_{1,n}[\nu] = \nu_1 + \cdots + \nu_n$.

The rv $\bar{\nu}$ is distributed as the number of heads after n tosses of a possibly "dishonest" coin that shows head with probability $p \geq \frac{1}{2}$. Consequently, it follows from known large deviation estimates for the binomial distribution (e.g., Bernstein's bound of Proposition 3.3.1) that $\mathbf{P}\left\{\bar{\nu} \leq \frac{1}{4}n\right\} \leq \exp\{-c'n\}$. Hence

$$\begin{aligned}\left|\mathbf{E}\exp\{i(t/n)|S_{1,n}[\xi]|\}\right| &\leq \exp\{-cn\} + \left|\mathbf{E}H(\bar{\nu})\mathbf{1}_{\{\bar{\nu}\geq n/4\}}\right|, \\ H(l) &= \mathbf{E}\exp\{i(t/n)|S_{1,[n/4]}[\xi^{(1)}] + Y_l|^2\}, \quad l \geq [n/4],\end{aligned}$$

where the RV Y_l is independent of $S_{1,[n/4]}[\xi^{(1)}]$, and its distribution is that of $S_{1,l-[n/4]}[\xi^{(1)}] + S_{1,n-l}[\xi^{(0)}]$.

To derive the estimate of the lemma, it remains to apply the symmetrization of Lemma 5.7.1 to eliminate Y_l. ○

5.7.3 Error Estimates for Higher-Order Expansions

This section contains theorems about higher-order Edgeworth expansions stated without proof. The theorems fit into the scheme developed in Section 5.3 — they refer to distributions of some regular functions of the normalized sum of vector "observations". They can be established using methods developed in the preceding sections. The tool used to evaluate precision of higher-order approximations to the distribution of the sum is usually the CF of the "smooth statistics" considered.

It is general knowledge that to reconstruct a function from its Fourier transform with very high degree of accuracy one has to know this latter in a very wide range of values of its argument ("frequencies" or "wave-vectors" in physical terminology). When the object reconstructed is, e.g., the difference of DF's one of which has smooth density, Lemma 5.3.1 and Lemma 5.3.2 show that Fourier transforms should be evaluated for "frequencies" up to $1/\varepsilon$ to get approximation with error $\mathcal{O}(\varepsilon)$.

Let $\zeta_n[\xi]$ be the normalized sum of i.i.d. RV's. In the setting typical of theorems on Edgeworth expansions, it is relatively easy to get for the CF $\mathbf{E}\exp\{itf(\zeta_n)\}$ the estimates that yield the bound $\mathcal{O}(1/\sqrt{n})$. The larger values of argument, $|t| \gg \sqrt{n}$, are much less tractable because Lemma 5.6.4 fails.

One has either to introduce some additional restrictions on the summand's distribution, or find some special problems in whose study one can get the necessary estimates by some new means. Neither the "true" restriction nor the natural class of special problems that can be treated in a "truly ∞D" style seem to be determined with any degree of finality, while all known results are technically quite demanding. For this reason the kind of commentary presented here seems not entirely out of place.

Modification of Cramér's condition. Below X is a separable real Hilbert space, and $\xi, \xi_1, \ldots \in X$ a sequence of i.i.d. RV's such that

$$\mathbf{E}\,|\xi|^2 < \infty, \quad \mathbf{E}\xi = 0, \quad \mathbf{cov}\big(\langle \xi, f\rangle, \langle \xi, g\rangle\big) = \langle \mathcal{V} f, g\rangle . \tag{5.7.9}$$

A standard condition used to evaluate error of higher-order Edgeworth expansions in one or several dimensions is as follows (in the formula $\|\cdot\| = \sqrt{(\cdot,\cdot)}$ are the norm and inner product in $\mathbf{R}^d$):

$$\theta(L) \stackrel{def}{=} \sup_{\|t\| \geq L} |\mathbf{E}\exp\{i(\xi, t)\}| < 1 \text{ for } L \geq L_0.$$

In finite dimensions this restriction is fairly natural. It is fulfilled, e.g., when the distribution of $\xi \in \mathbf{R}^d$ has a density.

A condition currently used in infinite-dimensional setting is a straightforward adaptation of the classical Cramér's condition. It is assumed that for some bounded operator $\mathcal{K} : X \to X$

$$\vartheta_{\mathcal{K}}(L) \stackrel{def}{=} \sup_{|\mathcal{K}y| \geq L} |\mathbf{E}\exp\{i\langle \xi, y\rangle\}| < 1 \text{ if } L \geq L_0. \tag{5.7.10}$$

By contrast with its finite-dimensional version, this condition is most difficult to verify.

Assume that the i.i.d. RV's $\xi, \xi_1, \ldots$ satisfy condition (5.7.9). The operator $\mathcal{V}$ is the covariance operator both of ξ and of the centered Gaussian RV η with the accompanying Gaussian law.

Theorem 5.7.1 *If condition* $\mathbf{E}\,|\xi|^{m+3} < \infty$ *is satisfied for some* $m > 2$ *as well as condition* (5.7.10) *with some operator* $\mathcal{K}$ *for which the subspace* $\mathcal{K}\mathcal{V}\mathcal{K}X$ *is infinite-dimensional, then the DF*

$$F_n(r, a) \stackrel{def}{=} \mathbf{P}\left\{|Z_n - a|^2 \leq r^2\right\}$$

and its Edgeworth approximation

$$G_{n,m}(r,a) \stackrel{def}{=} \mathbf{P}\left\{|\eta - a|^2 \le r^2\right\} + \sum_{k=1}^{m} U_k(r;a)$$

are related through the inequality

$$|F_n(r,a) - G_{n,m}(r,a)| \le c/n^{(m+1)/2}.$$

Distribution of ω_n^2-statistics. The special problem discussed below is classical in statistics.

Consider a sample from the uniform distribution on (0,1), i.e., the i.i.d. rv's $U, U_1, \ldots \in (0,\ 1)$ such that $\mathcal{U}(u) \stackrel{def}{=} \mathbf{P}\{U \le u\} = u$ for $u \in (0,\ 1)$. Its sample DF and ω_n^2-statistics are

$$S_n(u) \stackrel{def}{=} \tfrac{1}{n}\sum_{j=1}^{n} \mathbf{1}_{(0,u)}(U_j), \quad \omega_n^2 = n\int_0^1 \left[S_n(u) - u\right]^2 du.$$

The statistics considered fits into the infinite-dimensional setting adopted here. Indeed, the functions

$$\xi_j(u) = \mathbf{1}_{(0,u)}(U_j) - u$$

are i.i.d. RV's with values from the Hilbert space $X = L_2(0,\ 1)$, and $\omega_n^2 = |Z_n|^2$, $Z_n = \frac{1}{\sqrt{n}} S_{1,n}[\xi]$.

It is easy to see that $|\xi| \le 1$ a.s., so the moment conditions which ensure applicability of the central limit theorem and permit the construction of Edgeworth expansions of arbitrary order are satisfied.

Condition (5.7.10) has not been verified so far. Nevertheless, the following theorem has been established by methods that exploit the special construction used to produce the RV's $\xi_\bullet$. As before, $F_n(r) = \mathbf{P}\{|Z_n| \le r\}$ is the DF being approximated, and $G_{n,m}(r)$ is its Edgeworth approximation with the "smallest" term $\mathcal{O}(n^{m/2})$.

Theorem 5.7.2 *a) The CF of ω_n^2 and its derivatives admit the following estimates:*

$$\forall\, A \ge 0, k \in \mathbf{N}\ \exists\, \delta > 0 \quad \sup_{|t| \ge n^\delta} \left| \frac{d^k}{dt^k} \mathbf{E} \exp\left\{it\omega_n^2\right\} \right| \le c/n^A.$$

b) The bound on error term of the Edgeworth expansion for the DF is, for arbitrary order m, given by the inequality

$$|F_n(r) - G_{n,m}(r)| \le c/n^{(m+1)/2}.$$

When the bounds of part a) are already available, the estimate of part b) follows by calculations described in the preceding sections.

On the contrary, part a) was established (by R. Zitikis) using much more specific methods.

Chapter 6

Fine Asymptotics of Moderate Deviations

In this chapter X is a separable real Hilbert space with norm $|\cdot| = \sqrt{\langle\cdot,\cdot\rangle}$. The RV's $\xi, \xi_1, \ldots \in X$ are independent and, unless specially mentioned, identically distributed.

The RV ξ is supposed to satisfy the Cramér condition (3.6.2). It is centered, with covariance $\mathcal{V}$:

$$\exists A_1 > 0 \;\; A_2 = \mathbf{E} \exp\{A_1 |\xi|\} < \infty, \;\; \mathbf{E}\xi = 0, \;\; \langle \mathcal{V} f, f\rangle = \mathbf{E}\langle \xi, f\rangle^2, \;\; f \in X.$$

Notation for the eigenvalues and eigenvectors of $\mathcal{V}$, its trace, etc. reproduces that of (2.3.4)–(2.3.6). It will be convenient to normalize the distribution of ξ as in (2.3.7) assuming that the principal eigenvalue is $\sigma_1^2 = 1$.

The covariance operator is supposed to have infinite-dimensional range. A stronger nondegeneracy restriction is imposed for the sake of convenience:

$$\forall f \in X, \; f \neq 0 \;\; \langle \mathcal{V} f, f\rangle > 0. \tag{6.0.1}$$

Under this assumption, all eigenvalues of the covariance operator are positive, and there is an orthonormal basis of eigenvectors.

Notation η refers to a RV with the accompanying Gaussian law whose CF is

$$\mathbf{E} \exp\{i\langle \eta, t\rangle\} = \exp\left\{-\tfrac{1}{2}\langle \mathcal{V} t, t\rangle\right\}. \tag{6.0.2}$$

The problem treated below in most detail is as follows. Select a function $\Phi : X \to \mathbf{R}$ and consider the probability

$$\mathrm{p}_n(r) \stackrel{def}{=} \mathbf{P}\left\{\Phi\left(\tfrac{1}{r}\zeta_n\right) \geq 0\right\}, \;\; \zeta_n = \tfrac{1}{\sqrt{n}} S_{1,n}[\xi], \tag{6.0.3}$$

where $r = r_n \to \infty$ and $r/\sqrt{n} \to 0$ as $n \to \infty$. (Zero in the inequality $\Phi \geq 0$ could evidently be changed to one more parameter u.) The calculations of this chapter aim at finding asymptotic expressions for probabilities (6.0.3). Asymptotics of this kind are labelled "fine" to distinguish them from the "rough" asymptotic formulae of Chapter 4 for $\ln \mathbf{p}_n(r)$.

In the range of moderately large deviations, "rough" LD asymptotics can be described in terms of the "Gaussian" deviation functional (DF)

$$\Lambda_0(x) \stackrel{def}{=} \sup_{h\in X}\left[\langle x,h\rangle - \tfrac{1}{2}|\mathcal{V}h|^2\right] = \tfrac{1}{2}|\mathcal{V}^{-1/2}x|^2, \tag{6.0.4}$$

which is nonnegative and can assume the value $+\infty$. Under certain regularity assumptions (see Theorem 4.6.1) the logarithm of probability (6.0.3) admits the asymptotic representation

$$\ln \mathbf{P}\left\{\Phi\left(\tfrac{1}{r}\zeta_n\right)\geq 0\right\} = -[1+o(1)]\,\Lambda_\Phi, \quad \Lambda_\Phi = \inf_{\Phi(x)\geq 0}\Lambda_0(x). \tag{6.0.5}$$

Techniques developed in the present chapter permit to improve this approximation and obtain for the probability itself an asymptotic expression like

$$\mathbf{p}_n(r) = \text{const}\cdot r^A \exp\left\{-\Lambda_\Phi r^2\right\}[1+o(1)].$$

A closely related problem, not discussed here in detail, concerns exponential moments (cf. Theorem 4.4.2). A function $\varphi : X \to \mathbf{R}$ is selected, and approximations are sought to the expectation

$$M[\varphi, r_n] = \mathbf{E}\exp\left\{r_n^2\varphi(\zeta_n)\right\}$$

under appropriate assumptions concerning the growth of the function at infinity, the range of deviations, r_n, etc. In this case, the asymptotic expression is

$$\ln M[\varphi, r_n] = \ln \mathbf{E}\exp\left\{r_n^2\varphi\left(\tfrac{1}{r_n}\zeta_n\right)\right\} \sim r_n^2\Lambda[\varphi], \tag{6.0.6}$$

with $\Lambda[\varphi] = \sup_x\left\{\varphi(x) - \tfrac{1}{2}\left|\mathcal{V}^{-1/2}x\right|^2\right\}$. The improved "fine" asymptotics replace this expression by a product of polynomial and exponential as described after (6.0.5).

6.1 Fine LD Asymptotics in the Gaussian Case

6.1.1 Asymptotic Expression for Probability

For a centered Gaussian RV with the covariance operator $\mathcal{V}$, the DF is the quadratic form $\Lambda_0(x)$ of (6.0.4). It is finite on the subspace

$$X_0 = \mathcal{V}^{1/2}X = \left\{x\in X : |x|_+ \stackrel{def}{=} |\mathcal{V}^{-1/2}x| < \infty\right\}. \tag{6.1.1}$$

The covariance operator of a Gaussian RV has finite trace, so the embedding of X_0 in X is compact.

Lemma 6.1.1 *Consider an upper semicontinuous function $\Phi : X \to \mathbf{R}$.*

If the lower bound Λ_Φ of (6.0.5) is finite, then it is attained at some point of X_0:

$$\exists\, x_0 \in X_0 \;\; \tfrac{1}{2}|x_0|_+^2 = \Lambda_\Phi = \inf_{\Phi\geq 0}\tfrac{1}{2}|\mathcal{V}^{-1/2}x|^2, \;\; \Phi(x_0)\geq 0.$$

Proof of Lemma 6.1.1. Let $x_k \in X_0 \bigcap \{\Phi \geq 0\}$ be a sequence for which $\frac{1}{2}|x_k|_+^2 \to \Lambda_\Phi$. It contains a subsequence x_l' whose image $y_l = \mathcal{V}^{-1/2}x_l'$ converges weakly in X to some limit $y_0 \in X$. The convergence of $x_l' \to x_0 = \mathcal{V}^{1/2}y_0$ is strong as $\mathcal{V}$ has finite trace. By the assumption of the lemma

$$0 \leq \limsup_{k\to\infty} \Phi(x_k') \leq \Phi(x_0), \quad \Lambda_0(x_0) \geq \Lambda_\Phi.$$

At the same time,

$$|y_0|^2 = \lim_{l\to\infty}\langle y_l, y_0\rangle \leq \lim_{l\to\infty}\tfrac{1}{2}\left(|y_l|^2 + |y_0|^2\right) = \Lambda_\Phi + \tfrac{1}{2}|y_0|^2,$$

so $\Lambda_0(x_0) = \frac{1}{2}|x_0|_+^2 = \frac{1}{2}|y_0|^2 \leq \Lambda_\Phi$. This proves the lemma. ○

To calculate the fine LD asymptotics, it is necessary to impose some restrictions on the set where the extremum of Lemma 6.1.1 is attained and on the function Φ.

Restrictions on the set of extrema. Below, each extremum of the DF is supposed to lie on the boundary

$$H_\Phi = \{x : \Phi(x) = 0\}$$

of the set $\{\Phi \geq 0\}$ and to be the expectation of some distribution F_h^* which is the Cramér transform (3.6.4) of the "original" distribution $F = \mathcal{L}(\eta)$.

By (3.6.7), the expectation of the Cramér conjugate F_h^* is

$$m(h) = \nabla \ln \int_X e^{\langle x,h\rangle} F(dx).$$

For the Gaussian distribution $\mathcal{L}(\eta)$ the Cramér conjugate is itself Gaussian with same covariances and expectation $m(h) = \mathcal{V}h$: the CF of a RV η_h^* whose distribution is the Cramér conjugate to that of η is

$$\mathbf{E}\exp\{i\langle\eta_h^*, t\rangle\} = \exp\left\{i\langle\mathcal{V}h, t\rangle + \tfrac{1}{2}\langle\mathcal{V}t, t\rangle\right\}. \tag{6.1.2}$$

For this reason the restrictions on the set of extrema include the following condition:

There is some bounded set $F_0 \subset X$ such that $\Phi(\mathcal{V}h) = 0$ for $h \in F_0$ and

$$\mathcal{E}_\Phi \stackrel{def}{=} \{x \in X : \Phi(x) \geq 0, \ \Lambda_0(x) = \Lambda_\Phi\} \subset \mathcal{V}F_0. \tag{6.1.3}$$

The embedding $\mathcal{V}X \subset X$ is compact, so the set $\mathcal{V}F_0$ is at least compact.

Below, the functional Φ is smooth, and its gradient does not vanish in a neighborhood of the set where the DF attains its minimum. More precisely, it is assumed that Φ has continuous Fréchet derivatives of arbitrary order and that

$$\left|\nabla^{\otimes k}\Phi(x)\right| \leq \phi_k, \ |\nabla\Phi(x)| \geq \phi = \phi(\varepsilon_0) > 0 \text{ if } \operatorname{dist}(x, \mathcal{E}_\Phi) \leq \varepsilon_0. \tag{6.1.4}$$

Thus, the unit outward normal vector to H_Φ at x is well defined by the equality

$$\bar{n}(x) = |\nabla\Phi(x)|^{-1}\nabla\Phi(x)$$

in a neighborhood of each point of $\mathcal{E}_\Phi$.

If $x = \mathcal{V}h$, the value of the DF is $\frac{1}{2}|\mathcal{V}^{1/2}h|^2$. Thus its extrema on H_Φ are extrema of a smooth function of $h \in X$ under the restriction $\Phi(\mathcal{V}h) = 0$ and solve the problem

$$\tfrac{1}{2}|\mathcal{V}^{1/2}h|^2 \sim \min, \quad \Phi(\mathcal{V}h) = 0.$$

Consequently, they obey the well-known necessary condition: $\mathcal{V}h = \beta\mathcal{V}\bar{n}(x)$ for some $\beta \in \mathbf{R}$. The subspace $\mathcal{V}X$ being dense by (6.0.1), this condition can be written as

$$\exists\beta = \beta(x) \in \mathbf{R} : \; h = \beta\bar{n}(x) \text{ if } x = \mathcal{V}h \in \mathcal{V}F_0. \tag{6.1.5}$$

The factor $\beta = \beta(x)$ can also be expressed in terms of $\bar{n}(x)$ and the extremal value Λ_Φ. Indeed, multiplying (6.1.5) by $\mathcal{V}h$ and using this equality once again to express h through n, one arrives at the relations

$$\begin{aligned} \Lambda_\Phi &= \langle \mathcal{V}h, h\rangle = \langle \beta\bar{n}(x), \mathcal{V}h\rangle = \beta^2 \langle \mathcal{V}\bar{n}(x), \bar{n}(x)\rangle, \\ \beta^2 &= \langle \mathcal{V}\bar{n}(x), \bar{n}(x)\rangle^{-1}\Lambda_\Phi. \end{aligned} \tag{6.1.6}$$

Focusing condition. The restrictions listed below will make it possible to consider one minimum of the DF at a time (even if there is a fairly complicated set of extrema).

In addition to (6.1.3), the set of extrema is assumed to admit a special parameterization: there is a space $S = \{s\}$ carrying a measure σ on the σ-algebra of its measurable subsets $\boldsymbol{S}$, a measurable set $S_0 \in \boldsymbol{S}$, and a measurable mapping $h : S \to X$ such that

$$F_0 = \{h(s) : s \in S_0\}. \tag{6.1.7}$$

By (6.1.5), the vector $h(s)$ corresponding to a DF minimum $x(s) = \mathcal{V}h(s)$ has the direction of the outward normal to H_Φ. The corresponding one-dimensional projection is invariably denoted by

$$\mathcal{P}_s : x \mapsto \langle x, e_0\rangle e_0, \quad e_0 = e_0(s) = \bar{n}(s), \quad \bar{n}(s) \stackrel{def}{=} \bar{n}[x(s)]. \tag{6.1.8}$$

For each value of the parameter $s \in S_0$, projection (6.1.8) is supposed to be a part of the decomposition of the identity operator into a sum of orthogonal projections,

$$\mathcal{I} = \mathcal{P}_s + \mathcal{Q}_s + \mathcal{R}_s, \tag{6.1.9}$$

where the subspaces $\mathcal{P}_sX$, $\mathcal{Q}_sX$, and $\mathcal{R}_sX$ are orthogonal. Moreover, the subspace $\mathcal{Q}_sX$ is finite-dimensional:

$$d = d(s) \stackrel{def}{=} \dim(\mathcal{Q}_sX) < +\infty. \tag{6.1.10}$$

The unit normal $e_0 = e_0(s)$ is complemented by some unit vectors so as to make some orthonormal basis $(e_j, j = \overline{0,d})$ in the subspace $[\mathcal{P}_s + \mathcal{Q}_s]X$ (the basis is fixed later on for each s).

$$\mathcal{V}_s \stackrel{def}{=} ||\langle \mathcal{V}e_j, e_k\rangle||_{j,k=\overline{0,d}}. \tag{6.1.11}$$

is the matrix of the operator $[\mathcal{P}_s + \mathcal{Q}_s]\,\mathcal{V}\,[\mathcal{P}_s + \mathcal{Q}_s]$ in this basis.

One more object associated with parameterization (6.1.7) is the "focusing device" — a triple of functions $q_s(u)$, $u \in \mathbf{R}$, $\widehat{q}_s(x)$, $x \in X$, and $Q_r(x)$, $x \in X$, that serves to produce a specific decomposition of unity in a neighborhood of the extremal set.

All these functions are measurable in their arguments $s \in S_0$, $u \in \mathbf{R}_+$, $x \in X$. The last function of the triple is generated from the two first according to the formula

$$Q_r(x) \stackrel{def}{=} \int_{S_0} r^{D(s)}\, q_s\left(|r\mathcal{Q}_s[x - rx(s)]|\right) \widehat{q}_s\left(\tfrac{1}{r}x\right) d\sigma(s).$$

It is convenient to make $\widehat{q}_s\left(\frac{1}{r}x\right)$ and Q depend on one more parameter $\frac{1}{F} > 0$ (e.g., inserting in $\widehat{q}_s\left(\frac{1}{r}x\right)$ an additional cutoff factor whose part will become clear from the calculations below). This parameter is mostly dropped in the notation.

The "focusing device" obeys the following restrictions.

For a fixed s, the function q_s is bounded and vanishes outside a neighborhood of zero: $q_s(u) = 0$ if $u \geq 1$.

The function $\widehat{q}_s$ has continuous Fréchet derivatives of arbitrary order with respect to its argument $x \in X$. Moreover, it vanishes outside a neighborhood of the extremum selected:

$$\widehat{q}_s(y) = 0 \ \text{ if } \ |y - x(s)| \geq \tfrac{1}{F}, \ \ \tfrac{1}{F} > 0.$$

The integral $Q_r(x)$ is a bounded measurable function of $x \in X$.

The *focusing condition* stipulates that for large r the function $Q_r(x)$ approximates the indicator of a neighborhood of the extremal set for the DF. It reads:

There exist $c > 0$ and $\varepsilon = \varepsilon(F) > 0$ such that for $r > r_0$

$$\left|\mathbf{P}\left\{\Phi\left(\tfrac{1}{r}\eta\right) \geq 0\right\} - \mathbf{E}Q(\eta)\,\mathbf{1}_{\{\Phi \geq 0\}}\left(\tfrac{1}{r}\eta\right)\right| \leq c\exp\left\{-\left[\Lambda_\Phi + \varepsilon\right] r^2\right\}. \quad (6.1.12)$$

It may be convenient to permit the functions above depend on some additions parameters, e.g., on small parameters $\frac{1}{r} \ll 1$, $\tau \ll 1$. In this case, dependence on all parameters is supposed smooth (with a possible exception of $\frac{1}{F}$, which remains fixed in calculations), and inequalities in conditions imposed on them hold uniformly in sufficiently small values of all small parameters.

The construction above is employed to reduce the original problem in which the DF Λ_0 may have a manifold of minima to one with a single relevant extremum.

Under restriction (6.1.12), there is little loss of generality if the upper bound on the norm of Fréchet derivatives in (6.1.4) is assumed to hold in all X. Indeed, suppose that condition (6.1.12) holds. After calculation of the probability is reduced to computing $\mathbf{E}q_s\widehat{q}_s\mathbf{1}_{\{\Phi\geq 0\}}$ for separate values of s, one can replace Φ by a function that is better behaved outside the support of $q_s\widehat{q}_s$ with negligible error.

The calculations that follow are based on results of Sections 5.5–5.6 for the almost linear mappings generated by a smooth function. Recall that their form is (cf. (5.6.3))

$$\phi(x) = \phi_s(x) = \frac{r\Phi\left(\widehat{x}+\frac{1}{r}x\right)}{\left|\nabla\Phi\left(\widehat{x}\right)\right|}e_0 + \sum_{j=1}^{d}\langle x, e_j\rangle e_j, \quad \widehat{x} = \widehat{x}(s). \tag{6.1.13}$$

The point $\widehat{x}$ will later on coincide with (or at least be close to) one of the extrema $x(s)$ of the DF in (6.1.7). The mapping ϕ is supposed to satisfy the nondegeneracy conditions (5.6.6)–(5.6.7). (The constants in these conditions are supposed to be independent of the "position tag" $s \in S_0$ which selects the point $\widehat{x} = \widehat{s}$.)

Recall that condition (5.6.7) is essentially a restriction on the spectrum of the covariance operator of $\mathcal{V}^{1/2}\nabla^{\otimes 2}\Phi(\widehat{x}+y)\eta$, stipulating that its set of positive eigenvalues is infinite. The part played by averaging in η_2 is to ensure validity of the condition in "typical" points close to $\widehat{x}$.

The local properties of H_Φ near a point $x(s)$ are characterized by the operator

$$\ddot{\Phi} = \ddot{\Phi}(x(s)) \stackrel{def}{=} \frac{\beta(s)}{\left|\nabla\Phi\left[x(s)\right]\right|}\nabla^{\otimes 2}\Phi\left(x(s)\right), \quad \beta(s) = \beta(x(s)). \tag{6.1.14}$$

It will be assumed below that for each extremum $x(s)$ the quadratic form generated by $\ddot{\Phi}\left(x(s)\right)$ is dominated by a measurable seminorm $|\cdot|_*$ in X such that for some $\alpha > 0$

$$A_*(\alpha) = \mathbf{E}\exp\left\{\tfrac{1}{2}(1+\alpha)\left|\eta\right|_*^2\right\} < +\infty. \tag{6.1.15}$$

Namely, there exists a collection of positive constants such that

$$\langle\ddot{\Phi}(x(s)+\tfrac{1}{r}x)y, y\rangle \le c_1 + |y|_*^2 \tag{6.1.16}$$

whenever $\Phi\left(x(s)+\frac{1}{r}y\right) \ge 0$, $|r\mathcal{Q}_s y| \le c_2$, and $\widehat{q}_s\left(x(s)+\frac{1}{r}x\right) \ne 0$ for $\frac{1}{r} \le c_4$.

In the situation described above, LD asymptotics can be calculated using the following proposition.

Theorem 6.1.1 a) *If the set of DF extrema obeys conditions* (6.1.3) *and* (6.1.7) *as well as the focusing condition* (6.1.12) *with sufficiently small* $\frac{1}{F}$ *and projections* (6.1.9)–(6.1.10), *and the function* Φ *satisfies* (6.1.4), (6.1.15)–(6.1.16), *and the nondegeneracy conditions* (5.6.6)–(5.6.7), *then the following asymptotic representation is valid:*

$$\begin{aligned}\mathbf{p}(r) \stackrel{def}{=} \mathbf{P}\left\{\Phi\left(\tfrac{1}{r}\eta\right) \ge 0\right\} &= \exp\left\{-\Lambda_\Phi r^2\right\} \\ &\times \left[\int_{S_0} r^{D(s)}\mathbf{E}\,\mathcal{M}_s\left(\eta, \tfrac{1}{r}\right)d\sigma(s) + \mathcal{O}\left(e^{-\varepsilon r^2}\right)\right],\end{aligned} \tag{6.1.17}$$

where $\varepsilon = \varepsilon(F) > 0$ *and*

$$\begin{aligned}\mathcal{M}_s\left(y, \tfrac{1}{r}\right) &= \exp\left\{-r\langle h(s), y\rangle\right\}\mathbf{1}_{\{\Phi\ge 0\}}\left(x(s)+\tfrac{1}{r}y\right) \\ &\times q_s\left(|r\mathcal{Q}_s y|\right)\,\widehat{q}_s\left(x(s)+\tfrac{1}{r}y\right).\end{aligned} \tag{6.1.18}$$

b) *The expectation* $\mathbf{E}\mathcal{M}_s$ *in the right-hand side of the above equality admits,* *for each $s \in S_0$, an asymptotic expansion in powers of $\frac{1}{r}$: for every $M \in \mathbf{N}$*

$$\mathbf{E}\ \mathcal{M}_s\left(\eta, \tfrac{1}{r}\right) = \frac{1}{r^{d(e)+1}} \sum_{k=0}^{M} \frac{m_k}{r^k} + O\left(\frac{1}{r^{M+1}}\right),$$

where $d(s)$ is the number of dimensions of the subspace $\mathcal{Q}_s X$ in decomposition (6.1.9).

The first coefficient in the asymptotic expansion is (cf. (6.1.11)*)*

$$m_0 = C\frac{\widehat{q}_s(x(s);0)}{\sqrt{\det\left(2\pi\mathcal{V}_s\right)}}\mathbf{E}\left\{\exp\left\{\tfrac{1}{2}\langle\ddot{\Phi}\mathcal{R}_s\eta, \mathcal{R}_s\eta\rangle\right\}\Big|\, [\mathcal{P}_s + \mathcal{Q}_s]\,\eta = 0\right\},$$

$$C = \int_{\mathbf{R}^d} q_s\left(|y|\right) dy,$$

The proof of this theorem is postponed until a later subsection.

6.1.2 The Laplace Method: a Reminder

The proof of Theorem 6.1.1 below is a combination of two techniques. It uses first the "focusing" of the preceding subsection to concentrate attention on a single minimum of the DF, and the Cramér transform to separate the factor that is responsible for the rough LD asymptotics. The problem is reduced to calculation of an integral of a "localized" function which contains a large scale parameter. This latter is treated using the Laplace method.

It seems natural to preface the proof with a brief reminder of the technique.

Consider a function $f : \mathbf{R}^{d+1} \to \mathbf{R}$ which satisfies the following condition: for each $B > 0$ and $m \in \mathbf{Z}_+$

$$\int_{\mathbf{R}^{d+1}} [1 + \|y\|]^m |f(y)| dy = c'(m) < \infty,$$

$$\int_{\|y\| \geq L} [1 + \|y\|]^m\, |f(y)|\, dy \leq \frac{c''}{L^B}, \quad c'' = c''(m, B), \quad L \geq 1. \tag{6.1.19}$$

Let $g : \mathbf{R}^{d+1} \to \mathbf{R}$ be a bounded function having derivatives of arbitrary orders in a neighborhood of zero.

Proposition 6.1.1 *The asymptotic expansion*

$$J(r) \stackrel{def}{=} \int_{\mathbf{R}^{d+1}} f\left(ry\right) g\left(y\right) dy = \frac{1}{r^{d+1}} \sum_{\kappa=0}^{M} \frac{1}{r^\kappa} J_\kappa + \rho_M(r), \quad r \to +\infty, \tag{6.1.20}$$

holds for each natural number $M \geq 1$, with

$$J_\kappa = \frac{1}{\kappa!} \int_{\mathbf{R}^{d+1}} \left(y, \nabla\right)^\kappa g(0) f(y) dy, \quad |\rho_M(r)| \leq \frac{C(M)}{r^{M+1}}.$$

Notation $(y, \nabla)^k g$ for differentials (or "directional derivatives") agrees with that of Lemma 3.4.2 or (5.1.1).

Remark 6.1.1 If f and g depend on some parameter $u \in \mathbf{R}^p$, then the coefficients in the expansion of Proposition 6.1.1 also depend on this parameter.

Suppose that in a neighborhood of $u = 0$

$$|g(y;u)| \le G_0, \quad \|\nabla_y^{\otimes k} g(y;u)\| \le G_k \tag{6.1.21}$$

with same constants $G_\bullet$ for all $\|u\| \le c_1$ and $\|y\| \le c_2$. Then the expansion of Proposition 6.1.1 is valid for $\|u\| \le c_1$ with a uniform bound on the remainder term ρ_M.

If the dependence of f and g on the parameter u is smooth, and for all $k, l \in \mathbf{Z}_+$, $B > 0$

$$\int_{\mathbf{R}^{d+1}} [1 + \|y\|]^l \, \|\nabla_u^{\otimes k} f(y)\| \, dy \;\le\; c_1 < \infty, \quad c_1 = c_1(k, l),$$

$$\int_{\|y\| \ge L} [1 + \|y\|]^l \, \|\nabla_u^{\otimes k} f(y;u)\| \, dy \;\le\; \frac{c_2}{L^B}, \quad c_2 = c_2(k, l, B), \tag{6.1.22}$$

then the coefficients in the expansion of Proposition 6.1.1 have derivatives in u equal to

$$\nabla^{\otimes k} J_\kappa(u) \;=\; \int_{\mathbf{R}^{d+1}} \nabla_u^{\otimes k} \left[(y, \nabla_y)^\kappa g(0;u) f(y;u)\right] dy. \tag{6.1.23}$$

Proof of Proposition 6.1.1. The change to the scaled integration variable $y' = ry$ transforms the integral into

$$J(r) \;=\; \frac{1}{r^{d+1}} \int_{\mathbf{R}^{d+1}} f(y) g\left(\tfrac{1}{r} y\right) dy$$

(here and below primes are suppressed to simplify notation).

For arbitrary $B > 0$ and $L = \sqrt{r}$ by condition (6.1.19)

$$J(r) \;=\; \frac{1}{r^{d+1}} \int_{\|y\| \le L} f(y) g(\tfrac{1}{r} y) dy + O\left(r^{-(d+1)} L^{-B}\right).$$

The formula of the proposition now follows from the Taylor formula and condition (6.1.19): for some $\vartheta \in [0, 1]$,

$$g(y) \;=\; \sum_{\kappa=0}^{M} \frac{1}{r^\kappa} \frac{1}{\kappa!} (y, \nabla)^\kappa g(0) + \frac{1}{r^{M+1}} \frac{1}{(M+1)!} (y, \nabla)^{M+1} g\left(\tfrac{\vartheta}{r} y\right).$$

It remains to substitute the above expansion into the integral and use condition (6.1.19) to evaluate the error due to omission of the remainder term and the change of integration domain to the whole space. ○

A similar argument substantiates the assertions of Remark 6.1.1.

Proposition 6.1.1 will be used to calculate expectations containing smooth transforms of a Gaussian RV η by the mapping with finite-dimensional range

$$\phi(x; v) \in \mathbf{R}^{d+1}, \quad x \in X, \; v \in \mathbf{R}^p.$$

Later on, the mapping is that of (6.1.13), and one of the coordinates of the vector-valued small parameter $v \in \mathbf{R}^p$ is $\frac{1}{r}$. The Laplace method does not use any special properties due to near linearity of mapping (6.1.13) for very small values of $\frac{1}{r}$.

Proposition 6.1.2 *Assume that the measure*

$$M(B; v) = \mathbf{E}\mathbf{1}_B\left(\phi\left(\eta; v\right)\right)\Psi\left(\eta; v\right), \quad B \in \mathcal{B}\left(\mathbf{R}^{d+1}\right),$$

is absolutely continuous with respect to the Lebesgue measure in $\mathbf{R}^{d+1}$ *and its density*

$$\mu(y; v) \stackrel{def}{=} \frac{dM(\cdot, v)}{d\,\mathrm{mes}}(y), \quad y \in \mathbf{R}^{d+1}, \; v \in \mathbf{R}^p,$$

is bounded and has bounded derivatives of all orders in all its variables.

If the function f *satisfies conditions* (6.1.19) *and* (6.1.22) *uniformly in values of the small parameter* v *from a neighborhood of zero, then the expectation* $\mathbf{E}f\left(r\phi(\eta; v)\right)\Psi(\eta; v)$ *admits the asymptotic expansion of Proposition* 6.1.1 *in powers of* $\frac{1}{r}$, *and its coefficients are smooth functions of the small parameter* $v \to 0$. *They can be calculated using formulae of Proposition* 6.1.1 *with* $g = \mu$.

Proof of Proposition 6.1.2. Using the density of M, the expectation of the proposition can be written as

$$\mathbf{E}f\left(r\phi(\eta; v); v\right)\Psi\left(\eta; v\right) = \int_{\mathbf{R}^{d+1}} f(ry; v)\mu\left(y; v\right) dy.$$

The existence of an asymptotic expansion in powers of the small parameter $u = \frac{1}{r}$ becomes apparent, since conditions of Proposition 6.1.1 are fulfilled. The same proposition describes the properties of the coefficients as functions of the other small parameter. ○

Remark 6.1.2 If the distribution of ϕ has a density $p_v(y)$, $y \in \mathbf{R}^{d+1}$, $v \in \mathbf{R}^p$, that is a smooth function of both its arguments, and the conditional expectation

$$G_v(y) \stackrel{def}{=} \mathbf{E}\left\{\Psi(\eta; v)\,|\phi(\eta; v) = y\right\}$$

is also a smooth function of y and v, then the density of the measure in Proposition 6.1.2 is

$$\mu_v(y) = G_v(y)p_v(y), \tag{6.1.24}$$

and the coefficients of the asymptotic expansion for the expectation of Proposition 6.1.2 can be described in terms of these two functions. For instance, the principal term of the expansion is given by

$$\mathbf{E}f(r\phi(\eta; v)\Psi(\eta; v) = \frac{1}{r^{d+1}}\left[p_0(0)G_0(0) + o(1)\right]\int_{\mathbf{R}^{d+1}} f(y; 0)dy, \tag{6.1.25}$$

where $G_0(0) = \mathbf{E}\{\Psi(\eta;0)\,|\phi(\eta;0) = 0\}$.

Indeed, it follows from the definition of conditional expectation that the density of $M(\cdot;v)$ in this case can be represented as the product $\mu(y;v) = G_v(y)p_v(y)$.

The special form of coefficients (similar to (6.1.25)) in asymptotic expressions of Proposition 6.1.1 is convenient in the situation where the mapping ϕ degenerates into a linear one as $\frac{1}{r} \to 0$ and $v \to 0$. In this case, the Gaussian density $p_0(y)$ can be written down explicitly, as well as the conditional expectation $G_0(y)$, which becomes an integral with respect to a Gaussian distribution.

6.1.3 Derivation of LD Asymptotics

Proof of Theorem 6.1.1 consists of a few steps which occupy the rest of this section. For further reference, it is convenient to write down the results of some intermediate calculations as separate propositions.

a) Representation (6.1.17) follows from condition (6.1.12):

$$\mathbf{p}(r) = \mathbf{E}\int_{S_0} r^{D(s)} q_s\left(|r\mathcal{Q}_s\left[\eta - rx(s)\right]|\right)\widehat{q}_s\left(\tfrac{1}{r}\eta, \tfrac{1}{r}\right) d\sigma(s) + \rho_s(r),$$

where $\rho_s(r) = \mathcal{O}\left(\exp\left\{-(\Lambda_\Phi + \varepsilon) r^2\right\}\right)$. To prove the first assertion of the theorem it is necessary to change the order of integration and transform the expectation in the integrand.

To get a convenient representation for the expectation considered, the "gravity center" is shifted to the minimum $x = \mathcal{V}h(s)$ of the DF by the Cramér transform with parameter $rh(s)$. Formula (6.1.17) results from the subsequent use of the Cramér transform inversion formula (3.6.5). Indeed, for the vector parameter $h = rh(s)$ the Laplace transform equals $\varphi(h) = \mathbf{E}\exp\{\langle\eta, rh\rangle\} = \frac{1}{2}r^2\langle\mathcal{V}h, h\rangle$, and the expectation of the Cramér conjugate is $x_h^* = rx(s)$, so

$$\begin{aligned} \ln\varphi(h) - \langle x_h^*, h\rangle &= \tfrac{1}{2}r^2\langle\mathcal{V}h(s), h(s)\rangle - \langle\mathbf{E}x_{rh(s)}^*, rh(s)\rangle \\ &= -\tfrac{1}{2}r^2\langle\mathcal{V}h(s), h(s)\rangle = -\Lambda_\Phi r^2. \end{aligned}$$

This calculation yields the formula (cf. (6.1.18))

$$\mathbf{E}q_s\left(|r\mathcal{Q}_s\left[\xi - rx(e)\right]|\right)\widehat{q}_s\left(\tfrac{1}{r}\eta, \tfrac{1}{r}\right)\mathbf{1}_{\{\Phi\geq 0\}}\left(\tfrac{1}{r}\eta\right) = \exp\left\{-\Lambda_\Phi r^2\right\}\mathbf{E}\mathcal{M}_s(\eta).$$

b) The expression $\mathcal{M}_s(y)$ contains an exponential factor which can assume large values in the integration domain. It has to be transformed to a more tractable form.

It is convenient to do these calculations in a slightly more general setting and include a perturbation of the extremal pair $\{x(s), h(s)\}$ that appears in (6.1.18).

The perturbation considered here is determined by the functions

$$\widehat{x} = \widehat{x}(v,s) \in X, \quad \widehat{h} = \widehat{h}(v,s) \in \mathbf{X}, \quad \widehat{\beta} = \widehat{\beta}(v,s) \in \mathbf{R},$$

where the parameter s selects the extremum and $v \in \mathbf{R}^p$ is one more small parameter. Dependence of these functions on the new parameter is smooth, i.e., they have bounded derivatives of arbitrary order in v in a neighborhood of zero in $\mathbf{R}^p$. The functions $\widehat{h}, \widehat{x}, \widehat{\beta}$ are supposed to solve the equations

$$\Phi(\widehat{x}) = 0,\ \mathcal{Q}_s[\widehat{x} - x_0] = 0,\ [\mathcal{I} - \mathcal{Q}_s]\widehat{\beta}\bar{n}(\widehat{x}) = [\mathcal{I} - \mathcal{Q}_s]\widehat{h}, \tag{6.1.26}$$

with $\bar{n}(x)$, as usual, the unit normal vector to H_Φ at the point x. The "initial conditions" are

$$\widehat{h}(0,s) = h_0 \stackrel{def}{=} h(s),\ \widehat{x}(0,s) = x_0 \stackrel{def}{=} x(s),\ \widehat{\beta}(0,s) = \beta_0 \stackrel{def}{=} \beta(s). \tag{6.1.27}$$

Moreover, later on $n_0 = \bar{n}(x_0)$ is the unit vector of the outward normal at x_0.

Lemma 6.1.2 *Consider the function*

$$\begin{aligned}\widehat{\mathcal{M}}_s(x, \tfrac{1}{r}, v) &\stackrel{def}{=} \exp\{-r\langle\widehat{h}(s), x\rangle\}\mathbf{1}_{\{\Phi\geq 0\}}(\widehat{x}(s) + \tfrac{1}{r}x)\\ &\times\ q_s(|r\mathcal{Q}_s x|)\,\widehat{q}_s(\widehat{x}(s) + \tfrac{1}{r}x),\quad x \in X.\end{aligned}$$

If condition (6.1.16) *is satisfied as well as* (6.1.26)–(6.1.27), *then this function can be represented in the form* $\widehat{\mathcal{M}}_s(x, \frac{1}{r}, v) = g(r\phi(x))\Psi(x)$, *where the functions* $g(y)$, $y \in \mathbf{R}^{d+1}$, $\Psi(x)$, $x \in X$, *in the right-hand side are smooth in their arguments* y, x *and depend smoothly on the small parameters* $\frac{1}{r}$, v.

The mapping ϕ *is defined by formula* (6.1.13) *with* $\widehat{x} = \widehat{x}(v,s)$ *of* (6.1.26)–(6.1.27).

The function Ψ *satisfies condition* (5.4.15) *with* $\Gamma(x) = \exp\{\frac{1}{2}|x|_*^2\}$ *the exponential function of condition* (6.1.15).

Proof of Lemma 6.1.2. In the calculations to follow it is convenient to use a special coordinate system in X. The coordinates include the vector $\mathcal{R}_s x$ (see (6.1.9)) and the scalar functions $\phi(x)$ of (6.1.13).

It is easy to see that in a neighborhood of zero

$$\phi(x) \stackrel{def}{=} \sum_{j=0}^{d}\phi_j(x)e_j = \left[\widehat{\mathcal{P}}_s + \mathcal{Q}_s\right]x + \mathcal{O}\left(\tfrac{1}{r}|x|^2\right),$$

where $\widehat{\mathcal{P}}_s : x \mapsto \langle x, \widehat{n}\rangle e_0$, $\widehat{n} = \bar{n}(\widehat{x})$, and that the mapping $x \mapsto \phi(x) + \mathcal{R}_s x$ is smoothly invertible if v and $\frac{1}{r}$ are small. This mapping can be extended to all X in such a way that condition (5.4.13) would still hold, e.g., using the formula $\tilde{\phi} = \zeta(x)\phi(x) + [1 - \zeta(x)][\mathcal{P}_s + \mathcal{Q}_s]x$, with ζ some smooth cutoff function selecting the neighborhood of zero where the initial mapping has the desired properties.

In the rapidly varying exponential factor of $\widehat{\mathcal{M}}$, the exponent can be written in the form

$$-r\langle\widehat{h}, x\rangle = -r\langle\widehat{\beta}\widehat{n}, x\rangle + \langle\widehat{\beta}\widehat{n} - \widehat{h}, r\mathcal{Q}_s x\rangle$$

using (6.1.26). This expression can be transformed to read

$$\begin{aligned}-r\langle\widehat{h}, x\rangle = &- \frac{r^2\widehat{\beta}\Phi(\widehat{x} + \frac{1}{r}x)}{|\nabla\Phi(\widehat{x})|}\\ &+ \tfrac{1}{2}\langle\widehat{\Phi}''x, x\rangle + \langle\widehat{\beta}\widehat{n} - \widehat{h}, r\mathcal{Q}_s x\rangle + cr^2\widetilde{W}\left(\tfrac{1}{r}x\right),\end{aligned} \tag{6.1.28}$$

where $\widehat{\Phi}'' = |\nabla\Phi(\widehat{x})|^{-1}\,\widehat{\beta}\nabla^{\otimes 2}\Phi(\widehat{x})$ and

$$\widetilde{W}(y) \stackrel{def}{=} \frac{\widehat{\beta}}{|\nabla\Phi(\widehat{x})|}\left[\Phi(\widehat{x}+y) - \langle\nabla\Phi(\widehat{x}), y\rangle - \tfrac{1}{2}\langle\nabla^{\otimes 2}\Phi(\widehat{x})\,y, y\rangle\right]$$

is a smooth function of $y \in X$ which vanishes at $y = 0$ with its Fréchet derivatives of the first and second order.

Employing (6.1.28) and the coordinates $\{\phi,\ \mathcal{R}_s x\}$, the expression of the lemma can be transformed into

$$\begin{aligned}\widehat{\mathcal{M}}_s(x, \tfrac{1}{r}) &= \exp\{-r\widehat{\beta}\phi_0(x)\}\,\mathbf{1}_{\mathbf{R}_+}(\phi_0(x))\\ &\times\ \exp\{\langle\,\widehat{\beta}\widehat{n} - \widehat{h}, r\mathcal{Q}_s\phi(x)\rangle\}\\ &\times\ q_s(|r\mathcal{Q}_s\phi(x)|)\,\widehat{q}_s\left(\widehat{x} + \tfrac{1}{r}x\right)\\ &\times\ \exp\{\tfrac{1}{2}\langle\widehat{\Phi}''x, x\rangle + r^2\widetilde{W}(\tfrac{1}{r}x)\},\end{aligned} \tag{6.1.29}$$

where $\phi_0(y) = r\Phi(\widehat{x} + \frac{1}{r}y)/|\nabla\Phi(\widehat{x})|$ is the only coordinate of mapping (6.1.13) obtained by a nonlinear transformation of the argument.

The quadratic form $\langle\widehat{\Phi}''y, y\rangle$ satisfies condition (6.1.16) uniformly in $\frac{1}{r}+\|v\| \le c_0$. Indeed, the function Φ is smooth, so for small values of v and $\frac{1}{r}$

$$\langle\widehat{\Phi}''x, x\rangle \le \langle\ddot{\Phi}x, x\rangle + c\,|\widehat{x} - x_0|\,|x|^2 .$$

Moreover, for $|\frac{1}{r}x| \le \frac{1}{F}$ (see (6.1.12)) $r^2\widetilde{W}(\frac{1}{r}x) \le c\frac{1}{F}|x|^2$.

Hence $\widehat{\mathcal{M}}_s$ can be represented in the form stipulated in the lemma with

$$\Psi(x) = \widehat{q}_s(\widehat{x} + \tfrac{1}{r}x)\exp\left\{\tfrac{1}{2}\langle\widehat{\Phi}''x, x\rangle + r^2\widetilde{W}(\tfrac{1}{r}x)\right\},$$

having the desired properties if $\frac{1}{F}$ is sufficiently small. ○

c) Consider in $\mathbf{R}^d$ the measure

$$M(B) = M(B; u) \stackrel{def}{=} \mathbf{E}\mathbf{1}_B(\phi(\eta])\,\Psi(\eta), \tag{6.1.30}$$

where Ψ is the exponentially growing function of Lemma 6.1.2 (note that in the setting of the theorem there is no additional small parameter, so the only small parameter is $\frac{1}{r}$ and $\widehat{\beta}\widehat{n} - \widehat{h} = 0$).

The mapping ϕ and the function Ψ satisfy conditions (5.4.13)–(5.4.15) as well as (5.6.6)–(5.6.7). Thus Lemma 5.6.3 can be applied to verify that the Fourier transform of M is summable, and this measure has a density $\mu(y; \frac{1}{r})$, $y \in \mathbf{R}^{d+1}$. Moreover, the density has smooth derivatives in its variable y and is a smooth function of the small parameter in a neighborhood of zero. It follows from Lemma 6.1.2 that

$$\mathbf{E}\mathcal{M}_s(\eta, \tfrac{1}{r}) = \mathbf{E}f(r\phi\,(\eta))\Psi(\eta) = \int_{\mathbf{R}^d} f(ry)\mu(y)dy,$$

with

$$f\left(y^{(0)}, \ldots, y^{(d)}\right) = \exp\left\{-y^{(0)}\right\}\mathbf{1}_{[0,\infty)}\left(y^{(0)}\right) q_s\left(|\mathcal{Q}_s y|^2\right).$$

Thus, Proposition 6.1.2 is applicable, and $\mathbf{E}\mathcal{M}_s(\eta, \frac{1}{r})$ admits an asymptotic expansion in powers of the small parameter.

d) It remains to identify the principal term of the asymptotic expansion. The argument of part c) is valid if $\Psi \equiv 1$, so the mapping to finite dimensions produces a RV with smooth density. In the limit at $\frac{1}{r} \to \infty$ mapping (6.1.13) degenerates into a linear one:

$$\phi(x, 0) = \mathcal{P}_+ x, \quad \mathcal{P}_+ = [\mathcal{P}_s + \mathcal{Q}_s].$$

It is convenient to identify $\mathcal{P}_+ X$ with $\mathbf{R}^{d+1}$, and denote by $\mathcal{P}_+^*$ the corresponding embedding of $\mathbf{R}^{d+1}$ in X.

The RV $\mathcal{P}_+ \eta \in \mathbf{R}^{d+1}$ is centered Gaussian with covariance matrix (6.1.11) that is nondegenerate by condition (6.0.1). Its density does not vanish, so the density $p(y; \frac{1}{r})$ of $\phi\left(\eta, \frac{1}{r}\right)$ is also nonzero in a neighborhood of zero for small values of $\frac{1}{r}$. By (6.1.24)

$$\mu(y; \tfrac{1}{r}) = p(y; \tfrac{1}{r}) \mathbf{E}\left\{\Psi(\eta) \,|\phi(\eta) = y\right\}$$

This shows that the conditional expectation is a smooth function of y on the support of f, and one can compute the principal coefficient in the asymptotic expression of the theorem calculating the conditional expectation and the density p with $\frac{1}{r}$ replaced by zero.

In this limit, ϕ becomes the Gaussian RV $\mathcal{P}_+ \eta \in \mathbf{R}^{d+1}$ whose density at the origin is $p(0; 0) = 1/\sqrt{\det(2\pi \mathcal{V}_s)}$. This yields the formula of the theorem. ○

One can make the expression of the theorem slightly more explicit. Indeed, the conditional distribution of η given $\mathcal{P}_+ \eta = y$ is itself Gaussian, and an argument following the proof of Theorem 1.1.7 shows that its expectation and covariance operator are

$$m = \mathcal{V}\mathcal{P}_+^* \left(\mathcal{P}_+ \mathcal{V}\mathcal{P}_+^*\right)^{-1} y; \quad \tilde{\mathcal{V}} = \mathcal{V} - \mathcal{V}\mathcal{P}_+^* \left(\mathcal{P}_+ \mathcal{V}\mathcal{P}_+^*\right)^{-1} \mathcal{P}_+ \mathcal{V}.$$

6.2 Landing Outside a Sphere

Calculation of fine LD asymptotics in Section 6.1 postulated numerous properties of the set where DF attains its minimum. This section is dedicated to some more special problems, for which the "focusing device" of Section 6.1 can actually be constructed.

The first of these problems is to derive, using the new technique, the asymptotic expression of Theorem 2.3.1 for the "tail" of the norm of a Gaussian RV η with CF (6.0.2):

$$\begin{aligned} \mathrm{p}_0(r) &\stackrel{def}{=} \mathbf{P}\left\{|\eta| \geq r\right\} \\ &= \left[1 + \mathcal{O}\left(\tfrac{1}{r}\right)\right] \frac{\mathcal{K}}{\Gamma(\nu/2)} \left(\tfrac{1}{2} r^2\right)^{\nu/2-1} \exp\left\{-\tfrac{1}{2} r^2\right\}, \end{aligned} \tag{6.2.1}$$

where $\mathcal{K} = \mathbf{E} \exp\left\{\frac{1}{2} |\mathcal{R}\eta|^2\right\} = \prod_{j=1}^{\infty} 1/\sqrt{1 - \sigma_{\nu+j}^2}$.

Notation related to the covariance operator $\mathcal{V}$ is that of (2.3.5)–(2.3.6), so E is the principal eigenspace, $\nu = \dim(E)$ the multiplicity of the principal eigenvalue, etc. The covariance operator $\mathcal{V}$ is normalized as in (2.3.7). Orthogonal projection onto E is $\mathcal{P}$, and $\mathcal{R} = \mathcal{I} - \mathcal{P}$.

The probability in (6.2.1) equals $\mathbf{P}\left\{\Phi(\frac{1}{r}\eta) \geq 0\right\}$ if the function is $\Phi(x) = \frac{1}{2}(|x|^2 - 1)$.

It is fairly easy to verify that the minimum of the DF outside the unit ball is $\Lambda_\Phi = 1$ (recall normalization (2.3.7)). This minimum is attained at all points in the intersection of the unit sphere with the principal eigenspace, $F_0 = E\bigcap\{|x| = 1\}$. It is natural to parameterize F_0 by points e of the unit sphere $S = \{x \in \mathbf{R}^\nu : \|x\| = 1\}$.

To avoid trivial complications, it is convenient to identify E with $\mathbf{R}^\nu$, F_0 with S, and use e rather than s to denote the "label" of an extremum. Under this agreement, the outward unit normal at a point $x(e) = e \in F_0 = S$, coincides with e, and in decomposition (6.1.9) the projections $\mathcal{Q}_e$ are onto the tangent subspaces of F_0, which all have the same number of dimensions $d = \nu - 1$. Moreover, the only projection onto an infinite-dimensional subspace, $\mathcal{R} = \mathcal{I} - \mathcal{P}$, does not depend on $e \in S$.

No effort is spent on discerning between the norm and inner products in E and $\mathbf{R}^\nu$: these are $|\cdot| = \sqrt{\langle\cdot,\cdot\rangle}$ in both spaces.

In the calculations to follow, $d\sigma(\cdot)$ is the surface element of S.

6.2.1 Constructing the "Focusing Device"

For the special problem considered here, the focusing device of Section 6.1 can be constructed starting with a pair of smooth functions $\chi : \mathbf{R} \to [0, 1]$ and $q : \mathbf{R} \to [0, +\infty)$. They are assumed infinitely differentiable, although in a later subsection it will be more convenient to use less regular functions in a similar setting. For the time being, the principal eigenvalue is assumed multiple, $d = \nu - 1 \geq 1$.

The function $q(u)$ is even on $\mathbf{R}$. On the positive half-line it is nonincreasing and has compact support: $q(u) = 0$ for $u \geq 1$.

The function $\chi(u)$ vanishes for $u \leq -1$ and satisfies the identity

$$\chi(u) + \chi(-u) \equiv 1. \tag{6.2.2}$$

Otherwise the two functions are arbitrary.

Put

$$\tilde{q}(x, v) \stackrel{def}{=} v^{2(\nu-1)} \left(\frac{1}{2|S|} \int_S q\left(\left|v^{-2}\mathcal{Q}_e x\right|^2\right) d\sigma(e)\right)^{-1} \tag{6.2.3}$$

for $v > 0$ and $x \in X$, $\mathcal{P}x \neq 0$. As usual, $|S|$ is the surface area of the unit sphere in $\mathbf{R}^\nu$ (see (A.1.8) in Appendix A). This function actually depends on $\mathcal{P}x$ and v only.

Lemma 6.2.1 *The function $\tilde{q}$ satisfies the identity*

$$I(\gamma, \delta, \alpha) \stackrel{def}{=} \frac{(\gamma\delta)^{-(\nu-1)}}{|S|} \int_S q\left(\left|\tfrac{1}{\delta}\mathcal{Q}_e x\right|^2\right) \chi\left(\alpha\left\langle x, e\right\rangle\right) \tilde{q}\left(\gamma x, \sqrt{\gamma\delta}\right) d\sigma(e) = 1$$

for all values of $\gamma, \delta > 0$ *and* $\alpha \in \mathbf{R}$.

For each x *with* $\mathcal{P}x \neq 0$, $\widetilde{q}(x, v)$ *can be extended to a smooth function of* v *defined in a neighborhood of zero. It is nonincreasing in* $v > 0$ *and*

$$\lim_{v \searrow 0} \widetilde{q}(x, v) = |\mathcal{P}x|^{\nu-1}/\overline{q}, \quad \overline{q} = \frac{1}{|S|} \int_{|z| \leq 1} q(|z|^2) dz.$$

If $v^2 \leq \frac{1}{2}|\mathcal{P}x|$, *then* $\widetilde{q}(x, v) \leq c|\mathcal{P}x|^{\nu-1}$.

Proof of Lemma 6.2.1. By symmetry the integral in the identity of the lemma does not depend on the sign of α, so $I(\gamma, \delta, \alpha) = \frac{1}{2}[I(\gamma, \delta, \alpha) + I(\gamma, \delta, -\alpha)]$, and it follows from (6.2.2) that one can replace χ by $\frac{1}{2}$ in the integrand without changing the value of the integral. After this, the identity of the lemma becomes evident.

To prove the remaining assertions it is convenient to use in calculations coordinates of e in an orthonormal basis of E than includes the unit vector $e_0 = |\mathcal{P}x|^{-1}\mathcal{P}x$ and some unit vectors e_i, $i = \overline{1, \nu - 1}$. If the integration argument e is represented as $e = \langle e, e_0 \rangle e_0 + y$, $y \perp e_0$, then $|y|^2 = 1 - \langle e, e_0 \rangle^2 = |\mathcal{Q}_e e_0|^2$. The vector y varies in the unit ball of $\mathcal{Q}_{e_0}$, and the original surface integral can be represented as a volume integral with respect to the Lebesgue measure in this $(\nu - 1)$-dimensional Euclidean space: if $v^2/|\mathcal{P}x| < 1$, then

$$\begin{aligned} \widetilde{q}(x, v) &= v^{2(\nu-1)} \left(\frac{1}{|S|} \int_{|y| \leq 1} q\left(v^{-4}|\mathcal{P}x|^2|y|^2\right) \frac{dy}{\sqrt{1 - |y|^2}} \right)^{-1} \\ &= |\mathcal{P}x|^{\nu-1} \left(\frac{1}{|S|} \int_{|z| \leq 1} q\left(|z|^2\right) \frac{dz}{\sqrt{1 - (v^4/|\mathcal{P}x|^2)|z|^2}} \right)^{-1}. \end{aligned}$$

(The integral is transformed by changing to the variable $z = v^{-2}|\mathcal{P}x|y$, and the integration domain is the support of q.)

The assertions of the lemma follow from the above integral representation. ○.

It is supposed for the time being that the principal eigenvalue is multiple, $\nu > 1$. For $\nu = 1$ equation (6.2.3) is replaced by

$$\widetilde{q}(u) = \left[\tfrac{1}{2}\chi\left(\langle x, e\rangle\right) + \tfrac{1}{2}\chi\left(-\langle x, e\rangle\right)\right]^{-1} = 2. \qquad (6.2.4)$$

Calculation of Gaussian asymptotics. The preparations above suffice to derive (6.2.1).

However, some additional work is needed to show applicability of the method of Section 6.1.

Fix a small value of $\varepsilon > 0$.

It follows from Theorem 4.5.2 that

$$\mathbf{P}\{\,|\eta| \geq (1 + \sqrt{\varepsilon})r\} = \mathcal{O}\left(\exp\{-\tfrac{1}{2}r^2\left(1 + \sqrt{\varepsilon}\right)\}\right),$$

so

$$\mathbf{p}(r) = \mathbf{P}\left\{r \leq |\eta| \leq \left(1 + \sqrt{\varepsilon}\right) r\right\} + \mathcal{O}\left(\exp\left\{-\tfrac{1}{2}r^2\left(1 + \sqrt{\varepsilon}\right)\right\}\right).$$

In a similar way one can introduce an additional restriction on the component $\mathcal{R}\eta$. The DF of the RV $\eta_+ = \mathcal{P}\eta + \sqrt{1+\varepsilon}\mathcal{R}\eta$ has same minimum on $\{|x| \geq r\}$ as the DF of the original RV η for small $\varepsilon > 0$ because the principal eigenvalue of the covariance operator and the corresponding subspace do not change. Consequently,

$$\begin{aligned}\mathbf{P}\{|\eta| \geq r, |\mathcal{R}\eta| \geq \sqrt{\varepsilon}\, r\} &\leq \mathbf{P}\{|\eta_+|^2 \geq (1+\varepsilon^2)r^2\} \\ &= \mathcal{O}(\exp\{-\tfrac{1}{2}(1+\sqrt{\varepsilon})r^2\}).\end{aligned}$$

Thus, probability (6.2.1) can be represented as

$$\mathbf{p}(r) = \mathbf{E}\mathbf{1}_{[r,\infty)}(|\eta|)\,\overline{\chi}(\eta) + \mathcal{O}\left(\exp\{-\tfrac{1}{2}(1+\sqrt{\varepsilon})r^2\}\right), \tag{6.2.5}$$

where $\overline{\chi}$ is a smooth cutoff function which equals one at all points of interest and vanishes if $|\mathcal{R}x|$ or $|x| - r$ exceeds $r\sqrt{\varepsilon}$.

Multiply the expression under the expectation sign above by the identity of Lemma 6.2.1 (with $\gamma = \delta = \frac{1}{r}$) and integrate the result with respect to the distribution of η. The result is the equality

$$\mathbf{E}\mathbf{1}_{[r,\infty)}(|\eta|)\,\overline{\chi}(\eta) = \frac{r^{2(\nu-1)}}{|S|}\int_S \mathbf{E}\overline{\mathcal{M}}(\varepsilon, e)\, d\sigma(e) = r^{2(\nu-1)}\mathbf{E}\overline{\mathcal{M}}(\varepsilon, e_0),$$

where

$$\overline{\mathcal{M}}(x, e) = \mathbf{1}_{[r,\infty)}(|x|)\,\overline{\chi}(x) q\left(|r\mathcal{Q}_e[x - re]|^2\right)\ \widetilde{q}\left(\tfrac{1}{r}x, \tfrac{1}{r}\right)$$

and the vector $e_0 \in S$ can be chosen arbitrarily as the expectation of the integrand does not actually depend on its argument e by symmetry. On the same domain, $\overline{\chi} = 1$.

Cramér's transform reduces the expectation to the form

$$\mathbf{E}\overline{\mathcal{M}}(\eta, e) = \exp\{-\tfrac{1}{2}r^2\}\mathbf{E}\mathcal{M}^0(\eta), \tag{6.2.6}$$

where

$$\begin{aligned}\mathcal{M}^0(x) &= \exp\{-r\phi^{(0)}(x)\}\mathbf{1}_{[0,\infty}(r\phi^{(0)}(x))q(|r\mathcal{Q}_e x|^2) \\ &\times\ \widetilde{q}(e + \tfrac{1}{r}x, \tfrac{1}{r})\exp\{\tfrac{1}{2}|\mathcal{P}x|^2\} \\ &\times\ \exp\{\tfrac{1}{2}|\mathcal{R}x|^2\}\overline{\chi}(re + x),\end{aligned}$$

and $\phi(x)$ is mapping (6.1.13), which is quadratic in the special case considered here.

In the domain of integration, $\langle x, e\rangle$ can be represented as a function of ϕ and $\mathcal{R}x$. Indeed, $X = \langle x, e\rangle$ is the positive root of the quadratic equation

$$\phi^{(0)} = X + \frac{1}{2r}\left(X^2 + |\mathcal{Q}_e\phi|^2 + |\mathcal{R}x|^2\right),$$

so

$$\langle x, e\rangle = \frac{\phi^{(0)} - \frac{1}{2r}(|\mathcal{Q}_e\phi|^2 + |\mathcal{R}x|^2)}{\frac{1}{2} + \frac{1}{2}\sqrt{1 + \frac{1}{r}\phi^{(0)} - \frac{1}{r^2}[\,|\mathcal{Q}_e\phi|^2 + |\mathcal{R}x|^2\,]}}.$$

Additional restrictions of (6.2.5) that exclude large values of $|\eta|$ and $\mathcal{R}\eta$ ensure regularity of this solution as a function of ϕ and $\mathcal{R}x$, so in the integration domain

$$\exp\left\{\tfrac{1}{2}|x|^2 - r\phi^{(0)}(x)\right\} = g\left(r\phi(x), \tfrac{1}{r}\right)\exp\left\{\tfrac{1}{2}\left(1+\rho\left(x,\tfrac{1}{r}\right)\right)|\mathcal{R}x|^2\right\},$$

where g is bounded and smooth in its arguments, while $\rho\left(x, \frac{1}{r}\right)$ goes to zero uniformly in x as $\frac{1}{r} \to \infty$.

Thus, one can apply Theorem 6.1.1 to calculate $\mathbf{E}\mathcal{M}^0$, and the result is the equality

$$\mathbf{E}\mathcal{M}^0 = \frac{C}{r^\nu}\left(1 + \mathcal{O}\left(\tfrac{1}{r}\right)\right).$$

It remains to identify the constant in the right-hand side. After $\frac{1}{r}$ is replaced by 0, the mapping $\phi(\eta)$ degenerates into the projection $\mathcal{P}\eta$, whose distribution is the standard Gaussian law in $\mathbf{R}^\nu$, so

$$C = \int_{\mathbf{R}^{\nu-1}} q\left(|u|^2\right) dy \frac{\widetilde{q}\,(e,0)}{\sqrt{\det\left(2\pi\mathcal{P}\mathcal{V}\mathcal{P}\right)}} \mathbf{E}\left\{\exp\left\{\tfrac{1}{2}|\mathcal{R}\eta|^2\right\} | \mathcal{P}\eta = 0\right\}.$$

The projections $\mathcal{P}\eta$ and $\mathcal{R}\eta$ are independent. Consequently, Theorem 1.1.3 and Lemma 6.2.1 yield the equality

$$C = \frac{|S|}{(2\pi)^{\nu/2}} \prod_{j=1}^{\infty} 1/\sqrt{1-\sigma_{\nu+j}^2}$$

so (6.2.1) results from substitution of formula (A.1.8) for the surface area of the unit sphere.

6.2.2 A "Correct" Bernstein-Type Bound

The focusing of subsection 6.2.1 can be used to obtain one more Bernstein-type exponential bound on the probability $\mathbf{P}\left\{|\xi_1 + \cdots + \xi_n| \geq r\right\}$.

Inequalities of this kind were treated in some detail in subsections 3.3.4 and 3.4.1. The classical bound of S. N. Bernstein (see Proposition 3.3.1) has the important advantage of being almost as precise and much easier to prove than any later asymptotic expression for the probability that it evaluates. In the multi- or infinite-dimensional case the similarity persists.

As before, the object considered are sums of independent RV's $\xi_j \in X$, where X is a separable real Hilbert space. However, the summands are not supposed to have identical distributions. Instead, they obey the "collective" condition of Theorem 3.3.2:

$$\sum_{j=1}^{n} \mathbf{E}\,|\xi_j|^l \leq \tfrac{1}{2} l!\, B^2 H^{l-2}, \quad l = 2, 3, \ldots, \quad \mathbf{E}\xi_j = 0, \quad j = \overline{1, n}. \tag{6.2.7}$$

The constant $B > 0$ in this condition is used as the natural scale of the sum

$\xi_1 + \cdots + \xi_n$. Namely, it is assumed that

$$\forall f \in X \ \sum_{j=1}^{n} \mathbf{E} \langle \xi_j, f \rangle^2 \le B^2 \langle \mathcal{V} f, f \rangle , \tag{6.2.8}$$

where $\mathcal{V}$ is a nonnegative operator with trace $\mathrm{tr}(\mathcal{V}) \le 1$. Its eigenvalues are denoted σ_j^2 and ordered according to their value as in (2.3.4).

$$\sigma_1^2 = \cdots = \sigma_\nu^2 > \sigma_{\nu+1}^2 \ge \sigma_{\nu+2}^2 \ge \cdots$$

The natural number ν in this formula is the multiplicity of the principal eigenvalue of $\mathcal{V}$. In addition to eigenvalues, the calculations use upper bounds on the "trace residues":

$$\widehat{\sigma}_m^2 \ge \sum_{j=1}^{\infty} \sigma_{m+j}^2 = \mathrm{tr}(\mathcal{V}\mathcal{R}_m). \tag{6.2.9}$$

The sequence of eigenvalues σ_j^2 is associated with a fixed orthonormal basis of eigenvectors e_j of $\mathcal{V}$, so, for instance, $(e_j, j = \overline{1,\nu})$ is an orthonormal basis in the principal ν-dimensional eigenspace $E = \{x : \mathcal{V}x = \sigma_1^2 x\}$.

As usual, $\mathcal{P}_a$ is the orthogonal projection onto the line generated by $a \ne 0$, and $\mathcal{P}_j = \mathcal{P}_{e_j}$, $j = \overline{1,\nu}$, are the projections to basis axes in E. Special notation is reserved for certain other projections related to the eigenbasis:

$$\mathcal{P}_m^+ \stackrel{def}{=} \mathcal{P}_1 + \cdots + \mathcal{P}_m, \ \mathcal{R}_m \stackrel{def}{=} \mathcal{I} - \mathcal{P}_m^+, \tag{6.2.10}$$

where $\mathcal{I}$ is the identity operator. Moreover, special abbreviation will be used for projections acting in the principal eigenspace:

$$\mathcal{P} \stackrel{def}{=} \mathcal{P}_\nu^+; \ \mathcal{R} \stackrel{def}{=} \mathcal{R}_\nu; \ \mathcal{Q}_e \stackrel{def}{=} \mathcal{P} - \mathcal{P}_e, \ e \in E, \ |e| = 1. \tag{6.2.11}$$

The "correct" Bernstein-type exponential inequality for sums is given by

Theorem 6.2.1 *Assume that the independent RV's ξ_j obey conditions* (6.2.7) *and* (6.2.8). *Put*

$$\mu = \widehat{\sigma}_m^2/\sigma_1^2, \ \beta = \sigma_{\nu+1}^2/\sigma_1^2, \ \varepsilon = rH/(\sigma_1^3 B).$$

There exist constants $c_i > 0$ such that

$$\begin{aligned} \mathbf{p}_n(r) &\stackrel{def}{=} \mathbf{P}\{|\xi_1 + \cdots + \xi_n| \ge r\sigma_1 B\} \\ &\le c_1 (1+r)^{\nu-1} \exp\{-\tfrac{1}{2}r^2 + c_2\varepsilon r^2\} \end{aligned}$$

if for some natural m

$$\mu < (1-\beta)^2/(2-\beta). \tag{6.2.12}$$

The values of the constants are determined only by the parameters ν, m, β, and μ.

To simplify the calculations below, it is invariably supposed that the covariance operator $\mathcal{V}$ is normalized according to (2.3.7), $\sigma_1^2 = 1$.

Even though the proof of Theorem 6.2.1 is much easier than calculation of the principal term of "fine" asymptotics, it is quite long. For this reason some of its independent segments occupy several next subsections, and joining of the segments is postponed until the concluding one.

6.2.3 "Focusing" and Cramér's Transform

For the centered Gaussian law with covariance operator $\mathcal{V}$, the principal term of asymptotics is described in formula (6.2.1). By (6.2.5), the probability of landing outside a sphere of radius $r \gg 1$ differs little from that of landing in the εr-neighborhood of the set $E \bigcup \{|x| = r\}$ where the DF attains its minimum. This probability is "joined" from probabilities of landing near one of the minimum points of its DF.

It is natural to conjecture that similar behavior should be typical also of normalized sums of "small" independent RV's.

The bound of Theorem 6.2.1 is not quite as precise as a "fine" LD asymptotic formula, and the "focusing" employed to separate a single minimum of the DF need not be as sharp as in the proof of (6.2.1).

Lemma 6.2.2 *Let ζ be an arbitrary X-valued RV. If $\nu \geq 2$, then*

$$\mathbf{P}\{r \leq |\zeta| < R, |\mathcal{R}\zeta| < V\} \leq \frac{r^{\nu-1}}{|S|} \int_S \mathbf{E}\mathcal{N}_e(\zeta)\, d\sigma(e),$$

where

$$\begin{aligned}\mathcal{N}_e(x) &\stackrel{def}{=} q\left(|Q_e x|^2\right) \tilde{q}\left(\tfrac{1}{r}x, \tfrac{1}{\sqrt{r}}\right) \widehat{I}_s(x), \\ \widehat{I}_s(x) &= \mathbf{1}_{[r,R]}(|x|)\, \mathbf{1}_{[0,\infty)}(\langle x, e\rangle)\, \mathbf{1}_{[0,V)}(|\mathcal{R}x|),\end{aligned}$$

and q, $\tilde{q}$ are the functions of Lemma 6.2.1.

The inequality of the lemma is evident for $\nu = 1$ under the agreement that $q \equiv \frac{1}{2}\tilde{q} = 1$ and

$$\frac{1}{|S|} \int_S \varphi(e)\, ds(e) = \tfrac{1}{2}\varphi(e_1) + \tfrac{1}{2}\varphi(-e_1)\ . \tag{6.2.13}$$

Proof of Lemma 6.2.2. In the setting of the lemma, the function χ need not be smooth, and one can take $\chi(u) = \frac{1}{2}(1 + \operatorname{sign}(u))$ without changing $\tilde{q}$.

Abbreviate, for the time being, the notations of the lemma to

$$\begin{gathered}\widetilde{I} = \mathbf{1}_{[r,R]}(|x|)\, \mathbf{1}_{[0,V]}(|\mathcal{R}x|), \qquad \widehat{I} = \widehat{I}_e(x), \qquad \chi = \chi(\langle x, e\rangle), \\ q = q\left(|Q_s x|^2\right), \qquad \tilde{q} = \tilde{q}\left(\tfrac{1}{r}x, \tfrac{1}{\sqrt{r}}\right).\end{gathered}$$

Since $\widetilde{I}\chi \leq \widehat{I}$, it follows from the identity of Lemma 6.2.1 that

$$\widetilde{I} = \frac{1}{|S|} \int_S \widetilde{I}\chi\, q\, \tilde{q}\, d\sigma \leq \frac{1}{|S|} \int_S \widehat{I}\, q\, \tilde{q}\, d\sigma.$$

To derive the inequality of the lemma, it remains to put $x = \zeta$ and calculate the expectations of both sides of the expression obtained above. ○

The decomposition of Lemma 6.2.2 permits one to consider instead of the whole space E a "slice" of it. There is freedom in choosing the value of V (below $V = \infty$ will suffice).

The next step towards the proof of Theorem 6.2.1 is shifting the "gravity center" of the distribution to some point near re for calculation of the "slice mean" of Lemma 6.2.2. This is done using the Cramér transform (see Section 3.6) with appropriately chosen parameter. To get the inequality of the theorem, this choice need not be too careful.

Below, notation for the remainder in Taylor's expansion of exponentials is

$$\mathcal{E}_k(a) \stackrel{def}{=} e^a - \sum_{j=0}^{k} a^j/j!, \quad k \in \mathbf{N}, \quad \mathcal{E}_k(a) \le a^{k+1} e^a/(k+1)! \tag{6.2.14}$$

The calculations below use one more elementary identity and an estimate that it implies:

$$G_{k,m}(v) \stackrel{def}{=} \sum_{l=k}^{\infty} \frac{(l+m)!}{l!} v^l = \frac{d^m}{dv^m}\left[\frac{v^{k+m}}{1-v}\right], \quad 0 \le u < 1, \tag{6.2.15}$$

$$G_{k,m}(v) \le c_{k,m} v^k, \quad 0 \le v \le \tfrac{1}{2} \,. \tag{6.2.16}$$

Lemma 6.2.3 *Under condition* (6.2.7) *the Laplace transforms of all distributions* $\mathcal{L}(\xi_j)$ *are finite for* $|h|\,H < \frac{1}{2}$, *and the exponential moments of their Cramér conjugates admit the estimate*

$$A(u,h) \stackrel{def}{=} \sum_{j=1}^{n} \mathbf{E}\mathcal{E}_2\left(u\,|\xi|^*\right) \le cB^2 H u^3$$

if $u \ge 0$ *and* $(|h| + u)\,H < \frac{1}{2}$; *here* c *is an absolute constant.*

Proof of Lemma 6.2.3. By the Jensen inequality, $\varphi_j(h) \stackrel{def}{=} \mathbf{E}\exp\{(h,\xi_j)\} \ge 1$. For this reason, a combination of Taylor's expansion for exponentials and (6.2.14) implies the estimate

$$A \le \frac{1}{3!} \sum_{j=1}^{n} \mathbf{E} u^3 |\xi_j|^3 e^{[u+|h|]|\xi_j|} = \sum_{l=0}^{\infty} \frac{u^3 (u+|h|)^l}{3!\,l!} \sum_{j=1}^{n} \mathbf{E}\,|\xi_j|^{l+3} .$$

The inequality of the lemma now follows from (6.2.15) and (6.2.16):

$$A \le \tfrac{1}{12} B^2 H u^3 G_{0,3}([u+|h|]) \le cB^2 H u^3. \ \bigcirc$$

The moments of conjugate distributions admit the estimates of Lemma 3.6.3. One more estimate used below is an inequality for covariances. Let ξ_j^* be RV's whose distributions are Cramér's conjugates to distributions of ξ_j, and put $\bar{\xi}_j = \xi_j^* - \mathbf{E}\xi_j^*$.

Lemma 6.2.4 *Under the assumptions of Lemma* 3.6.3

$$\sum_{j=1}^{n} \left| \mathbf{E} \left| \mathcal{A}\bar{\xi}_j \right|^2 - \mathbf{E} \left| \mathcal{A}\xi_j \right|^2 \right| \le c \left| \mathcal{A} \right|^2 B^2 \left| h \right| H$$

for $|h|\, H < \frac{1}{4}$ *and arbitrary bounded operator* $\mathcal{A}$.

Proof of Lemma 6.2.4. The starting point is the inequality

$$\begin{aligned} \left|\mathbf{E}|\mathcal{A}\bar{\xi}_j|^2 - \mathbf{E}|\mathcal{A}\xi_j|^2\right| &\le |\mathcal{A}|^2\big[|\mathbf{E}\xi_j^*|^2 \\ &+ (\varphi_j - 1)\mathbf{E}|\xi_j|^2 + \mathbf{E}|\xi_j|^2(e^{|h||\xi_j|} - 1)\big]. \end{aligned}$$

Its verification follows the same lines as the proof of Lemma 6.2.3 or Lemma 3.6.3. After it is established, one can repeat the proof of the latter. ○

6.2.4 Rough Exponential Bounds for Cramér Transforms

In Theorem 6.2.1, the covariance operator of the sum is only evaluated from above. In its proof, it will be convenient to include in the sum an additional Gaussian summand ξ^0 independent of the rest, centered, and having the covariance operator

$$\langle \mathcal{U} f, g \rangle = \langle \mathcal{V} f, g \rangle - \frac{1}{B^2} \sum_{j=1}^{n} \mathbf{cov}\left(\langle \xi_j, f \rangle, \langle \xi_j, g \rangle \right).$$

For the resulting extended sum

$$\zeta = B^{-1}(\xi_1 + \cdots + \xi_n + \xi^0), \tag{6.2.17}$$

the covariance operator equals $\mathcal{V}$. Existence of a Gaussian distribution having the desired covariances is evident as $\mathcal{U}$ is nonnegative and trace class. (Recall that the Cramér transform translates Gaussian distributions without affecting covariances.)

Put

$$\varepsilon = rH/B, \quad h = (r/B)e, \quad e \in E, \quad |e| = 1. \tag{6.2.18}$$

The quantity ε will later play the part of a small parameter. Use h to define the centered Cramér conjugate sum

$$\begin{aligned} \bar{\zeta} &= \zeta^* - \mathbf{E}\zeta^*, \\ \zeta^* &= B^{-1}\left[\xi_1^* + \cdots + \xi_n^* + \mathcal{U}h + \xi^0\right], \end{aligned} \tag{6.2.19}$$

where all the summands are independent, the distribution of ξ_j^* is the Cramér transform (3.6.4) with the above vector parameter.

If $\varepsilon < \frac{1}{2}$, the expectation of the sum ζ^* in (6.2.19) satisfies the inequality

$$|\mathbf{E}\zeta^* - re| \le cr\varepsilon. \tag{6.2.20}$$

The calculations below make use of some rather special Bernstein-type exponential inequalities similar to those of subsection 3.4.2. They are based upon

Lemma 6.2.5 *Let $\mathcal{P}'$ be an orthogonal projection in X. There exists a set of constants $c_i > 0$ such that for $\sigma_1^2(\varepsilon+\delta) \le c_0$ and $\delta \overset{def}{=} v/B$*

$$\mathbf{E}\cosh\left(v\left|\mathcal{P}'\zeta\right|\right) \le \exp\left\{\tfrac{1}{2}v^2\mathbf{E}\left|\mathcal{P}'\zeta\right|^2 + c_1(\varepsilon+\delta)v^2\right\}$$

and, if the Cramér transform is applied with the vector parameter of (6.2.18), *then*

$$\mathbf{E}\cosh\left(v\left|\mathcal{P}'\bar{\zeta}\right|\right) \le \exp\left\{\tfrac{1}{2}v^2\mathbf{E}\left|\mathcal{P}'\zeta\right|^2 + c_1(\varepsilon+\delta)v^2\right\}.$$

Proof of Lemma 6.2.5. The Gaussian distribution is stable, so one can replace the additional summand of (6.2.17) or (6.2.19) by a normalized sum of arbitrarily large number N of independent copies of ξ^0 without changing the distribution of ζ or ζ^*. Indeed, $\mathcal{L}(\xi^0) = \mathcal{L}(\xi_N^{0+})$, $\xi_N^{0+} = \frac{1}{\sqrt{N}}\sum_{k=1}^N \xi_k^0$. It is easily verified that in notation of Lemma 6.2.3

$$\forall\, v>0 \quad \lim_{N\to\infty}\sum_{k=1}^N \mathbf{E}\,\mathcal{E}_2\left(v\left|\tfrac{1}{\sqrt{N}}\xi_k^0\right|\right) = \lim_{N\to\infty} N\mathbf{E}\,\mathcal{E}_2\left(v\left|\tfrac{1}{\sqrt{N}}\xi^0\right|\right) = 0,$$

so the desired estimate can be obtained by a passage to the limit from the rough "collective" estimates of Chapter 3.

For each natural l and $j = \overline{1,n}$

$$\mathbf{E}|\bar{\xi}_j|^l \le 2^{l-1}(\mathbf{E}|_j^*|^l + |\mathbf{E}\xi_j^*|^l) \le \mathbf{E}(2|\xi_j^*|)^l.$$

It follows that for large number of Gaussian "splinters" N there are inequalities

$$\sum_{j=1}^n \mathbf{E}\,\mathcal{E}_1(\alpha|\mathcal{P}'\bar{\xi}_j|) + \sum_{i=1}^N \mathbf{E}\mathcal{E}_1\left(\alpha\left|\tfrac{1}{\sqrt{N}}\xi^0\right|\right)$$

$$\le \tfrac{1}{2}v^2\mathbf{E}\left|\mathcal{P}'\bar{\zeta}\right|^2 + \sum_{j=1}^n \mathbf{E}\,\mathcal{E}_2(2\alpha|\xi_j^*|) + o(1), \quad \alpha = v/B.$$

At the same time, Lemma 3.6.3 and Lemma 6.2.4 provide the estimates

$$\mathbf{E}\left|\bar{\zeta}\right|^2 \le \mathbf{E}|\zeta|^2 + c\varepsilon, \quad \sum_{j=1}^n \mathbf{E}\,\mathcal{E}_2(2\alpha|\xi_j^*|) \le cv^2\delta \ \text{ if } \sigma_1^2(\varepsilon+\delta) \le c'.$$

The first inequality of the lemma now follows from the inequality of Theorem 3.3.4:

$$\mathbf{E}\cosh\left(\textstyle\sum_j \eta_j\right) \le \prod_j \mathbf{E}\left[e^{h|\eta|} - h|\eta|\right] \le \exp\left\{\textstyle\sum_j \mathbf{E}\mathcal{E}_1(|\eta_j|)\right\}.$$

The second estimate is derived similarly. ○

Lemma 6.2.6 *Put $\rho \overset{def}{=} \sqrt{\nu+\beta(m-\nu)+\mu}$ (see Theorem 6.2.1). There exist positive constants c_i such that*

$$\mathbf{P}\{|\zeta| \ge \rho r\} \le 2\exp\left\{-\tfrac{1}{2}r^2 + c_1\varepsilon r^2/\rho^3\right\} \quad \textit{if } \varepsilon/\rho \le c_0.$$

Proof of Lemma 6.2.6. Under the assumptions of Theorem 6.2.1 $\mathbf{E}\,|\zeta|^2 \le \rho^2$. The above estimate follows from Lemma 6.2.5 and Chebyshev's inequality in the form

$$\mathbf{P}\left\{|\zeta| \ge \rho r\right\} \;\le\; 2e^{-v\rho r}\mathbf{E}\cosh\left(v\,|\zeta|\right), \quad v = r/\rho. \; \bigcirc$$

The next lemma is a version of Lemma 3.4.1.

Lemma 6.2.7 *Assume that the conditions of Theorem* 6.2.1 *hold and* $R \ge r$, $\mu t^2 < 1$. *There exist constants* $c_i > 0$ *such that*

$$\mathbf{E}\exp\left\{\tfrac{1}{2}t^2|\mathcal{R}_m\bar{\zeta}|^2\right\}\mathbf{1}_{\{|\mathcal{R}_m\zeta^*|\le R\}} \;\le\; c_1\left(1 + \mu t^2/[1-\mu t^2]\right)$$

if $\varepsilon\left(1 + R/[\mu r]\right) \le c_0\left(1 - \mu t^2\right)$.

Proof of Lemma 6.2.7. By (6.2.20) the conditions of the lemma imply the inclusion $\{|\mathcal{R}_m\zeta^*| \le R\} \subset \{|\mathcal{R}_m\bar{\zeta}| \le 3R\}$. One can apply Lemma 6.2.5 and the Chebyshev inequality in the form

$$\mathbf{P}\{|\mathcal{R}_m\bar{\zeta}|^2 \ge u\} \;\le\; 2e^{-v\sqrt{u}}\mathbf{E}\cosh(v|\mathcal{R}_m\bar{\zeta}|)$$

to obtain, after putting $v = \frac{1}{\mu}\sqrt{u}$, the estimate

$$\mathbf{P}\left\{|\mathcal{R}_m\bar{\zeta}|^2 \ge u\right\} \;\le\; 2\exp\left\{-\tfrac{1}{2}u\mu^{-1}(1 - c'\varepsilon\mu^{-1}(1 + R/[\mu r]))\right\},$$

which is valid if $u \le (3R)^2$ and $\varepsilon(1 + R/[\mu r]) \le c''$. It remains to integrate by parts to obtain an upper bound in terms of the quantity

$$\begin{aligned}\int_0^{3R} e^{\frac{1}{2}t^2 u}\, d\left(-\mathbf{P}\{|\bar{\zeta}| \ge u\}\right) \;&\le\; 1 + \tfrac{1}{2}t^2 \int_0^{3R} e^{\frac{1}{2}t^2 u}\mathbf{P}\{|\bar{\zeta}| \ge u\}du \\ &\le\; 1 + \tfrac{1}{2}t^2\left[(2\mu)^{-1} - \tfrac{1}{2}t^2 - c'\varepsilon\mu^{-1}(1 + R/[\mu r])\right].\end{aligned}$$

This proves the lemma. $\bigcirc$

The calculations below need a version of Lemma 3.4.2. Recall that the subspace $(\mathcal{I} - \mathcal{R}_m)X$ is m-dimensional.

Lemma 6.2.8 *Suppose that the conditions of Theorem* 6.2.1 *are satisfied and*

$$m \ge \nu, \quad R \ge r, \quad \kappa \overset{def}{=} \max\left\{s^2, \beta t^2\right\} < 1.$$

There is a constant c_0, *which depends on* m *only, such that if*

$$\varepsilon\left[1 + (s+t)^2\,(R/r)\right] \;\le\; c_0\min\left[1, (1-\kappa)(s+t)^{-2}\right],$$

then

$$\mathbf{E}\exp\left\{\tfrac{1}{2}\left[s^2|\mathcal{P}\bar{\zeta}|^2 + t^2|(\mathcal{R} - \mathcal{R}_m\bar{\zeta})|^2\right]\right\}\mathbf{1}_{\{|\zeta^*|\le R\}} \;\le\; c_1\,(1-\kappa)^{-m/2}.$$

Proof of Lemma 6.2.8. Put $\tilde{\zeta} = s\mathcal{P}\bar{\zeta} + t\,(\mathcal{R} - \mathcal{R}_m)\bar{\zeta}$. If the constant c_0 is chosen small enough, it follows from (6.2.20) that

$$\left\{|\zeta^*| \le R\right\} \subset \left\{|\bar{\zeta}| \le 3R\right\} \subset \left\{|\tilde{\zeta}| \le 3\,(s+t)\,R\right\}.$$

Below the subspace $X_m = (\mathcal{P} + [\mathcal{R} - \mathcal{R}_m] X)$ with some fixed orthonormal basis e_j, $j = \overline{1, m}$, is identified with $\mathbf{R}^m$.

Consider a standard Gaussian RV $\eta = \sum_{j=1}^m Y_j e_j \in \mathbf{R}^m$ whose coordinates $Y_j \in \mathbf{R}$ are independent of each other (with standard normal distribution) and $\bar{\zeta}$. Denote by $\mathbf{M}$ the conditional expectation with respect to distribution of η given fixed values of all other rv's, and use $\mathbf{E}$ as notation for conditional expectations given the value of η. In this notation

$$\exp\left\{\tfrac{1}{2}\left[s^2|\mathcal{P}\bar{\zeta}|^2 + t^2|(\mathcal{R} - \mathcal{R}_m)\bar{\zeta}|^2\right]\right\} = \mathbf{M}\exp\left\{\langle\tilde{\zeta}, \eta\rangle\right\}.$$

If $|\tilde{\zeta}| \leq 3\,(s+t)\,R$, then known asymptotic expressions for incomplete Gamma-functions show that

$$\mathbf{M}\exp\left\{\langle\tilde{\zeta}, \eta\rangle\right\}\mathbf{1}_{\{|\eta| \geq 9(s+t)R\}} \leq \exp\left\{\tfrac{1}{2}|\tilde{\zeta}|^2\right\}\mathbf{M}\mathbf{1}_{\left\{|\eta| > 2|\tilde{\zeta}|\right\}} \leq c(m).$$

It remains to evaluate

$$z = \mathbf{E}\mathbf{M}\exp\left\{\langle\tilde{\zeta}, \eta\rangle\right\}\mathbf{1}_{\{|\eta| \leq 9(s+t)R\}}\mathbf{1}_{\left\{|\tilde{\zeta}| \leq 3(s+t)R\right\}}.$$

The sum $\tilde{\zeta}$ consists of RV's which satisfy the inequality

$$|s\mathcal{P}\bar{\xi}_j + t(\mathcal{R} - \mathcal{R}_m\bar{\xi}_j)| \leq (s+t)\,|\bar{\xi}_j|.$$

One can repeat the argument from the proof of Lemma 6.2.5 and use this circumstance to obtain the inequality

$$\begin{aligned} \mathbf{E}\exp\left\{\langle\tilde{\zeta}, \eta\rangle\right\} &\leq \exp\left\{\tfrac{1}{2}\left(s^2|\mathcal{P}\eta|^2 + t^2\beta|(\mathcal{R} - \mathcal{R}_m)\eta|^2\right)\right\} \\ &\times \exp\left\{c'\left(\varepsilon|\eta|^2[s+t]^2 + |\eta|^3[s+t]^3 H/B\right)\right\} \end{aligned}$$

which is valid in the whole domain of integration with respect to the distribution of η if $\varepsilon\left(1 + [s+t]^2 R/r\right) \leq c''$ (recall that there $|\eta| \leq 9\,(s+t)\,R$). Under the stronger restriction of the lemma, the estimate assumes the form

$$\mathbf{E}\exp\left\{\langle\tilde{\zeta}, \eta\rangle\right\} \leq \exp\left\{\tfrac{1}{2}|\eta|^2\left[1 - \tfrac{1}{2}(1-\kappa)\right]\right\}.$$

To complete the proof, one can use the known explicit formula for the CF of a Gamma-distribution, which yields the inequality $z \leq c'''(m)\,[1-\kappa]^{-m/2}$ and the bound of the lemma. ○

6.2.5 Derivation of "Correct" Bernstein-Type Bound

In the calculations that follow, notation $c, c_i, \bar{c}$, etc. is used for the constants whose values are determined by the dimension numbers ν, m, and ratios of variances β, μ only. These constants are enumerated locally, and only if this is necessary to avoid confusion. No attempt is made at writing out their explicit numerical values. In different parts of the proof, same notation can refer to constants with different numerical values.

The parameter $\varepsilon = rH/B$ is invariably assumed small, viz., $\varepsilon \leq c_0$.

Proof of Theorem 6.2.1. (a) The probability considered in the theorem admits the estimate

$$\mathbf{p}_n(r) \leq 2\mathbf{P}\{|\zeta| \geq r\}, \tag{6.2.21}$$

where the additional centered Gaussian summand ξ^0 is included into the sum ζ of (6.2.17) to trim its covariances. Indeed, whatever the choice of $a \in X$

$$\mathbf{P}\{|a+\xi^0| \geq |a|\} \geq \tfrac{1}{2}\mathbf{P}\{\max[|a+\xi^0|, |a-\xi^0|] \geq |a|\} = \tfrac{1}{2}$$

by the symmetry of $\mathcal{L}(\xi^0)$ and the trianle inequality.

By Lemma 6.2.6, one can concentrate on the probability of landing in the layer between the original sphere and one of radius ρr, $\rho = \sqrt{\nu + \beta(m-\nu) + \mu}$. It yields the estimate

$$\begin{aligned} \mathbf{p}_n(r) &\leq 2\overline{\mathbf{p}}_n(r) + 4\exp\{-\tfrac{1}{2}r^2 + c\varepsilon r^2\}, \\ \overline{\mathbf{p}}_n(r) &= \mathbf{P}\{r \leq |\zeta| < \rho r\}. \end{aligned} \tag{6.2.22}$$

Thus, a sufficiently precise bound on $\overline{\mathbf{p}}_n(r)$ is necessary to prove the theorem.

(b) An application of Lemma 6.2.2 (with $V = \infty$ and $R = \rho r$) reduces the task to considering the probabilities of landing in "slices" of the spherical ring:

$$\overline{\mathbf{p}}_n(r) \leq \frac{r^{\nu-1}}{|S|}\int_S \mathbf{E}\mathcal{N}_e(\zeta)\, d\sigma(e), \tag{6.2.23}$$

where

$$\mathcal{N}_e(x) = q\left(|Q_e x|^2\right) \mathbf{1}_{[0,\infty)}(\langle x, e\rangle)\mathbf{1}_{[r,\rho]}(|x|)\ \tilde{q}\left(\tfrac{1}{r}x, \tfrac{1}{\sqrt{r}}\right).$$

While calculating the expectation in (6.2.23), it is convenient to use the Cramér transform with the vector parameter (6.2.18). This is done to to shift the expectation of the transformed sum ζ^* near the point re of the principal eigenspace, which selects the "slice". (In the Gaussian case this point would be an extremum of the DF.) Lemma 3.6.3 shows that the expectation of the transformed sum is indeed not too far from the chosen point.

All higher-order semi-invariants of Gaussian laws are zero. Lemma 6.2.3 and (6.2.20) yield the estimate

$$\begin{aligned} \mathbf{E}\mathcal{N}_e(\zeta) &= \exp\{\ln \mathbf{E}e^{\langle Bh,\zeta\rangle} - \langle Bh, \mathbf{E}\zeta^*\rangle\} \\ &\times\ \mathbf{E}e^{-\langle Bh,\bar{\zeta}\rangle}\mathcal{N}_e(\zeta^*) \leq \exp\{-\tfrac{1}{2}r^2 + c\varepsilon r^2\}\mathbf{m}(e), \end{aligned} \tag{6.2.24}$$

where

$$\mathbf{m}(e) \stackrel{def}{=} \mathbf{E}\, e^{-\langle re,\bar{\zeta}\rangle}\mathcal{N}_e(\mathbf{E}\zeta^* + \bar{\zeta}). \tag{6.2.25}$$

(c) In the right-hand side of (6.2.25), the exponent may take on large values for some values of $\bar{\zeta}$. This makes the evaluation of $\mathbf{m}(e)$ rather laborious.

In the conditions of the theorem one can choose two numbers, $p > 1$ and $\gamma \in (0,1)$, so as to satisfy the restrictions

$$(1-\gamma^2)p < 1,\quad (1+\gamma)\beta p < 1,\quad \frac{\mu(1+\gamma)}{1-1/p} < 1. \tag{6.2.26}$$

The range of possible values of these new parameters is determined by that of

μ, β. In the rest of the proof the values of γ, p are fixed. The part played by restrictions (6.2.26) will become clear later on.

The calculations use the RV

$$\widehat{\zeta} \overset{def}{=} \mathbf{E}\zeta^* - re + \bar{\zeta}. \tag{6.2.27}$$

It follows from (6.2.20) that $-r\langle e, \bar{\zeta}\rangle \le -r\langle e, \widehat{\zeta}\rangle + c\varepsilon r^2$, so

$$\mathbf{m}(e) \le \exp\{c\varepsilon r^2\}\mathbf{E}\mathcal{N}_e^+(\widehat{\zeta}), \quad \mathcal{N}_e^+(x) \overset{def}{=} e^{-\langle re, x\rangle}\mathcal{N}_e(re + x). \tag{6.2.28}$$

It is convenient to analyze the behavior of $\mathcal{N}_e^+$ using in X the "coordinates"

$$\widehat{\xi} = \langle x, e\rangle, \quad Y = |\mathcal{Q}_s x|, \quad Z = |\mathcal{R}x|. \tag{6.2.29}$$

In the domain where $\mathcal{N}_e^+ \neq 0$,

$$\widehat{\xi} \ge -r, \quad 0 \le Y \le 1, r^2 \le |re + x|^2 \equiv \left(\widehat{\xi} + r\right)^2 + Y^2 + Z^2 \le \rho^2 r^2. \tag{6.2.30}$$

The argument of the exponential in the expression $\mathcal{N}_e^+$ can be written in the form

$$-\langle re, x\rangle = -r\widehat{\xi} = \tfrac{1}{2}\left[r^2 - |re + x|^2\right] + \tfrac{1}{2}\left(\widehat{\xi}^2 + Y^2 + Z^2\right). \tag{6.2.31}$$

The choice of domain imposes restriction on the "transversal" coordinate Y. For this reason the projection of $re + x$ on the principal eigenspace E admits the bound

$$|\mathcal{P}(re + x)|^2 = Y^2 + \left(\widehat{\xi} + r\right)^2 \le \left(1 + \widehat{\xi} + r\right)^2. \tag{6.2.32}$$

To proceed with the evaluations, it is convenient to subdivide into a few subdomains the set $\{\mathcal{N}_e^+ \neq 0\}$.

(c1) If $\widehat{\xi} \ge 0$ and $r^2 \ge \nu - 1$, then by (6.2.32) and Lemma 6.2.1

$$\mathcal{N}_e^+(x) \le ce^{-r\widehat{\xi}} \le c. \tag{6.2.33}$$

The constant in the final bound is determined by the choice of the function q in the "focusing device" and the dimension number ν.

(c2) Consider the domain selected by restrictions (6.2.30). For its points satisfying the additional restriction

$$\widehat{\xi} < 0, \quad Y^2 + Z^2 \ge \gamma r^2, \tag{6.2.34}$$

where γ is one of parameters (6.2.26), there is inequality $\widehat{\xi}^2 \le r^2 \le \frac{1}{\gamma}\left(Y^2 + Z^2\right)$. Consequently, the exponent (6.2.31) admits the estimate

$$-\langle x, e\rangle = -r\widehat{\xi} \le \tfrac{1}{2}\left(\widehat{\xi} + Y^2 + Z^2\right) \le \tfrac{1}{2}(1 - \gamma^2)\widehat{\xi}^2 + \tfrac{1}{2}(1 + \gamma)Z^2 + 1.$$

Besides, Lemma 6.2.1 and (6.2.32) show that $q\widehat{q} \le c$. It follows that restrictions (6.2.30) and (6.2.34) imply validity of the bound

$$\mathcal{N}_e^+(x) \le c' \exp\left\{\tfrac{1}{2}(1 - \gamma^2)\widehat{\xi}^2 + \tfrac{1}{2}(1 + \gamma)Z^2\right\} \tag{6.2.35}$$

(the constant being of the above described type).

(c3) If, contrary to the assumption of (c2),

$$\widehat{\xi} < 0, \quad Y^2 + Z^2 < \gamma r^2, \tag{6.2.36}$$

then

$$-r\widehat{\xi} \leq r\left(r - \sqrt{r^2 - Y^2 - Z^2}\right) \leq \frac{Y^2 + Z^2}{1 + \sqrt{1-\gamma}} \leq \tfrac{1}{2}(1+\gamma)\left(Y^2 + Z^2\right).$$

Hence

$$\mathcal{N}_e^+(x) \leq c'' \exp\left\{\tfrac{1}{2}(1+\gamma)Z^2\right\}. \tag{6.2.37}$$

Finally, inequalities (6.2.33), (6.2.35), and (6.2.37) provide the bound that is valid for all values of the argument of $\mathcal{N}_e^+$:

$$\mathcal{N}_e^+(x) \leq \tilde{c} \exp\left\{\tfrac{1}{2}(1-\gamma^2)\widehat{\xi}^2 + \tfrac{1}{2}(1+\gamma)Z^2\right\} \mathbf{1}_{\{|re+x| \leq \rho r\}}. \tag{6.2.38}$$

(d) Inequality (6.2.38) reduces the task to evaluation of the expectation of the rv

$$W \stackrel{def}{=} \exp\left\{\tfrac{1}{2}(1-\gamma^2)|\mathcal{P}\widehat{\zeta}|^2 + \tfrac{1}{2}(1+\gamma)|\mathcal{R}\widehat{\zeta}|^2\right\}\widehat{I}, \quad \widehat{I} = \mathbf{1}_{\{|\zeta^\bullet| \leq \rho r\}}$$

(recall that $\widehat{\zeta}$ was defined in (6.2.27)).

Inequality (6.2.20) can be used to verify that for small enough values of ε and each bounded linear operator $\mathcal{A}$

$$|\mathcal{A}\zeta|^2 \leq \left[|\mathcal{A}\widehat{\zeta}| + cr\varepsilon|\mathcal{A}|\right]^2 \leq (1+\varepsilon)\,|\mathcal{A}\widehat{\zeta}|^2 + cr^2\varepsilon\,|\mathcal{A}|^2$$

(below $\mathcal{A}$ will be a projection, so $|\mathcal{A}| = 1$). Hence for small ε, $\varepsilon \leq \widehat{c}_0$,

$$W \leq \widehat{c}_1 \widehat{I} \exp\left\{\widehat{c}_2 r^2\varepsilon + \tfrac{1}{p}V_1 + \left(1 - \tfrac{1}{p}\right)V_2\right\},$$

where

$$\begin{aligned} V_1 &= \tfrac{1}{2}p\,(1+\varepsilon)\left[\left(1-\gamma^2\right)|\mathcal{P}\bar{\zeta}|^2 + (1+\gamma)\,|(\mathcal{R} - \mathcal{R}_m)\,\bar{\zeta}|^2\right], \\ V_2 &= \frac{1+\gamma}{2(1-1/p)}|\mathcal{R}_m\bar{\zeta}|^2. \end{aligned}$$

The estimate

$$\mathbf{E}\,W \leq \check{c}_1 \exp\left\{\check{c}_2\, r^2\,\varepsilon\right\}, \quad \varepsilon \leq \check{c}_0, \tag{6.2.39}$$

now follows from the Hölder inequality, Lemma 6.2.7, and Lemma 6.2.8.

The assertion of the theorem results from estimates (6.2.21)–(6.2.39). ○

6.3 Moderately Large Deviations for Sums

In this section $\xi, \xi_1, \ldots$ are i.i.d. RV's satisfying the Cramér condition (3.6.2), and the aim of calculations is an asymptotic formula for probability (6.0.1),

$$\mathbf{p}_n(r_n) \stackrel{def}{=} \mathbf{P}\{\Phi(\tfrac{1}{r_n}\zeta_n) \geq 0\}, \quad \zeta_n = \tfrac{1}{\sqrt{n}}(\xi_1 + \cdots + \xi_n).$$

The RV's have zero expectation and covariance operator $\mathcal{V}$ (see the introduction to this chapter for notation).

In calculation of LD asymptotics, the principal object is DF (4.1.2). For the normalized sum ζ_n the DF is

$$\Lambda(a; \zeta_n) = \sup_{h \in X} [\langle a, h\rangle - \ln \mathbf{E} \exp\{\zeta_n, h\}],$$

and it is easily seen that

$$\ln \mathbf{E} \exp\{\zeta_n, h\} = n \ln \varphi\left(\tfrac{1}{\sqrt{n}} h\right) = \tfrac{1}{2}\langle \mathcal{V}h, h\rangle + \mathcal{O}(\tfrac{1}{\sqrt{n}}), \tag{6.3.1}$$

where $\varphi(h) = \mathbf{E} \exp\{\langle \xi, h\rangle\}$. The principal term on the right-hand side of the last formula is the logarithm of the Laplace transform for the accompanying Gaussian law. This makes plausible the similarity between LD asymptotics of normalized sums and Gaussian RV's (see Section 6.1 and subsection 6.2.1). For "rough" LD asymptotics this similarity is established in Theorem 4.6.1. The corresponding result for "fine" asymptotics of moderately large deviations is Theorem 6.3.1 below.

The new feature in "fine" theorems on large deviations is the appearance of a special correction, intermediate in its order of magnitude between the quadratic exponential and polynomial.

For distributions of real-valued sums, this correction is described by the so-called Cramér series, and this name will be used to refer to it also in the case of vector-valued RV's. Its construction is quite laborious.

The main result of the section is Theorem 6.3.1 stated in subsection 6.3.3. Its proof and the necessary preparations occupy the next subsections.

6.3.1 Focusing and Choosing Cramér Transform

To state Theorem 6.3.1, it is necessary to introduce numerous conditions. Its proof follows the pattern of Section 6.1, and most restrictions and notation are borrowed from it.

The small parameters below are

$$0 < 1/\sqrt{n} \ll 1, \quad 0 < \tau = r/\sqrt{n} \ll 1, \quad 1/r \ll 1. \tag{6.3.2}$$

The function Φ obeys restrictions (6.1.4) and (6.1.15)–(6.1.16).

It is assumed that the set where the Gaussian DF (6.0.4) attains its minimum on $\{\Phi \geq 0\}$ admits the parameterization described in (6.1.3)–(6.1.7).

Once again, it is assumed that this parameterization can be used to focus on one minimum of Λ_0 at a time applying specific decomposition of unity. The "focusing device" includes the functions

$$q_s(|x|) \geq 0, \quad \widehat{q}_s(x; v) \geq 0, \quad s \in S_0, \quad x \in X, v \in [0, 1],$$

a natural-valued "effective dimension number" $D(s)$, and one more function

$$Q\left(x; \tfrac{1}{r}\right) = \int_{S_0} r^{D(s)} q_s\left(|rQ_s[x - x(s)]|\right) \widehat{q}_s\left(x, \tfrac{1}{r}\right) d\sigma(s) \tag{6.3.3}$$

(with $d\sigma(s)$ being the measure of (6.1.7)).

All the above functions are supposed to comply with obvious measurability requirements and some additional restrictions. These latter seem to be quite natural although no attempt is made here to make them minimal. It is supposed that the restrictions listed below are uniform in the "tag" $s \in S_0$.

The "focusing condition" (6.1.12) used in the Gaussian case is modified. As before, the function $q_s(u) \geq 0$ is even and decreases on the positive half-line. Moreover,

$$q_s(u) = 0, \quad |u| \geq 1, \quad \left|\frac{d^k}{du^k} q_s(u)\right| \leq \Theta_k, \quad k = 0, 1, \ldots \tag{6.3.4}$$

It is convenient to include in the function $\widehat{q}$ one more smooth cutoff factor to make it vanish outside a neighborhood of the selected extremum $x_0(s)$. Namely, $\widehat{q}_s(x; v)$, $x \in X$, $v \in [0, 1]$, vanishes outside a neighborhood of the extremum selected by s:

$$\widehat{q}_s(x; v) = 0 \text{ if } |x - x(s)| \geq \tfrac{1}{F} > 0. \tag{6.3.5}$$

As a consequence, $\widehat{q}_s$ (and through it also the function Q of (6.1.12)) is supposed to depend on $v = \frac{1}{r}$, and a new parameter $F > 0$. (To avoid encumbering notation it is not mentioned explicitly.) Dependence on the "spatial" variable $x \in X$ and the small parameter $v = \frac{1}{r}$ is smooth if F is fixed.

The function $\widehat{q}$ and its derivatives admit the estimates

$$\left|\nabla_x^{\otimes k}(\partial^l/\partial v^l)\widehat{q}_s(x; v)\right| \leq \widehat{\Theta}_{k,l}. \tag{6.3.6}$$

The new focusing condition is as follows: for $\tau \leq c_0$ and any $F > 0$

$$\begin{aligned} \left|\mathbf{P}\left\{\Phi\left(\tfrac{1}{r}\zeta_n\right) \geq 0\right\}\right. &- \left.\mathbf{E}Q_r(\zeta_n)\mathbf{1}_{\{\Phi \geq 0\}}\left(\tfrac{1}{r}\zeta_n\right)\right| \\ &\leq c_1(F)\exp\left\{-\left[\Lambda_\Phi + \delta(F)\right]r^2\right\}, \end{aligned} \tag{6.3.7}$$

where $\delta(F) > 0$ depends on the parameter selecting the neighborhood of $x(s)$ in (6.3.5).

Verification of the focusing condition may prove difficult, but its use seems justified by the fact that there are situations of interest where it holds (cf. Section 6.2).

The following lemma shows that it is possible to localize the calculations and consider only a neighborhood of a single extremum of the Gaussian DF at a time.

Lemma 6.3.1 *Under condition* (6.3.7) *there is representation*

$$\mathbf{P}\left\{\ \Phi\left(\tfrac{1}{r}\zeta_n\right)\ \geq 0\right\} = \int_{S_0} r^{D(s)}\mathbf{E}\widetilde{\mathcal{M}}_s\left(\zeta_n\right)d\sigma\left(s\right) + \rho_{\mathrm{f}}\left(r,n\right),$$

where

$$\widetilde{\mathcal{M}}_s(x) \stackrel{def}{=} q_s\left(|rQ_s(x - rx(s))|\right)\mathbf{1}_{\{\Phi\geq 0\}}\left(\tfrac{1}{r}x\right)\widehat{q}_s\left(\tfrac{1}{r}x,\tfrac{1}{r}\right),$$

and the order of the remainder is determined by the choice of F *in* (6.3.5)*:*

$$|\rho_{\mathrm{f}}\left(r,n\right)| \leq c\exp\left\{-[\Lambda_\Phi + \delta(F)]r^2\right\}.$$

Proof of Lemma 6.3.1 is a straightforward application of (6.3.3) and condition (6.3.7). ○

A special Cramér transform. While calculating the "slice" expectation in the right-hand side of the equation of Lemma 6.3.1, it is natural to separate the effects which determine exponential decay as $r \to \infty$ from those responsible for the finer details of asymptotic behavior. To this end, the distribution of ζ_n is submitted to the Cramér transform. Its aim is to bring the expectation of the resulting distribution $\mathcal{L}(\zeta_n^*)$ as close as possible to $rx(s)$.

For a Gaussian RV with DF (6.0.4), the possibility to shift the expectation to $rx(s)$ applying the Cramér transform to $\mathcal{L}(\zeta_n)$ is postulated in (6.1.3) and (6.1.7). Unfortunately, the DF of ζ_n differs from Λ_0, and this makes the choice of parameter more difficult. It may not be possible to match $\mathbf{E}\zeta_n^*$ and $e(s)$ exactly. The next best thing is to satisfy conditions (6.1.26) (see Lemma 6.1.2).

Select an extremum of the Gaussian DF: fix a point $s \in S_0$ and thus determine h_0, x_0, n_0, and β_0 of (6.1.27). In the calculations below the "position tag" $\cdot_s$ is usually skipped: $Q = Q_s$, etc.

To shift the expectation of a centered Gaussian RV η with covariance operator $\mathcal{V}$ to $rx(s)$ by Cramér's transform, the vector parameter of the latter should equal $rh(s)$. By the stability of Gaussian laws, the RV η is distributed as the normalized sum of its independent copies: $\eta \sim \frac{1}{\sqrt{n}}\sum_{j=1}^{n}\eta_j$. The Cramér conjugate to this Gaussian sum has expectation $rx(s)$ if the distributions of individual summands are each changed to its Cramér conjugate with the parameter $\tau h(s)$.

For the summands in the non-Gaussian sum ζ_n, the parameter of the Cramér transform is sought in the form $\tau\widehat{h}$ with $\widehat{h}$ a perturbation of its "Gaussian" value (cf. (6.1.7)):

$$\widehat{h}\left(\tau,s\right) = h\left(s\right) + \tau g\left(\tau,s\right). \tag{6.3.8}$$

(The function g will be smooth in τ later on.) For this choice of parameter, the Cramér transform $F^*_{\tau\widehat{h}}$ of the distribution $F = \mathcal{L}\left(\xi\right)$ has expectation $\tau\widehat{x}$ with

$$\widehat{x}\left(\tau,s\right) = \frac{1}{\tau}\nabla\ln\mathbf{E}e^{\langle\xi,H\rangle}\Big|_{H=\tau\widehat{h}} = \frac{1}{\tau}\left(\mathbf{E}e^{\langle\xi,\tau\widehat{h}\rangle}\right)^{-1}\mathbf{E}e^{\langle\xi,\tau\widehat{h}\rangle}\xi. \tag{6.3.9}$$

The unit outward normal vector to the hypersurface $\{\Phi(x) = \Phi\left(\widehat{x}\right)\}$ at this point is

$$\widehat{n}\left(\tau,s\right) = \mathbf{n}\left(\widehat{x}\right),\ \ \mathbf{n}\left(x\right) \stackrel{def}{=} \left|\nabla\Phi\left(x\right)\right|^{-1}\nabla\Phi\left(x\right). \tag{6.3.10}$$

Below, the vector parameter $\widehat{h}$ is chosen so as to satisfy equations (6.1.26)–(6.1.27) with τ playing the part of the additional small parameter v. Recal that these equations read

$$\Phi(\widehat{x}) = 0,\ Q_s[\widehat{x} - x_0] = 0,\ [\mathcal{I} - Q_s]\widehat{\beta}\widehat{n} = [\mathcal{I} - Q_s]\widehat{h}, \tag{6.3.11}$$

$$\widehat{h}(0, s) = h_0 \stackrel{def}{=} h(s),\ \widehat{x}(0, s) = x_0,\ \widehat{\beta}(0, s) = \beta_0, \tag{6.3.12}$$

where $x_0 \stackrel{def}{=} x(s)$ and $\widehat{\beta}(\tau, s)$ is a positive function. As before (see (6.1.5)–(6.1.6)),

$$h_0 = \beta_0 n_0,\ \beta_0 = \sqrt{\Lambda_\Phi / \langle \mathcal{V} n_0, n_0 \rangle}$$

with $n_0 \stackrel{def}{=} \mathbf{n}(x_0)$. The corresponding initial value of the normal is $\widehat{n}(0, s) = n_0$.

Equations (6.3.11)–(6.3.12) select two functions, $\tau \mapsto g(\tau, s) \in \mathbf{X}$ and $\tau \mapsto \widehat{\beta}(\tau) \in \mathbf{R}$.

In other terms, system (6.3.11) specifies the choice of parameter in the Cramér transform which shifts the mean value of the transformed distribution to a point on $H_\Phi = \{\Phi = 0\}$ whose "transversal" projection QX is that of $x(s)$, and where the normal to H_Φ inherits from n_0 the direction of its projection $[\mathcal{I} - Q]\,\bar{n}(\widehat{x})$. The possibility to solve (6.3.11) is discussed in Lemma 6.3.3 in a later subsection.

6.3.2 The Cramér Series

Fix a value of the "position tag" s in the decomposition of Lemma 6.3.1. It will usually be dropped in notation of (6.3.8)–(6.3.12).

Suppose that $\widehat{h} = \widehat{h}(\tau, s)$ is a smooth function of τ satisfying (6.3.11). The vector $\tau\widehat{h}$ is used as the parameter of the Cramér transform (3.6.4) applied to the distribution of an individual summand $\mathcal{L}(\xi)$. The resulting distribution is

$$F^*_{\tau\widehat{h}}(dx) = \frac{e^{\langle x, \tau\widehat{h}\rangle}\mathbf{P}\{\xi \in dx\}}{\mathbf{E}e^{\langle \xi, \tau\widehat{h}\rangle}}. \tag{6.3.13}$$

Notation $\xi^*, \xi_1^*, \ldots$ is used for i.i.d. RV's whose distribution is $F_{\tau\widehat{h}}$. The normalized sum $\bar{\zeta}_n = \frac{1}{\sqrt{n}}\left(\bar{\xi}_1 + \cdots \bar{\xi}_n\right)$ consists of the i.i.d. RV's $\bar{\xi}_1, \ldots$ that are copies of the centered RV $\bar{\xi} = \xi^* - \mathbf{E}\xi^*$.

Denote by

$$U(x) = \frac{r\left[\Phi\left(\widehat{x} + \frac{1}{r}x\right) - \Phi(\widehat{x})\right]}{|\nabla\Phi(\widehat{x})|}$$

the only "nonlinear" coordinate of mapping (6.1.13) associated with $\widehat{x}(\tau, s)$.

Lemma 6.3.2 *If the parameter* $\frac{1}{F} > 0$ *in* (6.3.6) *is small enough and the vector parameter* $\tau\widehat{h}$ *of the Cramér transform* (6.3.13) *satisfies* (6.3.11), *then the expectation on the right-hand side of the equation of Lemma* 6.3.1 *can be written down as*

$$\mathbf{E}\, q_s(|rQ_s\zeta_n|)\widehat{q}_s\left(\tfrac{1}{r}\zeta_n - x_0, \tfrac{1}{r}\right)\mathbf{1}_{\{\Phi \geq 0\}}\left(\tfrac{1}{r}\zeta_n\right) \tag{6.3.14}$$

$$= \exp\left\{-\Lambda_\Phi r^2 + nC(\tau, s)\right\}\mathbf{E}\mathcal{E}\left(\overline{\zeta}_n; s, \tfrac{1}{r}, \tau\right),$$

where the "Cramér series" is

$$C(\tau, s) = \tfrac{1}{2}\mathbf{E}\left\langle \xi, \tau h_0\right\rangle^2 + \ln \mathbf{E}e^{\langle \xi, \tau\widehat{h}\rangle} - \left(\mathbf{E}e^{\langle \xi, \tau\widehat{h}\rangle}\right)^{-1}\mathbf{E}\langle \xi, \tau\widehat{h}\rangle e^{\langle \xi, \tau\widehat{h}\rangle}, \quad (6.3.15)$$

and

$$\begin{aligned}\mathcal{E}\left(x; s, \tfrac{1}{r}, \tau\right) &\stackrel{def}{=} \exp\left\{-\widehat{\beta}rU(x) + \tfrac{1}{2}\widehat{\beta}U^2(x)\right\}\mathbf{1}_{[0, c_1 r/F]}\left(U\left(x\right)\right)\\ &\times \quad q\left(\left|rQ_s x\right|\right)\exp\left\{\langle\widehat{\beta}\widehat{n} - \widehat{h}, rQ_s x\rangle\right\}\Psi\left(x\right),\end{aligned}$$

with

$$\begin{aligned}\Psi\left(x\right) &\stackrel{def}{=} \widehat{q}_s\left(x_0 + \left[\widehat{x} - x_0\right] + \tfrac{1}{r}x; \tfrac{1}{r}, \tfrac{1}{F}\right)\exp\left\{\tfrac{1}{2}\langle\widehat{\Phi}'' x, x\rangle + \rho_\Phi(x)\right\},\\ \langle\widehat{\Phi}'' x, x\rangle &= \widehat{\beta}\frac{\langle \mathcal{R}_s x, \nabla\rangle^2 \Phi\left(\widehat{x}\right)}{\left|\nabla\Phi\left(\widehat{x}\right)\right|}.\end{aligned}$$

In the part of space where $\mathcal{E}$ does not vanish,

$$\left|\langle\widehat{\beta}\widehat{n} - \widehat{h}, rQ_s x\rangle\right| \le c_1\tau,$$

and

$$\left|\rho_\Phi\left(x\right)\right| \le c_2\left(\min\{\tfrac{1}{r}\left|x\right|^3, \tfrac{1}{F}\left|x\right|^2\} + \tfrac{1}{r} + \left(\tfrac{1}{\sqrt{F}} + \tfrac{1}{r}\right)\left|\mathcal{R}x\right|^2 + \sqrt{F}U^2\left(x\right)\right).$$

The derivatives of the factor Ψ in x and $v = \frac{1}{r}$ admit the estimates

$$\left|\nabla^{\otimes k}\frac{\partial^l}{\partial v^l}\Psi\left(x\right)\right| \le c\left(k, l\right)\left[1 + \left|x\right|\right]^{B_{k,l}}\exp\left\{\tfrac{1}{2}\langle\widehat{\Phi}''\mathcal{R}x, \mathcal{R}x\rangle + \rho_\Phi\right\}.$$

For small values of $\tau \ge 0$ the "Cramér series" nC has the usual order: $nC\left(\tau, e\right) = \mathcal{O}\left(n\tau^3\right)$.

The exponential factor containing $\langle\widehat{\Phi}'' x, x\rangle$ can be large in the neighborhood of $\widehat{x}$ cut out by $\widehat{q}_s$. To avoid complications it is postulated below that this quadratic form satisfies the following summability condition: there exist constants $\alpha_1, \alpha_2 > 0$ and $A > 0$ such that for all k, n, u, v

$$\mathbf{E}\exp\left\{\tfrac{1}{2}(1 + \alpha_1)[\langle\widehat{\Phi}''\overline{W}, \overline{W}\rangle + \alpha_2|\overline{W}|^2]\right\} \le A, \quad (6.3.16)$$

where

$$\overline{W} = \overline{W}_{k,n}\left[u, v\right] = \tfrac{1}{\sqrt{n}}\left(\textstyle\sum_{j<k}\overline{\xi}_j + u\overline{\xi}_k + v\eta_k + \sum_{j>k}\eta\right)$$

is any "hybrid" sum (cf. (5.4.9)) of Gaussian and centered Cramér conjugate RV's.

Proof of Lemma 6.3.2. By the inversion formula (3.6.5) for the Cramér transform with parameter H, the following formula holds whatever the choice of the function Q:

$$\mathbf{E}Q\left(\zeta_n\right) = \exp\left\{n\left[\ln \mathbf{E}e^{\langle \xi, H\rangle} - \mathbf{E}\left\langle \xi^*, H\right\rangle\right]\right\}\mathbf{E}e^{-\langle\sqrt{n}H, \overline{\zeta}_n\rangle}Q\left(\sqrt{n}\mathbf{E}\xi^* + \overline{\zeta}_n\right).$$

For $H = \tau\widehat{h}$, the factor preceding the expectation is that of the lemma.

It remains to write down the exponential under the expectation sign so as to make tractable its exponent $-\langle\sqrt{n}H, \overline{\zeta}_n\rangle = -\left\langle r\widehat{h}, \overline{\zeta}_n\right\rangle$, which is large on a part of the integration domain. This is where (6.3.11) plays a crucial part (cf. the proof of Lemma 6.1.2). By this condition

$$\begin{aligned} r\langle \widehat{h}, \overline{\zeta}_n \rangle &= r\langle \widehat{h}, \mathcal{Q}_s\overline{\zeta}_n \rangle + r\widehat{\beta}\langle (\mathcal{I} - \mathcal{Q}_s)\widehat{n}, \overline{\zeta}_n \rangle \\ &= \langle \widehat{h} - \widehat{\beta}\widehat{n}, r\mathcal{Q}_s\overline{\zeta}_n \rangle + r\widehat{\beta}\langle \widehat{n}, \overline{\zeta}_n \rangle. \end{aligned} \tag{6.3.17}$$

In the right-hand side, the first summand is small in the integration domain. Indeed, for small τ there is relation $\widehat{h} - \widehat{\beta}\widehat{n} = \mathcal{O}(\tau)$ since $h_0 = \beta_0 h_0$ and $\left|r\mathcal{Q}_s\overline{\zeta}_n\right| \le 1$ (see (6.3.4)).

The function Φ vanishes at $\widehat{x}$, so in the right-hand side of (6.3.17)

$$\begin{aligned} r\widehat{\beta}\langle \widehat{n}, \overline{\zeta}_n \rangle &= r\widehat{\beta}\frac{\langle \overline{\zeta}_n, \nabla\rangle\Phi(\widehat{x})}{|\nabla\Phi(\widehat{x})|} = r\widehat{\beta}U(\overline{\zeta}_n) - r^2\widehat{\beta}\widetilde{\Phi}(\overline{\zeta}_n), \qquad (6.3.18) \\ \widetilde{\Phi}(x) &= \frac{\Phi(\widehat{x} + \frac{1}{r}x) - \Phi(\widehat{x}) - \frac{1}{r}\langle x, \nabla\rangle\Phi(\widehat{x})}{|\nabla\Phi(\widehat{x})|}, \end{aligned}$$

where U was introduced before the lemma, and

$$\begin{aligned} \widehat{\beta}r^2\widetilde{\Phi}(x) &= \frac{\widehat{\beta}}{2}\frac{\langle x, \nabla\rangle^2\Phi(\widehat{x})}{|\nabla\Phi(\widehat{x})|} + \rho_\Phi^{(1)}(x), \qquad (6.3.19) \\ \rho_\Phi^{(1)}(x) &= \frac{\widehat{\beta}}{2!r}\int_0^1 \frac{\langle x, \nabla\rangle^3\Phi(\widehat{x} + \frac{\vartheta}{r}x)}{\widehat{\beta}\,|\nabla\Phi(\widehat{x})|}(1-\vartheta)^2\, d\vartheta. \end{aligned}$$

By (6.3.5) the only values of x that matter are from a (r/F)-neighborhood of $\widehat{x}$, so the remainder term in (6.3.19) is evaluated by

$$\left|\rho_\Phi^{(1)}(x)\right| \le c\min\left[\tfrac{1}{F}|x|^2, \tfrac{1}{r}|x|^3\right] \le \tfrac{1}{F}|x|^2.$$

It remains to transform the quadratic term in (6.3.19). For the purpose, it is convenient to change from the coordinates $X = \langle x, \widehat{n}\rangle, Y = \mathcal{Q}_s x, Z = \mathcal{R}_s x$ to $U = U(x)$, Y, Z. The change of variables is possible in the $\frac{1}{F}$-neighborhood of $\widehat{x}$ selected by focusing provided that $\frac{1}{F}$ is small enough. The alteration of coordinates is small:

$$|X - U| \le c\min\left\{\tfrac{1}{F}|U|, \tfrac{1}{r}\left[U^2 + |Y|^2 + |Z|^2\right]\right\}.$$

It follows that

$$\frac{\widehat{\beta}}{2}\frac{\langle x, \nabla\rangle^2\Phi(\widehat{x})}{|\nabla\Phi(\widehat{x})|} = \frac{\widehat{\beta}}{2}U^2\frac{\langle \widehat{n}, \nabla\rangle^2\Phi(\widehat{x})}{|\nabla\Phi(\widehat{x})|} + \frac{\widehat{\beta}}{2}\frac{\langle Z, \nabla\rangle^2\Phi(\widehat{x})}{|\nabla\Phi(\widehat{x})|} + \rho_\Phi^{(2)}.$$

The Cauchy inequality and the restriction $|\mathcal{Q}_s x| \le \frac{1}{r}$ can now be used to verify the relation

$$\left|\rho_\Phi^{(2)}\right| \le c\left[\tfrac{1}{r} + \sqrt{F}U^2 + \tfrac{1}{\sqrt{F}}|Z|^2\right].$$

This proves the lemma. ○

6.3.3 The Normal Approximation

It is shown below that the expectation of Lemma 6.3.2 can be approximated by one where the normalized sum $\bar{\zeta}_n$ is replaced by a centered Gaussian RV.

It might seem more natural to attempt constructing a higher-order asymptotic expansion for this quantity in the spirit of Chapter 5. However, the obstacle to such an improved approximations is the lack of reasonable substitutes for the Cramér condition on CF's (see subsection 5.7.3). To undertake laborious calculations using conditions that have so far proved unrealistic seems unpractical, so only the simplest approximation to $\mathcal{L}\left(\bar{\zeta}_n\right)$ is considered in any detail — the centered Gaussian law with the covariance operator $\mathcal{V}$. The relative order of approximation proves to have the order $\mathcal{O}(\tau)$.

Theorem 6.3.1 *Assume that for a small enough* $\frac{1}{F} > 0$ *there is a focusing triple which satisfies conditions* (6.3.4)-(6.3.7), *and system* (6.1.26)-(6.1.27) *can be used to define the Cramér transform and seriies as in Lemma* 6.3.2.

Assume, moreover, that the function Φ *obeys restriction* (6.1.4), *and mapping* (6.1.13) *associated with the expectation of* (6.3.13) *satisfies the summability condition* (6.3.16), *and the nondegeneracy conditions* (5.6.6) *and* (5.6.35).

Then the following equality holds if $\frac{1}{r} + \tau \leq c_0$, $\tau \overset{def}{=} \frac{r}{\sqrt{n}}$:

$$\mathbf{p}_n \overset{def}{=} \mathbf{P}\{\Phi(\tfrac{1}{r}\zeta_n) \geq 0\} = \exp\{-\Lambda_\Phi r^2\}$$
$$\times \left[\int_{S_0} (1+\gamma)\, e^{nC(\tau,s)} r^{D(s)-d(s)} \mathbf{E}\, \mathcal{E}(\eta; s, \tfrac{1}{r}, \tau) d\sigma(s) + \rho_{\mathbf{f}}\right].$$

In the expectation under the integral η *is a centered Gaussian RV with the covariance operator of* ξ *and* $\Lambda_\Phi = \min\left\{\frac{1}{2}|\mathcal{V}^{-1/2}x|^2 : \Phi(x) \geq 0\right\}$. *The "Cramér series"* $C(\tau, e)$ *is defined in* (6.3.15), *and the functions* $\mathcal{E}$, Ψ *in Lemma* 6.3.2. *The error terms satisfy the inequalities*

$$|\gamma| \leq c_3\tau, \quad \rho_{\mathbf{f}} \leq c_4 e^{-\delta(F)r^2}, \quad \delta(F) > 0.$$

If compared to Theorem 6.1.1, Theorem 6.3.1 seems incomplete. The expectation under the integral sign is basically that of Theorem 6.1.1. One can obtain for it asymptotic representations similar to those of Theorem 6.1.1. The difference between $\mathcal{E}$ and the expression $\mathcal{M}$ of Theorem 6.1.1 is in certain correction terms. It seems preferable to dispose of these latter after the random argument is replaced by a Gaussian RV with covariances independent of the small parameters.

Proof of Theorem 6.3.1. Application of Lemma 6.3.1 and Lemma 6.3.2 reduces the task to calculating the expectation $\mathbf{E}\,\mathcal{E}\left(\bar{\zeta}_n\right)$. The function under the expectation sign can be written as

$$\mathcal{E}(x) = g \circ \phi(x)\, \Psi(x),$$

where $x \mapsto \phi(x) = U(x)e + Q_s x$ is a smooth mapping with finite-dimensional range, and the last factor Ψ satisfies (6.3.16). The finite-dimensional function is, in notation of (6.19),

$$\begin{aligned} g(u,y) &= \exp\{-r\widehat{\beta}u + \tfrac{1}{2}\langle\widehat{\Phi}''\widehat{n},\widehat{n}\rangle u^2\}\mathbf{1}_{[0,cr/F]}(u) \\ &\quad \times q(|ry|)\exp\{\langle\widehat{\beta}\widehat{n} - \widehat{h}, ry\rangle\}, \quad u \in \mathbf{R}, \quad y \in \mathbf{R}^d, \end{aligned}$$

(the subspace $Q_s X$ is here identified with $\mathbf{R}^d$).

The setting here is that of Theorem 5.6.1, so the Gaussian approximation can be obtained using this proposition. What remains is to derive the necessary bounds on the L_1-norm of g and its oscillation.

If $\frac{1}{F}$ is small, the quadratic term in the exponent is dominated by the linear one, so for the pertinent values of the argument

$$\exp\left\{-r\widehat{\beta}u + \tfrac{1}{2}\langle\widehat{\Phi}''\widehat{n},\widehat{n}\rangle u^2\right\} \le \exp\{-\tfrac{1}{2}r\widehat{\beta}u\}$$

and

$$\frac{c'}{r^{d+1}} \le \int_0^{\bar{c}/r} du \int_{\|y\|\le\bar{c}/r} g(u,y)\,dy \le \int_0^\infty du \int_{\mathbf{R}} g(u,y)\,dy \le \frac{c''}{r^{d+1}}. \tag{6.3.20}$$

Calculation of oscillation has to take into account the presence of a jump in u. To do this, it is convenient to write the difference as

$$\begin{aligned} g(u+\Delta u, y+\Delta y) - g(u,y) &= g(u+\Delta u, y) - g(u,y) \\ &+ g(u+\Delta u, y+\Delta y) - g(u+\Delta u, y). \end{aligned}$$

If εr is small (and later on $\varepsilon = \frac{1}{\sqrt{n}} \ll \frac{1}{r}$), then

$$\begin{aligned} &\int_{\mathbf{R}} \sup_{|\Delta u|\le\varepsilon} |g(u+\Delta u,y) - g(u,y)|\,du \\ &\le \left(2\int_0^\varepsilon e^{-cru}du + \int_0^{cr} e^{-ru/2}\left[1 - e^{-\tilde{c}\varepsilon r}\right]du\right)\mathbf{1}_{\{\|y\|\le 1/r\}} \le c\varepsilon\mathbf{1}_{\{\|y\|\le 1/r\}}. \end{aligned}$$

(Recall that g vanishes if $\|y\| > 1$.) Similarly,

$$\int_{\mathbf{R}} \sup_{\|\Delta y\|\le\varepsilon} |g(u+\Delta u, y+\Delta y) - g(u+\Delta u, y)|\,du \le c\varepsilon\mathbf{1}_{\{\|y\|\le\varepsilon+1/r\}},$$

so the final estimate is

$$\int_{\mathbf{R}} du \int_{\mathbf{R}^d} \operatorname{osc}_\varepsilon g(u,y)\,dy \le \frac{c\varepsilon}{r^d}. \tag{6.3.21}$$

It was established in the proof of Theorem 6.1.1, which applies to the Gaussian LD asymptotics, that the measure $G(B) = \mathbf{E}\mathbf{1}_B(\phi(\eta))\Psi(\eta)$ has a smooth bounded density which does not vanish at the origin. Consequently, it follows from Proposition 6.1.1, (6.3.20), and (6.3.21) that

$$\mathbf{E}\mathcal{E}(\eta) \ge c/r^{d+1}, \quad \mathbf{E}\operatorname{osc}_\varepsilon g\circ\phi(\eta) \le c\varepsilon/r^d.$$

For $\varepsilon = \frac{1}{\sqrt{n}}$, Theorem 5.6.1 now yields the relation

$$\mathbf{E}\,\mathcal{E}\left(\overline{\zeta}_n\right) = \mathbf{E}\,\mathcal{E}\left(\eta\right) + \mathcal{O}\left(\frac{\varepsilon}{r^d}\right) = \left[1+\mathcal{O}\left(\tau\right)\right]\mathbf{E}\,\mathcal{E}\left(\eta\right), \quad \tau = r\varepsilon.$$

This proves the theorem. ○

6.3.4 Solvability of Equations for Cramér Parameter

The properties of the hypersurface H_Φ near $x(s)$ are reflected in the operator $\ddot{\Phi}(x(s))$ of (6.1.14). For small values of τ, the vector $\widehat{x}$ and the unit normal one at this point can be written as

$$\frac{1}{\tau}\left(\widehat{x} - x_0\right) = \mathcal{V}g + \mu + \mathcal{O}(\tau), \quad \frac{\beta_0}{\tau}\left[\widehat{n} - n_0\right] = \ddot{\Phi}\left(\mathcal{V}g + \mu\right) + \mathcal{O}(\tau) \qquad (6.3.22)$$

with $\mu = \mathbf{E}\langle\xi, h_0\rangle^2\xi$. Recall that the centered RV ξ satisfies the Cramér condition (3.6.2) and its covariance operator $\mathcal{V}$ obeys (6.0.1).

Lemma 6.3.3 *If the linear operator*

$$\mathcal{R}_s - \mathcal{R}_s\ddot{\Phi}\widetilde{\mathcal{V}}^{1/2}\left[\mathcal{I} - \widetilde{\mathcal{Q}}\right]\widetilde{\mathcal{V}}^{1/2}\mathcal{R}_s \;:\; \mathcal{R}_sX \to \mathcal{R}_sX$$

has bounded inverse, then equations (6.3.11)–(6.3.12) *have a unique solution* $\left\{g, \widehat{\beta}\right\}$ *which is a smooth function of* τ.

In the above, $\widetilde{\mathcal{V}} = \mathcal{V}^{1/2}\left[\mathcal{I} - \mathcal{P}_{\widetilde{n}}\right]\mathcal{V}^{1/2}$*, where* $\mathcal{P}_{\widetilde{n}} : x \mapsto \langle x, \widetilde{n}\rangle\widetilde{n}$ *is the projection onto the direction of* $\widetilde{n} = \left|\mathcal{V}^{1/2}n_0\right|^{-1}\mathcal{V}^{1/2}n_0$ *and*

$$\widetilde{\mathcal{Q}} = \widetilde{\mathcal{V}}^{1/2}\mathcal{Q}_s\left[\mathcal{Q}_s\widetilde{\mathcal{V}}\mathcal{Q}_s\right]^{-1}\mathcal{Q}_s\widetilde{\mathcal{V}}^{1/2}$$

is one more projection operator.

Remark 6.3.1 The operator $\mathcal{Q}_s\widetilde{\mathcal{V}}\mathcal{Q}_s : \mathcal{Q}_s\mathbf{X} \to \mathcal{Q}_sX$ has bounded inverse because of condition (6.0.1). Indeed, the space where it acts is finite-dimensional, so it is sufficient to verify that

$$\mathcal{Q}_sy \neq 0 \implies \langle\mathcal{Q}_s\widetilde{\mathcal{V}}\mathcal{Q}_sy, y\rangle = |[\mathcal{I} - \mathcal{P}_{\tilde{n}}]\,\mathcal{V}^{1/2}\mathcal{Q}_sy|^2 > 0.$$

One can argue by the rule of contraries. Since $\widetilde{n}$ has the direction of $\mathcal{V}^{1/2}e_0$, the vector in the above inequality can be represented as

$$[\mathcal{I} - \mathcal{P}_{\tilde{n}}]\,\mathcal{V}^{1/2}\mathcal{Q}_sy = \mathcal{V}^{1/2}z, \quad z = \alpha e_0 - \sum_{j=1}^{d}\beta_je_j,$$

where the coefficients β_j all vanish only if $\mathcal{Q}_sy = 0$. By (6.0.1), equality $\mathcal{V}^{1/2}z = 0$ is possible only if $z = 0$, and this would contradict the assumption $\mathcal{Q}_sy \neq 0$.

Thus only the invertibility of the second operator of the lemma is a restrictive condition as long as uniformity in s is unessential.

If the parameterization of extrema is by points of a smooth manifold and the norms of the inverse operators in the lemma are uniformly bounded, then the dependence of $\widehat{h}$ and $\widehat{\beta}$ on the parameter s selecting the extremum is also smooth.

Proof of Lemma 6.3.3. Introduce a new unknown $B \stackrel{def}{=} \frac{1}{\tau}\left(\widehat{\beta} - \beta_0\right)$ to transform (6.3.11):

$$\frac{1}{\tau}\left[\Phi(\widehat{x}) - \Phi(x_0)\right] = 0, \quad \frac{1}{\tau}\mathcal{Q}\left[\widehat{x} - x_0\right] = 0, \tag{6.3.23}$$

$$[\mathcal{I} - \mathcal{Q}]\left(B\widehat{n} + \frac{\beta_0}{\tau}[\widehat{n} - n_0] - g\right) = 0. \tag{6.3.24}$$

This is done with a view to use the fact that for $\tau = 0$ the "unperturbed" values $B = 0$ and h_0 solve the system.

The scalar unknown B can be expressed explicitly through the vector-valued one g: to do so it suffices to multiply scalarly (6.3.24) by n_0 and note that $\langle\widehat{n}, n_0\rangle \neq 0$ for small τ. To eliminate this unknown, substitute the result into equation (6.3.24). It assumes the form

$$[\mathcal{I} - \mathcal{Q}_s][\mathcal{I} - \widehat{\mathcal{P}}]\left(\frac{\beta_0}{\tau}[\widehat{n} - n_0] - g\right) = 0, \tag{6.3.25}$$

where $\widehat{\mathcal{P}}$ is the operator $\widehat{\mathcal{P}} : x \mapsto \langle n_0, x\rangle\widehat{n}$.

Evidently, $[\mathcal{I} - \mathcal{Q}_s][\mathcal{I} - \mathcal{P}_s] = \mathcal{R}_s$ and $\widehat{\mathcal{P}} \to \mathcal{P}_s$ as $\tau \to 0$. Using these relations, (6.3.22), and the operator $\ddot{\Phi}$ of (6.1.14), equations (6.3.23)–(6.3.24) can be written in the form

$$\begin{aligned} \langle n_0, \mathcal{V}g\rangle &= -\langle n_0, \mu\rangle + \Psi_0, \\ \mathcal{Q}_s\mathcal{V}g &= -\mathcal{Q}_s\mu + \Psi_1, \\ \left[\mathcal{R}_s - \mathcal{R}_s\ddot{\Phi}\mathcal{V}\right]g &= \mathcal{R}_s\ddot{\Phi}\mu + \Psi_2, \end{aligned} \tag{6.3.26}$$

where $\Psi_i(g, \tau)$ are smooth functions which vanish if $\tau = 0$.

It remains to verify that the linear system with the left-hand side of (6.3.26) defines a bounded operator to apply the theorem on implicitly defined functions to prove the lemma. Introduce the new set of unknowns

$$\alpha = \langle g, n_0\rangle \in \mathbf{R}, \quad y = \mathcal{Q}_s g, \quad z = \mathcal{R}_s g.$$

Combining equations (6.3.26) to make the system "triangular", one can reduce it to the following form:

$$\begin{aligned} |\widetilde{n}|^2\alpha + \langle\widetilde{n}, \mathcal{V}^{1/2}\mathcal{Q}_s y\rangle + \langle\widetilde{n}, \mathcal{V}^{1/2}\mathcal{R}_s z\rangle &= A_0, \\ \mathcal{Q}_s\widetilde{\mathcal{V}}\mathcal{Q}_s y + \mathcal{Q}_s\widetilde{\mathcal{V}}\mathcal{R}_s z &= A_1, \\ \left[\mathcal{R}_s - \mathcal{R}_s\ddot{\Phi}\widetilde{\mathcal{V}}^{1/2}\left(\mathcal{I} - \widetilde{\mathcal{Q}}\right)\widetilde{\mathcal{V}}^{1/2}\mathcal{R}_s\right] z &= A_2, \end{aligned}$$

where the operators are those from the statement of the lemma, and the functions on the right-hand side are combinations of those on the right-hand side in the original system (6.3.26).

It is evident that the last system defines a bounded linear operator, so the assertion of the lemma follows from the theorem on implicit functions. ○

The conditions of Lemma 6.3.3 above are quite natural if at the extrema the normal to $\{\Phi = 0\}$ belongs to the subspace $\mathcal{V}^{1/2}X$ and the projections of (6.1.9) commute with operator (6.1.14), while the parameterization of (6.1.7) is smooth.

Consider the points on the surface $H_\Phi = \{\Phi = 0\}$ which belong to the subspace $\mathcal{V}^{1/2}X$ and lie close to the extremum $x(s)$. To select these it is convenient to use an explicit formula to represent the surface. In what follows, the independent variable is $y \in T_s$, $T_s \stackrel{def}{=} \{y \in X : \langle y, \mathcal{V}^{1/2}n_0\rangle = 0\}$. The points of $H_\Phi \bigcap \mathcal{V}^{1/2}X$ are represented in the form $x = \mathbf{X}(y) \stackrel{def}{=} \mathcal{V}^{1/2}y + \zeta(y)n_0$, with $n_0 = \overline{n}(s)$ the unit normal to H_Φ. In this formula $\zeta(y)$ solves the equation

$$\Phi\left(x_0 + \mathcal{V}^{1/2}y + \zeta(y)n_0\right) = 0.$$

By the implicit function theorem, the function ζ is well defined and smooth in a neighborhood of zero. Calculating its derivatives for $y = 0$ and using Taylor's formula, one arrives at the equality

$$\zeta(y) = -(2\beta_0)^{-1}\langle \mathcal{V}^{1/2}\ddot{\Phi}(x_0)\mathcal{V}^{1/2}y, y\rangle + \mathcal{O}\left(|y|^3\right),$$

where notation is that of (6.1.14).

The DF Λ_0 (see (6.0.4)) is finite on $\mathcal{V}^{1/2}X$. For the points lying of H_Φ its value is

$$\Lambda_0\left(x_0 + \mathcal{V}^{1/2}y + \zeta(y)n_0\right) = \tfrac{1}{2}\left|\mathcal{V}^{-1/2}x_0 + y + \zeta(y)\mathcal{V}^{-1/2}n_0\right|^2 \quad (6.3.27)$$
$$= \Lambda_\Phi + \tfrac{1}{2}|y|^2 + \langle \mathcal{V}^{-1/2}x_0, y\rangle + \langle \mathcal{V}^{-1/2}x_0, \zeta(y)\mathcal{V}^{-1/2}n_0\rangle + \mathcal{O}\left(|y|^3\right).$$

It follows by (6.1.5) and the choice of coordinates that $\langle \mathcal{V}^{-1/2}x_0, \mathcal{V}^{-1/2}n_0\rangle = \beta(e)$ and $\langle \mathcal{V}^{-1/2}x_0, y\rangle = 0$. Thus (6.3.27) reduces to

$$\Lambda_0[\mathbf{x}(y)] = \Lambda_\Phi + \tfrac{1}{2}|y|^2 - \tfrac{1}{2}\langle\ddot{\Phi}y, y\rangle + \mathcal{O}\left(|y|^3\right).$$

On the set where Λ_0 attains its maximal value, this functional is constant. Hence the directions tangent to this set at point x_0 are selected by the projection $\mathcal{Q}_s$. Projection to the subspace orthogoal both to the normal and tangent directions is denoted by $\mathcal{R}_s$. These projections satisfy the conditions

$$\mathcal{Q}_s\left[\mathcal{I} - \mathcal{V}^{1/2}\ddot{\Phi}\mathcal{V}^{1/2}\right]\mathcal{Q}_s = 0, \quad \mathcal{R}_s\left[\mathcal{I} - \mathcal{V}^{1/2}\ddot{\Phi}\mathcal{V}^{1/2}\right]\mathcal{R}_s \geq 0.$$

Under natural additional restrictions the latter operator is strictly positive definite. The subtracted term in its expression is a finite trace operator (not necessarily positive), and the assumption that its spectrum is separated from zero need not be unrealistic. Note that in a situation where $\ddot{\Phi}$ commutes with $\mathcal{V}$ and the projections, this amounts to the condition of Lemma 6.3.3.

Appendix A

Auxiliary Material

A.1 Euclidean Space

A.1.1 Vectors and Sets

Vectors of the Euclidean space $\mathbf{R}^k$, $k \geq 1$, are, as a rule, defined in terms of their coordinates in an implicitly specified orthonormal basis. Whenever matrix notation is used, the vector $x \in \mathbf{R}^k$ is identified with the column of its coordinates $(x^{(j)}, j = \overline{1,k})$. The inner product in $\mathbf{R}^k$ and the corresponding norm are

$$(x, y) = x^{(1)}y^{(1)} + \cdots + x^{(k)}y^{(k)}, \ |x| = (x, x)^{1/2}.$$

Inequalities between vectors are coordinate-wise: for instance,

$$x < y \iff x^{(j)} < y^{(j)}, \ j = \overline{1,k}.$$

Same notation is used for operators in $\mathbf{R}^k$ and the corresponding matrices if this does not lead to misunderstanding. The matrix M^T is transposed to M. The determinant of a square $k \times k$ matrix M is $\det(M) = (M_{ij})$, and $\operatorname{tr}(M) = M_{11} + \cdots + M_{kk}$ is its trace. The inequality $M > 0$ means that the matrix is positive definite, and the sign $\geq$ implies that it is nonnegative definite. The unit matrix is invariably $I = (\delta_{ij})$, where $\delta_{ij} = \begin{cases} 1, & i = j, \\ 0, & i \neq j, \end{cases}$ is the Kronecker symbol.

Algebraic operations over subsets of vector spaces are defined by the usual equalities

$$\alpha A + \beta B = \{\alpha x + \beta y : x \in A, \ y \in B\},$$

where the coefficients are either real numbers or operators. Notation does not discern between a vector and the one-element set which contains it: $a + A = \{a + x : x \in A\}$ for $a \in \mathbf{R}^k$.

The indicator function of a set is always $\mathbf{1}_A(x) = \begin{cases} 1, & x \in A, \\ 0, & x \notin A. \end{cases}$

The distance between a point and a set, or between two sets, in a Euclidean space is

$$\operatorname{dist}(x,A)=\inf_{y\in A}|x-y|\,,\quad \operatorname{dist}(A,B)=\inf_{x\in A}\operatorname{dist}(x,B)\,. \tag{A.1.1}$$

Notations $\bar{A} = \{x : \operatorname{dist}(x,A)=0\}$, $A^c = \mathbf{R}^k \setminus A$ refer to the closure and complement of a set. Same notations are applied for sets in metric spaces, where the distance between sets is defined using the distance in this space, e.g., $\rho(x,y)$, instead of the Euclidean norm $|\cdot|$ above.

For a positive number ε,

$$A^\varepsilon = A+\varepsilon B_1 = \{x : \operatorname{dist}(x,A)<\varepsilon\}\,;\; A^{-\varepsilon} = ((A^c)^\varepsilon)^c\,. \tag{A.1.2}$$

Same notation is used for sets in metric spaces.

A.1.2 Volume Integrals

Measures in $\mathbf{R}^k$ are defined on the Borel σ-algebra $\mathcal{B}\left(\mathbf{R}^k\right)$. Distributions are measures which satisfy the normalization condition $\mu\left(\mathbf{R}^k\right) = 1$. The Lebesgue measure in $\mathbf{R}^k$ and its subspaces is denoted by dx, $dx^{(1)}\cdots dx^{(k)}$, $\operatorname{mes}(dx)$, etc.

If the integration domain of a volume integral is not specified, it coincides with $\mathbf{R}^k$:

$$\int \phi(x)dx = \int_{\mathbf{R}^k} \phi(x)dx.$$

A.1.3 Surface Integrals

It is convenient to remind here the main features of the construction used to define ***surface integrals*** over hypersurfaces in Euclidean spaces (see references in Appendix B for details).

The surface integral over a hypersurface M is most easily described when this latter can be defined by the equation

$$(x,e) = f(x'),\; x' = x-(x,e)e \in B,$$

for a suitably chosen unit vector $e \in \mathbf{R}^k$, $|e| = 1$, and a set B. Obviously, (x,e) is the projection of $x \in \mathbf{R}^k$ onto the direction of e, and $x' = x-(x,e)e$ is its projection onto the $(k-1)$-dimensional subspace $E = \{x : (x,e)=0\}$, so $x = (x,e)e + x'$. The set B which parameterizes the hypersurface is its projection onto E. In this parameterization, all points of the hypersurface are given by the equation

$$x = X(x') \equiv x' + f(x')e,\; x' \in B. \tag{A.1.3}$$

If both the integrand and the hypersurface are sufficiently regular, the surface integral is defined by the formula

$$\begin{aligned}\int_M \phi(x)ds(x) &= \int_B \left.\frac{\phi(x)}{|(e,n(x))|}\right|_{x=X(x')} \operatorname{mes}_E(dx') \qquad \text{(A.1.4)}\\ &= \int_B \phi(X(x'))\left(1+|\nabla f(x')|^2\right)^{1/2}\operatorname{mes}_E(dx'),\end{aligned}$$

where mes_E is the Lebesgue measure in E (with the metric induced by that of $\mathbf{R}^k$), and $n(x)$ is the normal unit vector of M at the point x.

If the hypersurface admits decomposition $M = \bigcup_{\kappa=1}^{\nu} M_\kappa$ into parts that can be represented as in (A.1.3), the integral over it is

$$\int_M \phi(x)ds(x) = \sum_{\kappa=1}^{\nu} \int_{M_\kappa} \phi(x)ds(x). \qquad \text{(A.1.5)}$$

The integral of $\phi(x) \equiv 1$ is the *area* of the hypersurface:

$$|M| = \int_M 1\, ds(x). \qquad \text{(A.1.6)}$$

It does not change if the hypersurface is translated. Consider, for instance, the open ball of radius r centered at the origin and the sphere which bounds it:

$$B_r = \{x : |x| < r\}, \quad S_r = \{x : |x| = r\} = \partial B_r. \qquad \text{(A.1.7)}$$

The surface area of any translate of the sphere is

$$|a + S_r| = r^{k-1} 2\pi^{k/2}/\Gamma(k/2). \qquad \text{(A.1.8)}$$

One can also apply the Minkowski definition of the surface area in the important special case where the hypersurface is the boundary of a "solid" set. This definition is based on the calculation of the Lebesgue measure of an infinitesimal "boundary layer" of the set considered. This definition does not contradict the above described one: whenever both are applicable,

$$\lim_{\epsilon \searrow 0} \frac{1}{\epsilon} (\text{mes}\,(A^\epsilon) - \text{mes}(A)) = \int_{\partial A} ds(x).$$

The surface area has an important monotonicity property. If the sets C_+ and C_- are convex and $C_+ \supset C_-$, their surface areas are related through the inequality

$$|\partial C_+| \geq |\partial C_-|. \qquad \text{(A.1.9)}$$

Thus, if a convex set C is contained in a ball $a + B_r$, then

$$|\partial C| \leq |\partial B_r| = |S_r| \leq r^{k-1} 2\,\pi^{k/2}/\Gamma(k/2). \qquad \text{(A.1.10)}$$

A.2 Infinite-Dimensional Random Elements

This section is a reminder of some facts concerning infinite-dimensional distributions which are essential for the exposition. It lays no claim to being either exhaustive or complete — most propositions are mentioned without proof. Appendix B provides the necessary references.

A.2.1 Random Elements in a Measurable Space

Let E be some set and $\mathcal{E}$ a σ-algebra of its subsets. A ***random element*** (RE) with values in the measurable space $\langle E, \mathcal{E}\rangle$ defined on the probability space $\langle \Omega, \mathcal{A}, \mathbf{P}\rangle$ is a $\mathcal{A}$-$\mathcal{E}$-measurable mapping $\xi : \Omega \to E$, i.e., one for which the pre-image $\xi^{-1}(B) = \{\omega : \xi(\omega) \in B\}$ belongs to $\mathcal{A}$ for each $B \in \mathcal{E}$. When E is a metric space, the measurable sets are usually those from its Borel σ-algebra: $\mathcal{E} = \mathcal{B}(E)$.

A ***distribution*** (or ***probability, probability measure***) on $\mathcal{E}$ is an arbitrary countably additive measure m satisfying the normalization condition $m(E) = 1$. In particular, the distribution of the RE $\xi \in E$ is the measure

$$F_\xi(B) = \mathbf{P}\{\xi \in B\}, \ B \in \mathcal{E}. \tag{A.2.1}$$

If $\xi : \Omega \to E_j$ are RE's, then for each finite $m \in Z_+$ the collection $\xi = (\xi_j, j = \overline{1,m})$ can be considered as a RV in the product space $E = E_1 \times \cdots \times E_m$ whose measurable sets belong to the σ-algebra $\mathcal{E} = \mathcal{E}_1 \times \cdots \times \mathcal{E}_m$, the smallest one which contains all the "rectangles" — Cartesian products $B = B_1 \times \cdots \times B_m$, $B_j \in \mathcal{E}_j$, $j = \overline{1,m}$.

The RE's ξ_j, $j = \overline{1,m}$, are ***independent*** if there is equality

$$\forall B_j \in \mathcal{E}_j \quad \mathbf{P}\left(\bigcap_{j=1}^m \{\xi_j \in B_j\}\right) = \prod_{j=1}^m \mathbf{P}\{\xi_j \in B_j\}. \tag{A.2.2}$$

The RE's from a finite (or infinite) family $\xi_a : \Omega \to E_a$, $a \in A$, are independent if the RE's $\xi_{a(j)}$ are independent whatever the choice of m and $a(j) \in A$, $j = \overline{1,m}$.

Consider an arbitrary finite family of distributions $F_j : \mathcal{E}_j \to [0, 1]$, $j = \overline{1,m}$, in arbitrary measurable spaces $\langle E_j, \mathcal{E}_j\rangle$. There is always a probability space Ω supporting a family of independent RE's $\xi_j : \Omega \to E$ with these distributions: $\mathbf{P}\{\xi_j \in \cdot\} = F_j(\cdot)$. The requirement is met, e.g., by the RE's $\xi : \omega = \langle e_1, \ldots, e_m\rangle \mapsto e_j$ defined on the product space $\Omega = E_1 \times \cdots \times E_m$ with the σ-algebra $\mathcal{A} = \mathcal{E}_1 \times \cdots \times \mathcal{E}_m$.

If ξ is a RE, a family of independent RE's with the same distribution is called a family of its ***independent copies.***

More particularly, the sequences of E-valued RE's $\Xi^{(\alpha)} = \left(\xi_j^{(\alpha)}, j \in J \subseteq \mathbf{Z}\right)$, $\alpha \in \Upsilon$, are independent copies of $\Xi = (\xi_j)$ if for each finite set $I \subset J$ the RE's $\Xi_\alpha^I = \left(\xi_j^{(\alpha)}, j \in I\right)$ with values in X^I have the distribution of $\Xi = (\xi_j, j \in I)$ and if, moreover, for each choice of $m \in \mathbf{N}$ and $\alpha(l) \in \Upsilon, l = \overline{1,m}$, the RV's $\Xi_{\alpha(l)}$ are independent.

It is always possible to obtain a finite number of independent copies of any family of random variables defined on some probability space $\langle \Omega, \mathcal{A}, \mathbf{P}\rangle$ choosing for a new probability space the product $\langle \Omega^n, \mathcal{A}^n, \mathbf{P}^n\rangle$ and "duplicating" the original random variables on Ω by means of the formula

$$\xi_j^{(\alpha)}(\omega) = \xi(\omega_\alpha), \quad \omega = \left(\omega_{\alpha'}, {\alpha'=1,\ldots,n}\right) \in \Omega^n.$$

A.2.2 Distributions in Polish Spaces

Let E be a Polish (i.e., separable complete metric) space. The distributions in a Polish space are measures on its Borel sets. Each distribution is ***tight***, i.e., for each $\varepsilon > 0$ there is a compact set K_ε such that

$$F(E \setminus K_\varepsilon) < \varepsilon. \tag{A.2.3}$$

It is also a ***Radon measure***: one can approximate each Borel set B by a compact one $K_{\varepsilon,B} \subset B$ so that (A.2.3) would remain valid with E replaced by B.

A sequence of measures $\mu_n : \mathcal{B}(E) \to \mathbf{R}_+$ ***weakly converges*** to the measure μ_∞ if

$$\int_E \varphi(x)\,\mu_n(dx) \;\to\; \int_E \varphi(x)\,\mu_\infty(dx) \tag{A.2.4}$$

for each bounded continuous functional φ on E.

A family of measures is ***weakly compact*** if each sequence of its elements contains a weakly convergent subsequence. One of the most commonly used criteria of weak compactness is the following theorem.

Proposition A.2.1 *A sequence of measures $(\mu_n,\ n \in \mathbf{N})$ is weakly compact if and only if*

a) *it is bounded: $\forall n \in \mathbf{N}\ \ \mu_n(E) \le c < \infty$;*

b) *it is tight: for each $\varepsilon > 0$ there is a compact set $K_\varepsilon \subset E$ such that*

$$\forall n \in \mathbf{N}\ \ \mu_n(E \setminus K_\varepsilon) \le \varepsilon.$$

A.2.3 Measurable Linear Spaces

A real linear space X with a σ-algebra $\mathcal{X}$ of its subsets is a ***measurable linear space*** if the following two operations are measurable: that of addition

$$\langle x, y\rangle \mapsto x + y,$$

acting from $\langle X \times X, \mathcal{X} \times \mathcal{X}\rangle$ to $\langle X, \mathcal{X}\rangle$, and multiplication by a real number

$$\langle \alpha, x\rangle \mapsto \alpha x \in X,$$

from $\langle \mathbf{R} \times X, \mathcal{B}(\mathbf{R}) \times \mathcal{X}\rangle$ to $\langle X, \mathcal{X}\rangle$. A RE with values in a linear measurable space is called below a ***random vector*** (RV) with values in X.

If for some $m \in \mathbf{N}$ and $j = \overline{1,m}$ the mappings $\alpha_j : \Omega \to \mathbf{R}$ are rv's and $\xi_j \in X$ are RV's, then $\eta = \alpha_1\xi_1 + \cdots + \alpha_m\xi_m$ defines one more RV. Measurability of this mapping follows from the fact that it is a finite superposition of measurable operations.

The distribution of a sum of independent RV's can be expressed in terms of distributions of summands (see (A.2.1)) using the ***convolution formula***

$$\mathbf{P}\{\xi + \eta \in B\} = \int_X F_\xi(B - x)\,F_\eta(dy). \tag{A.2.5}$$

Indeed, by independence, the joint distribution of ξ and η in $X \times X$ is the product of their distributions, and (A.2.5) follows by applying the Fubini theorem to the measurable function $X \times X \ni \langle x, y\rangle \mapsto \mathbf{1}_B(x+y) \in \mathbf{R}$.

A.2.4 Cylinder Sets

Measurable linear spaces can be constructed as follows.

Let X be a linear space and $T = \{t\}$ some family of linear functionals $x \mapsto \langle x, t\rangle \in \mathbf{R}$ on this space.

Associate with a finite set $\bar{t} = \left(t_j, j = \overline{1,k}\right) \subset T$ the mapping

$$\mathcal{P}_{\bar{t}} : X \ni x \mapsto \left(\langle x, t_j\rangle, j = \overline{1,k}\right) \in \mathbf{R}^k. \tag{A.2.6}$$

In what follows, it is convenient to label by the linear functionals t_j coordinates of vectors from the Euclidean space $\mathbf{R}^{\bar{t}} = \mathcal{P}_{\bar{t}}X$. In this latter, the inner product of $x = (x_s) \in \mathbf{R}^{\bar{t}}$, $y = (y_s) \in \mathbf{R}^{\bar{t}}$ is $(x,y)_{\bar{t}} = \sum_{s \in \bar{t}} x_s y_s$, and the corresponding norm is $|\cdot|_{\bar{t}} = \sqrt{(\cdot,\cdot)_{\bar{t}}}$.

A subset $A \subset X$ is *T-cylindrical* if for some finite collection of linear functionals $\bar{t}$ it belongs to the class $\mathcal{C}_{\bar{t}}$ defined by the formula

$$A \in \mathcal{C}_{\bar{t}} \iff A = \{x \in X : \mathcal{P}_{\bar{t}}x \in B\}, \quad B \in \mathcal{B}\left(\mathbf{R}^{\bar{t}}\right). \tag{A.2.7}$$

Each class $\mathcal{C}_{\bar{t}}$ is a σ-algebra. Put

$$\mathcal{C}_0(T) = \bigcup_{\bar{t} \subset T} \mathcal{C}_{\bar{t}}, \quad \mathcal{C}(T) = \sigma(\mathcal{C}_0(T)). \tag{A.2.8}$$

The class of T-cylindrical sets $\mathcal{C}_0(T)$ is an algebra, but not necessarily a σ-algebra. The smallest σ-algebra of subsets of X containing $\mathcal{C}_0(T)$ is $\mathcal{C}(T)$, the *T-cylindrical σ-algebra*.

Proposition A.2.2 *For each family T of linear functionals, the measurable space $\langle X, \mathcal{C}(T)\rangle$ is a measurable linear space.*

Proof of Proposition A.2.2. Let $\mathcal{K}$ be such a class of subsets of X that the set of pairs $\{[x,y] \in X \times X : x + y \in K\}$ belongs to $\mathcal{C}(T) \times \mathcal{C}(T)$ if $K \in \mathcal{K}$. This class is a σ-algebra, because $\mathcal{C}(T) \times \mathcal{C}(T)$ is one. Moreover, $\mathcal{K}$ contains the sets $\{x : \langle x, t\rangle < \alpha\}$ for all $\alpha \in \mathbf{R}$ and $t \in T$. This follows from the representation

$$\{[x,y] : \langle x + y, t\rangle < \alpha\} = \bigcup_{n=1}^{\infty} \bigcup_{k \in \mathbf{Z}} \left\{[x,y] : \langle x,t\rangle < \tfrac{k}{n}, \langle y, t\rangle < \alpha - \tfrac{k+1}{n}\right\}.$$

For this reason, the inclusion $\mathcal{K} \supset \mathcal{C}_{\bar{t}}$ is evident for each finite set $\bar{t}$. Hence $\mathcal{K} \supseteq \mathcal{C}(T)$.

A similar reasoning shows that multiplication by a real number is also a measurable operation. ○

Let $\mu : \mathcal{C}(T) \to [0, 1]$ be a distribution on the T-cylindrical σ-algebra of the linear space X, and $\hat{\mu}(v) = \int_X e^{i\langle y,v\rangle}\mu(dy)$ its characteristic functional (CF).

The ***finite-dimensional distributions*** of μ are distributions in Euclidean spaces $\mathbf{R}^{\bar{t}}$ associated with finite collections of elements of T. They are defined by the formula

$$\mu_{\bar{t}}(B) = \mu\{x : \mathcal{P}_{\bar{t}}x \in B\}, \quad B \in \mathcal{B}\left(\mathbf{R}^{\bar{t}}\right). \tag{A.2.9}$$

The correspondence between the characteristic function (CF) of a finite-dimensional distribution and the CF of the infinite-dimensional one which generates it is established by the identity

$$\hat{\mu}_{\bar{t}}(u) = \int_{\mathbf{R}^{\bar{t}}} \exp\{i(u,y)_{\bar{t}}\}\,\mu_{\bar{t}}(dy) = \hat{\mu}\left(\textstyle\sum_{s\in\bar{t}} u^{(s)}s\right), \tag{A.2.10}$$

where $u \in \mathbf{R}^{\bar{t}}$.

The family of all finite-dimensional distributions of an arbitrary distribution μ satisfies a natural *consistency condition.* Namely, if $\bar{s}$ and $\bar{t}$ are two collections of linear functionals such that $\bar{s} \subset \bar{t}$, the CF's of $\mathcal{P}_{\bar{s}}$ and $\mathcal{P}_{\bar{t}}$ should satisfy the identity

$$\hat{\mu}_{\bar{s}}\left[\left(u^{(s)}, s \in \bar{s}\right)\right] = \hat{\mu}_{\bar{t}}\left[\left(v^{(t)}, t \in \bar{t}\right)\right] \text{ if } \sum_{s\in\bar{s}} u^{(s)}s = \sum_{t\in\bar{t}} v^{(t)}t. \tag{A.2.11}$$

It is easily seen that the CF $\mu_{\bar{t}}(v)$ is continuous in its argument $v \in \mathbf{R}^{\bar{t}}$. The CF of the infinite-dimensional distribution, $\overline{\mu}(t) = \hat{\mu}_{\{t\}}(1)$, $t \in T$, is also continuous in the weak topology generated by the duality $\langle X, T\rangle$.

Proposition A.2.3 *Consider a family of finite-dimensional distributions $\mu_{\bar{s}} : \mathcal{B}(\mathbf{R}^{\bar{s}}) \to [0, 1]$, where $\bar{s}$ are all finite collections of elements of T. If it satisfies the consistency condition* (A.2.11)*, then this family is related to some finitely additive set function μ defined on the cylindric algebra $\mathcal{C}_0$ by formula* (A.2.9)*.*

If $\bar{t}$ is a finite collection of elements of T, the restriction of this set function μ to the σ-algebra $\mathcal{C}_{\bar{t}}$ is a countably additive distribution.

Proof of Proposition A.2.3. It follows from (A.2.11) that, whenever $\bar{t} \supset \bar{s}$, the distributions $\mu_{\bar{t}}$ and $\mu_{\bar{s}}$ obey the consistency condition

$$\mu_{\bar{s}}(B) = \mu_{\bar{t}}\left(B \times \mathbf{R}^{\bar{t}\setminus\bar{s}}\right), \quad B \in \mathcal{B}\left(\mathbf{R}^{\bar{s}}\right).$$

Indeed, consider the distribution in the "smaller" space defined as $\mu^0(B) \equiv \mu_{\bar{t}}\left(B \times \mathbf{R}^{\bar{t}\setminus\bar{s}}\right)$. In accordance with (A.2.11) its CF is that of $\mu_{\bar{s}}$:

$$\int_{\mathbf{R}^{s}} \exp\{i(u,y)_{\bar{s}}\}\,\mu(dy) = \hat{\mu}_{\bar{s}}(u) = \hat{\mu}_{\bar{t}}(v),$$

if $u \in \mathbf{R}^{\bar{s}}$ and $v \in \mathbf{R}^{\bar{t}}$ has coordinates $v^{(t)} = \begin{cases} u^{(t)}, & t \in \bar{s}, \\ 0, & t \in \bar{t} \setminus \bar{s}. \end{cases}$

Thus, $\mu^0 = \mu_{\bar{s}}$, and the function μ is well defined on the algebra $\mathcal{C}_0(T)$ by the formula

$$\mu\{x \in X : \mathcal{P}_{\bar{t}}x \in B\} = \mu_{\bar{t}}(B), \quad B \in \mathcal{B}\left(\mathbf{R}^{\bar{t}}\right). \tag{A.2.12}$$

Finite additivity of this set function follows from the fact that a finite collection of sets from $\mathcal{C}_0(T)$ can always be considered as a collection of sets from one and same σ-algebra $\mathcal{C}_{\bar{t}}$ for a sufficiently rich $\bar{t}$.

The fact that each restriction of μ to $\mathcal{C}_{\bar{t}}$ is σ-additive is as evident — this restriction is the finite-dimensional distribution $\mu_{\bar{t}}$. ○

When finite-dimensional distributions are considered, it may prove useful to interpret them as countably additive distributions on all the cylindrical σ-algebra $\mathcal{C}(T)$ using (A.2.6). The possibility of such interpretation is established by

Proposition A.2.4 *Assume that $\mu_{\bar{s}} : \mathcal{B}(\mathbf{R}^{\bar{s}}) \to [0, 1]$ is a countably additive distribution. There exists a countably additive distribution $\tilde{\mu}$ on the cylindric σ-algebra $\mathcal{C}(T)$ such that its restriction to $\mathcal{C}_{\bar{s}}$ is related to the finite-dimensional distribution $\mu_{\bar{s}}$ by the formula*

$$\tilde{\mu}_{\bar{s}}\{x \in X : \mathcal{P}_{\bar{s}}x \in B\} = \mu_{\bar{s}}(B), \; B \in \mathcal{B}(\mathbf{R}^{\bar{s}}).$$

Proof of Proposition A.2.4. Let $\bar{s}^0$ be some maximal subsystem of linearly independent vectors in $\bar{s}$. It is well known that, given a finite set $\bar{s}^0$ of linearly independent functionals on X, one can select a system of vectors $x_s \in X$, $s \in \bar{s}^0$, which is biorthogonal to it:

$$\langle x_{s'}, s''\rangle = \delta_{s's''}, \; s', s'' \in \bar{s}^0.$$

Let $\eta = \left(\eta^{(s)}, s \in \bar{s}^0\right)$ be a RV with the distribution $\mu_{\bar{s}^0}$ in the Euclidean space $\mathbf{R}^{\bar{s}^0}$. Formula

$$\xi = \sum_{s \in \bar{s}} \eta^{(s)} x_s$$

clearly defines a X-valued RV. Its distribution $\tilde{\mu}_{\bar{s}}$ is a countably additive measure on the cylinder σ-algebra $\mathcal{C}(T)$.

It is convenient to have recourse to CF's to verify that the restriction of $\tilde{\mu}_{\bar{s}}$ to $\mathcal{C}_{\bar{s}^0}$ does indeed have the properties stipulated in the proposition. The CF's of the corresponding finite-dimensional distributions are $\hat{\mu}_{\bar{s}}$ and $\hat{\mu}_{\bar{s}^0}$.

By the choice of the set $\bar{s}^0$ and the elements of the biorthogonal system x_s, the following equalities hold for an arbitrary $u = \left(u^{(s)}, s \in \bar{s}\right)$:

$$w \equiv \sum_{s \in \bar{s}} u^{(s)} s = \sum_{s' \in \bar{s}^0} u_0^{(s')} s', \quad u_0^{(s')} = \sum_{s \in \bar{s}} u^{(s)} \langle x_{s'}, s\rangle,$$

where $u^{(s)}$ and $u_0^{(s')}$ are the coordinates of the respective projections $\mathcal{P}_{\bar{s}}w$ and $\mathcal{P}_{\bar{s}^0}w$.

It follows immediately from the definition of the RV ξ and (A.2.11) that

$$\begin{aligned}
\mathbf{E}e^{i\langle\xi, w\rangle} &= \mathbf{E}\exp\left\{i\langle\xi, \sum_{s\in\bar{s}} u^{(s)} \sum_{s'\in\bar{s}^0}\langle x_{s'}, s\rangle s'\rangle\right\} \\
&= \mathbf{E}\exp\left\{i\sum_{s'\in\bar{s}^0} \eta_{s'} u_0^{(s')}\right\} = \mathbf{E}\exp\{i(\eta, u_0)_{\bar{s}^0}\} \\
&= \hat{\mu}_{\bar{s}^0}(u_0) = \hat{\mu}_{\bar{s}}(u) = \hat{\mu}_{\bar{s}}\left(\mathcal{P}^{\bar{s}}w\right).
\end{aligned}$$

These equalities prove the proposition. ○

The question whether it is possible to extend a finitely additive set function of Proposition A.2.3 to a countably additive distribution defined on all the cylindrical σ-algebra is much less trivial. It is often difficult to verify the necessary and sufficient condition for the existence of such an extension,

$$\mu(A_\infty) \leq \sum_{i=1}^{\infty} \mu(A_i) \text{ if } A_\infty \subseteq \bigcup_{i=1}^{\infty} A_i,\ A_i \in \mathcal{C}_0(T). \tag{A.2.13}$$

A.2.5 Distributions in Banach Spaces

To discern between finitely and countably additive set functions without introducing new names, the additivity property will sometimes be specified explicitly, although normally the terms "measure" and "distribution" refer to countably additive positive set functions.

Let X be a real Banach space and X^* its conjugate. Denote by $\langle x, x' \rangle$ the value of the linear functional $x' \in X^*$ on the vector $x \in X$ and by $|\cdot|$ the norms in X and X^*.

The *expectation* of a X-valued RV ξ is always understood as a Bochner integral:

$$\mathbf{E}\xi = \int_X x \mathbf{P}\{\xi \in dx\}.$$

Tools used to treat Bochner integrals include the well-known bound $|\mathbf{E}\xi| \leq \mathbf{E}|\xi|$ and the Jensen inequality $\varphi(\mathbf{E}\xi) \leq \mathbf{E}\varphi(\xi)$ for convex functionals φ.

In the more special setting of a Banach space X, one can state a number of additional results concerning measurability of sets and the possibility of using finite-dimensional distributions to define a *bona fide* distribution in X.

Proposition A.2.5 *If X is separable, then its Borel σ-algebra $\mathcal{B}(X)$ coincides with the X^*-cylindric σ-algebra $\mathcal{C}(X^*)$.*

Proof of Proposition A.2.5. Suppose that the sequence (x_j) is dense everywhere in X while the linear functionals $t_j \in X^*$ satisfy the conditions $|t_j| \leq 1$ and $\langle x_j, t_j \rangle = |x_j|$. Then the following inequality holds for each $x \in X$:

$$|x| = \sup_{k \in \mathbf{N}} \langle x, t_k \rangle.$$

Indeed, if $x_j \to x$, then there is convergence $\langle x_j, t_k \rangle \to \langle x, t_k \rangle$ for each k, so

$$\sup_{k \in \mathbf{N}} \langle x, t_k \rangle \leq \liminf_{j \to \infty} \sup_{k \in \mathbf{N}} \langle x_j, t_k \rangle.$$

Besides, $\langle x_j, t_j \rangle = |x_j| \to |x|$, so the converse inequality holds as well.

Thus all the balls $\{x : |x - a| < \varepsilon\}$ in X are $\mathcal{C}(X^*)$-measurable sets, which implies measurability of all Borel sets.

On the other hand, the half-spaces $H = \{x : \langle x, t \rangle < \alpha\}$ are open (consequently, $\mathcal{B}(X)$-measurable) sets for all $\alpha \in \mathbf{R}$ and $t \in X^*$. The set of these half-spaces generates $\mathcal{C}(X^*)$. ○

It is general knowledge that a Banach space is separable if its conjugate is separable. However, a separable Banach space may have a nonseparable conjugate — an example is the space of continuous functions on the unit interval $\mathbf{C}[0,1]$ with the supremum norm: its conjugate is the nonseparable space of signed measures on the Borel sets of $[0, 1]$.

Proposition A.2.6 *Assume that the Banach space X is separable and reflexive. Let*

$$\mu_{\bar{t}} : \mathcal{B}\left(\mathbf{R}^{\bar{t}}\right) \to [0,\ 1],\ \ \bar{t} \subset X^*,\ \ \mathrm{card}(\bar{t}) < \infty,$$

be a family of finite-dimensional distributions corresponding to arbitrary finite subsets of X^* which obeys the consistency condition (A.2.11).

It is sufficient for the existence of a countably additive distribution on $\mathcal{B}(X)$ related to the family $(\mu_{\bar{t}}, \bar{t} \subset X^*)$ by (A.2.9) *that the following conditions be satisfied:*

a) *The CF $\hat{\mu}(t) \equiv \hat{\mu}_{\{t\}}(1)$ is continuous, and $\hat{\mu}_{\{0\}}(1) = 1$;*

b) *there is a sequence $(t_j, j \in \mathbf{N})$ which is dense on the unit sphere of X^* such that*

$$\forall\, r \geq r_0\ \ \lim_{n\to\infty} \mu_{\bar{t}(n)} \left\{ x \in \mathbf{R}^{\bar{t}} :\ \max_{j=\overline{1,n}} |\langle x, t_j\rangle| \geq r \right\} \leq \varphi(r),\ \ \lim_{r\to\infty} \varphi(r) = 0,$$

where $\bar{t}(n) = \left(t_j, j = \overline{1,n}\right)$.

Proof of Proposition A.2.6. Under the assumptions of the proposition, the closed balls $B_r = B_{r,0} = \{x : |x| \leq r\}$ are compact subsets of the space X endowed with the weaker topology $\sigma(X, X^*)$. On each ball B_r this topology can be metrized; to do so, define the distance by the equality

$$\rho(x, y) = \sum_{k=1}^{\infty} \frac{1}{2^k} |\langle x - y, t_k\rangle| .$$

The neighborhoods $U_q(a) \equiv \{x :\ \rho(x, a) < q\}$, $q > 0$, are convex sets.

For each ball B_r, $r > 0$, the class of closed convex subsets defined using the original topology of the Banach space coincides with the class of closed convex sets for the topology $\sigma(X, X^*)$. Hence the Borel σ-algebras of the metric spaces $\langle B_r, |\cdot|\rangle$ and $\langle B_r, \rho\rangle$ coincide.

By Proposition A.2.4, each finite-dimensional distribution $\mu_{\bar{t}(n)}$ generates a distribution μ_n on $\mathcal{B}(X)$ which coincides, on $\mathcal{C}_{\bar{t}(n)}$, with the finitely additive set function of Proposition A.2.3 (see (A.2.12)). The measures $m_{n,r}(A) \equiv \mu_n\left(A \bigcap B_r\right)$ are concentrated in the compact metric space $\langle B_r, \rho\rangle$. It follows by Proposition A.2.1 that, for a fixed value of r, the sequence of these measures is weakly compact since the measure of the ball does not exceed one for each of them.

One can employ a version of the well-known diagonalization process to select a subsequence $n' = n'(k) \nearrow \infty$ such that on each compactum $\langle B_r, \rho\rangle$, $r \in \mathbf{N}$, there is weak convergence $m_{n',r} \overset{weak}{\longrightarrow} m_{\infty,r}$ (with the limit measure $m_{\infty,r}$ being possibly different for each ball). However, these limit measures have to agree in a natural way: it follows from a similar inequality for the distributions with finite numbers n' that

$$\int_{B_{r''}} f(x)\mathbf{1}_{B_{r''}}(x)m_{\infty,r''}(dx) \geq \int_{B_{r'}} f(x)\mathbf{1}_{B_{r'}}(x)m_{\infty,r'}(dx)$$

if $r'' > r'$ for each bounded continuous functional $f : X \to \mathbf{R}$. Consequently, for $r'' > r'$ the difference

$$\nu_{r',r''}(A) = m_{\infty,r''}\left(A\bigcap B_{r''}\right) - m_{\infty,r'}\left(A\bigcap B_{r'}\right)$$

defines a nonnegative countably additive measure on $\mathcal{B}(X)$, while the measure

$$\nu(A) = \sum_{r=1}^{\infty} \nu_{r,r+1}(A)$$

is nonnegative (and countably additive) on $\mathcal{B}(X)$. Condition b) implies that this new measure satisfies the inequality $\nu\{x : |x| \geq r\} \leq \varphi(r)$.

It remains to ascertain that the finite-dimensional distributions of ν are those of the finitely additive measure considered in Proposition A.2.3.

For $j \leq n$ the one-dimensional CF $\mu_{\{t_j\}}$ coincides with the CF of the "pre-limit" measure μ_n on $\mathcal{B}(X)$: $\hat{\mu}_{\{t_j\}}(\lambda) = \hat{\mu}_n(\lambda t_j)$. This is a simple consequence of the inclusion $t_j \in \bar{t}(n)$. For this reason, the condition of the proposition leads to the estimate

$$\left|\int_X \exp\{i\langle x, \lambda t_j\rangle\}\nu(dx) - \int_{\mathbf{R}^{\bar{t}(n)}} \exp\{i\lambda y_j\}\mu_{\bar{t}(n)}(dy)\right| \tag{A.2.14}$$

$$\leq 2\varphi(r) + \left|\int_{B_r} \exp\{i\langle x, \lambda t_j\rangle\}\nu(dx) - \int_{B_r} \exp\{i\langle x, \lambda t_j\rangle\}\mu_{n,r}(dy)\right|,$$

where y_j is the coordinate of y which corresponds to t_j, and the function φ is that of condition b).

The exponential in the integrand is bounded and continuous on the compact metric space $\langle B_r, \rho\rangle$ in the topology generated by the distance ρ. Hence the passage to the limit (along the above-described subsequence n') shows that, for each $r > 0$, there is convergence of integrals:

$$\int_{B_r} e^{i\langle x,\lambda t_j\rangle}\mu_{n',r}(dy) \to \int_{B_r} e^{i\langle x,\lambda t_j\rangle}m_{\infty,r}(dy) = \int_{B_r} e^{i\langle x,\lambda t_j\rangle}\nu(dy).$$

Thus, the fact that $\hat{\mu}$ and $\hat{\mu}_{\bar{t}(n)}$ coincide on the set $\bar{t}(n)$ and (A.2.14) yield the inequality

$$\left|\int_X e^{i\langle x,\lambda t_j\rangle}\nu(dx) - \hat{\mu}(\lambda t_j)\right| \leq 2\varphi(r),$$

where r is an arbitrary integer (or positive) number. The inequality shows that the CF's ν and μ agree on the set $\{\lambda t_j,\ j \in \mathbf{N}\}$. The latter is everywhere dense in X^* by the choice of the sequence $\{t_j, j \in \mathbf{N}\}$. Since $\hat{\mu}$ is a continuous functional by the assumption, this proves that $\hat{\mu}$ coincides with $\hat{\nu}$ everywhere in X^*: indeed, the fact that ν is also continuous follows from the estimate

$$\left|\int_X e^{i\langle x,t'\rangle}\nu(dx) - \int_X e^{i\langle x,t''\rangle}\nu(dx)\right| \le r\,|t'-t''| + \nu\{x:\ |x|\ge r\}. \bigcirc \quad (A.2.15)$$

Banach spaces with special properties. A Banach space X is said to *contain a basis* $(e_j, j\in\mathbf{N})$ if each of its elements admits of the unique norm-convergent decomposition

$$x = \sum_{j=1}^{\infty} \mathrm{x}_j e_k, \quad \mathrm{x}_k \in \mathbf{R}.$$

The coefficients have the form $\mathrm{x}_j = \langle x, f_j\rangle$, where $(f_j, j\in\mathbf{N})$ is a sequence of elements from the conjugate space X^* which is biorthogonal to (e_j): $\langle e_j, f_k\rangle = \begin{cases} 1, & j=k, \\ 0, & j\ne k. \end{cases}$ The biorthogonal sequence is uniquely determined by the choice of basis.

The projection operators

$$\mathcal{P}_N : x \mapsto \sum_{j=1}^{N} \langle x, f_j\rangle\, e_j, \quad \mathcal{R}_N : x \mapsto x - \mathcal{P}_N x,$$

are uniformly bounded:

$$|\mathcal{P}_N x| \le U\,|x|, \quad |\mathcal{R}_N x| \le U\,|x|,$$

where the constant $U>0$ does not depend on N.

A Banach space X is said to be a *space of type* 2 if for each sequence of independent centered RV's ξ_j, $j=\overline{1,n}$, the squared norm of the sum $\xi_1+\cdots+\xi_n$ satisfies the inequality

$$\mathbf{E}\,|\xi_1+\cdots+\xi_n|^2 \le c\left(\mathbf{E}\,|\xi|^2 + \cdots + \mathbf{E}\,|\xi_n|^2\right), \quad \mathbf{E}\xi_j = 0,\ j=\overline{1,n},$$

where the constant c does not depend on n or the choice of the summands, i.e., is determined by the properties of the norm in X.

Examples of type 2 spaces are the space of summable sequences l_p, $p\ge 2$, or the corresponding spaces of summable functions L_p over some measurable space.

A.2.6 Distributions in Hilbert Space

The finite-dimensional distributions of Proposition A.2.3 determine the CF μ which is positive-definite:

$$\forall n\in\mathbf{N}, t_j\in X^*, \varsigma_j\in\mathbf{C} \quad \sum_{j,k=1}^{n} \varsigma_j\bar{\varsigma}_k \mu(t_j - t_k) \ge 0.$$

Consider distributions in a separable Hilbert space X with inner product $(\cdot,\cdot)$ and norm $|\cdot| = \sqrt{(\cdot,\cdot)}$.

Necessary and sufficient conditions which ensure that a given family of finite-dimensional distributions corresponds to a countably additive distribution, or a family of distributions is weakly compact can be stated as restrictions on continuity of the CF.

Define *S-operators* as symmetric positive definite operators $A : X \to X$ with finite *trace* $\operatorname{tr}(A) = \sum_{k=1}^{\infty} (Ae_k, e_k)$, where $\{e_k\}$ is some orthonormal basis. (The definition of trace does not actually depend upon the choice of the basis.)

Define *S-topology* as the topology whose fundamental system of neighborhoods of zero consists of the sets $\{x : (Ax, x) < 1\}$, where A is an arbitrary S-operator.

Proposition A.2.7 *A functional $\hat{\mu} : X \to \mathbf{C}$ is the CF of some countably additive distribution defined on the Borel σ-algebra of the real separable Hilbert space X if and only if this functional is nonnegative definite, continuous with respect to the S-topology, and normalized to satisfy $\hat{\mu}(0) = 1$.*

Proposition A.2.8 *Let $\hat{\mu}_a$ be the CFs of distributions on the Borel σ-algebra of a separable real Hilbert space belonging to the family μ_a, $a \in \mathcal{A}$.*

The family of distributions μ_a, $a \in \mathcal{A}$, is weakly compact if the corresponding family of CFs is equicontinuous at the origin in the S-topology.

A.3 Deviation Function of Real Random Variable

A.3.1 Large Deviations and Deviation Function

Many results in the theory of large deviations are the offspring of the following result for sums of real-valued rv's.

Define the *deviation function* of the distribution $\mathcal{L}(Y)$ by the formula

$$\Lambda(v) = \Lambda_Y(v) = \sup_{t \geq 0} [vt - \ln \mathbf{E} \exp\{tY\}] . \tag{A.3.1}$$

This function need not be finite, and no assumptions are made concerning the existence of finite exponential moments or centering.

Proposition A.3.1 *Let $Y, Y_1, \ldots \in \mathbf{R}$ be i.i.d. rv's. The limit*

$$l(v) \stackrel{def}{=} \lim_{n \to \infty} \frac{1}{n} \ln 1/\mathbf{P}\left\{ \tfrac{1}{n}(Y_1 + \cdots + Y_n) > v \right\}$$

exists for each $v \in \mathbf{R}$. If $\mathbf{P}\{Y \leq v\} \neq 1$, then $l(v) = \Lambda_Y(v)$.

Note that the limit in the lemma is not necessarily finite.

The restriction in the statement of the proposition is essential.

Indeed, if $\mathbf{P}\{Y = v\} = 1$, then $p_n = \mathbf{P}\{\frac{1}{n}(Y_1 + \cdots + Y_n) > v\} = 0$ for all n, and $\limsup \frac{1}{n} \ln 1/p_n = +\infty$. At the same time,

$$\sup_{t \geq 0} \left[vt - \ln \mathbf{E} e^{tY}\right] = \sup_{t \geq 0} [vt - vt] = 0,$$

so the deviation function cannot be equal to the limit.

The proof of Proposition A.3.1 is sectioned into a number of steps summarized in separate lemmas.

It is sufficient to establish the assertion of the proposition for the special case of $v = 0$. To obtain the assertion in the general case, it suffices to change the rv's Y_j to $Y_j' = Y_j - v$.

Lemma A.3.1 *Under the assumptions of Proposition* A.3.1

$$l^- = \liminf_{n\to\infty} \frac{1}{n} \ln 1/p_n \geq \Lambda_Y,$$

where

$$\Lambda_Y \stackrel{def}{=} \sup_{t\geq 0} \{-\ln \psi(t)\}, \quad \psi(t) \stackrel{def}{=} \mathbf{E} \exp\{tY\}, \; p_n = \mathbf{P}\{Y_1 + \cdots + Y_n > 0\}.$$

Proof of Lemma A.3.1. By the Chebyshev inequality

$$p_n \leq \mathbf{E} \exp\{t(Y_1 + \cdots + Y_n)\} = \exp\{n \ln \psi(t)\}$$

for each $t \geq 0$. Hence $l^- \geq \Lambda_Y = \sup_{t\geq 0}[-\ln\psi(t)]$. ○

It is much more difficult to derive the upper bound for the sequence $\frac{1}{n}\ln 1/p_n$. It is natural to consider separately a number of special cases.

Lemma A.3.2 *If* $\mathbf{P}\{Y \geq 0\} = 1$ *and* $\mathbf{P}\{Y > 0\} \neq 0$, *then (cf. Lemma* A.3.1*)*

$$l^+ \equiv \limsup_{n\to\infty} \frac{1}{n} \ln 1/p_n = l^- = \Lambda = 0.$$

Proof of Lemma A.3.2. Under the restrictions of the lemma, $\mathbf{P}\{Y = 0\} < 1$, so

$$p_n = 1 - (\mathbf{P}\{Y = 0\})^n \to 1$$

as $n \to \infty$. Thus, both partial limits $l^\pm$ are equal to zero. Moreover, in this case $\psi(t) > 1$ for $t > 0$, so the upper bound in the definition of Λ_Y is attained at $t = 0$, and $\Lambda_Y = 0$. This proves the lemma. ○

The rest of the argument considers the remaining case, where

$$\mathbf{P}\{Y > 0\} \neq 0, \quad \mathbf{P}\{Y < 0\} \neq 0. \tag{A.3.2}$$

The crucial argument concerns rv's which have a finite number of distinct values.

Lemma A.3.3 *If the sequence* $Y, Y_1, \ldots$ *satisfies the condition*

$$\exists\, m \in \mathbf{N},\; y_k \in \mathbf{R}, q_k > 0,\; k = \overline{1,m} \quad \sum_{k=1}^{m} q_k = 1, \; q_k = \mathbf{P}\{Y = y_k\}, \tag{A.3.3}$$

then

$$l^+ = \limsup_{n\to\infty} \frac{1}{n} \ln 1/p_n \leq \Lambda_Y.$$

Proof of Lemma A.3.3. By the assumption, there are both negative and positive values among y_k:

$$\min_{1\le k\le m} y_k < 0 < \max_{1\le k\le m} y_k. \tag{A.3.4}$$

Put $\#_k \stackrel{def}{=} \sum_{j=1}^n \mathbf{1}_{\{Y_j=y_k\}}$. The probability that all these rv's assume prescribed values l_k is

$$\mathbf{P}(\mathbf{l}) = \mathbf{P}\left\{\#_k = l_k, k=\overline{1,m}\right\} = \frac{n!}{l_1!\dots l_m!}[q_1]^{l_1}\cdots[q_m]^{l_m}, \quad \mathbf{l} = \left(l_j, j=\overline{1,m}\right).$$

Consequently

$$p_n = \mathbf{P}\{Y_1+\dots+Y_n>0\} \ge \textstyle\sum_+ \mathbf{P}(\mathbf{l}) \ge \max_+ \mathbf{P}(\mathbf{l}),$$

where both the sum and the maximum are over the set of integer-valued vectors satisfying the two conditions

$$l_1+\dots+l_m = n, \quad y_1l_1+\dots+y_ml_m > 0. \tag{A.3.5}$$

The summands in the expression for p_n can be evaluated from below using Stirling's formula for the factorial:

$$\mathbf{P}(\mathbf{l}) \ge \frac{c}{n^{m/2}}\exp\left\{n\sum_{k=1}^m \pi_k \ln\frac{q_k}{\pi_k}\right\}, \quad \pi_k = \frac{l_k}{n}, \quad \sum_{k=1}^m \pi_k = 1. \tag{A.3.6}$$

Assume first that

$$\textstyle\sum_{k=1}^m y_kq_k > 0. \tag{A.3.7}$$

Then it is possible to choose natural sequences $l_k = l_k(n) = n[q_k+o(1)]$ so as to satisfy conditions (A.3.5). It follows from (A.3.6) that

$$\liminf_{n\to\infty}\frac{1}{n}\ln p_n \ge 0, \quad l^+ \le 0. \tag{A.3.8}$$

Besides, $\Lambda_Y = 0$. Indeed, under restriction (A.3.7) the function

$$\psi_m(t) = \textstyle\sum_{k=1}^m q_k e^{y_kt}$$

attains its minimum on the positive half-line at $t=0$: it is convex with positive derivative at the origin.

Assume now that condition (A.3.7) is violated:

$$\sum_{k=1}^m y_kq_k \le 0. \tag{A.3.9}$$

In this case the vector of "frequencies" $\pi_k = l_k/n$ satisfying (A.3.5) can approach only the points $(\hat\pi_k)$ for which

$$\sum_{k=1}^m \hat\pi_k = 1, \quad \sum_{k=1}^m y_k\hat\pi_k = 0, \quad \hat\pi_k \ge 0, \tag{A.3.10}$$

and the lower estimate for $\frac{1}{n}\ln p_n$ which follows from (A.3.6) is

$$\liminf_{n\to\infty}\frac{1}{n}\ln p_n \geq \max{}_* \sum_{k=1}^{m}\pi_k \ln\frac{q_k}{\pi_k},$$

where notation $\max_*$ refers to the maximum over the set of "frequencies" (π_k) selected by (A.3.10).

It follows from the Lagrange rule that the extremum is attained at the point with strictly positive coordinates

$$\pi_k^0(\tau) = \frac{q_k e^{\tau y_k}}{\psi_m(\tau)},\ l=\overline{1,m},\quad \psi(\tau)\equiv\sum_{k=1}^{m} e^{\tau y_k} q_k.$$

Such a point does exist if there is a value of the parameter τ which solves the equation

$$\sum_{k=1}^{m} y_k \pi_k^0(\tau) = \frac{d}{d\tau}\ln\psi_m(\tau) = 0.$$

Not all values y_k are zero. Consequently, the function $\psi(\tau)$ is strictly convex by (A.3.4); by (A.3.9) it is nonincreasing in a neighbourhood of zero. Hence the extremal point does exist, and it corresponds to the value of the parameter $\tau \geq 0$, which is the minimum of ψ on the positive half-line (viz., a maximum of $-\ln\psi(\tau)$).

By the choice of the "limit frequencies" $\pi_k^0(\tau)$,

$$\sum_{k=1}^{m}\pi_k^0(\tau)\ln\frac{q_k}{\pi_k^0(\tau)} = \ln\psi_m(\tau) = -\sup_{t\geq 0}\left[-\ln\psi(\tau)\right] = -\Lambda_Y.$$

The second equality is a consequence of the fact that τ is minimum of ψ.

Since all the $\pi_k^0(\tau)$ are strictly positive, for large values of n one can choose natural numbers $l_k = n\left[\pi_k^0 + o(1)\right]$ so that $\sum_{k=1}^{m} y_k l_k > 0$. Hence also in the case now considered

$$\liminf_{n\to\infty}\ \tfrac{1}{n}\ln p_n \geq \inf_{\tau\geq 0}\left[\ln\psi(\tau)\right],\ l^+ \leq \sup_{t\geq 0}\left[-\ln\psi(\tau)\right] = \Lambda_Y. \tag{A.3.11}$$

This proves the lemma. ○

The next step is to obtain the estimate of Lemmas A.3.2–A.3.3 for arbitrary discrete distributions.

Lemma A.3.4 *Assume that the i.i.d. rv's $Y, Y_1, \ldots$ satisfy condition* (A.3.2) *and have a discrete distribution:*

$$\exists\, y_k, k\in\mathbf{N}: \sum_{k=1}^{\infty} q_k = 1,\ q_k = \mathbf{P}\{Y = y_k\}. \tag{A.3.12}$$

Then the inequality of Lemma A.3.3 *is true.*

Note that, contrary to the assumption of the preceding lemma, the set of k for which $q_k > 0$ is no longer necessarily finite.

Proof of Lemma A.3.4. The assertion of the lemma is established by an argument which reduces the problem to one considered in Lemma A.3.3.

Select a number m so large that condition (A.3.4) is satisfied. This is possible because the distribution of Y has positive masses both on the positive and negative half-lines.

If $q_k = 0$ for all $k > m$, the distribution of Y has a finite set of possible values, and Lemma A.3.3 applies to it. In the converse case, the distribution of Y can be decomposed into a mixture of two distributions, of which one satisfies the requirements of Lemma A.3.3:

$$\begin{aligned} \mathbf{P}\{Y \in B\} &= \alpha_m G_m(B) + (1-\alpha_m)G'(B), \quad \alpha_m = \sum_{k=1}^{m} q_k, \\ G_m(B) &= \sum_{k=1}^{m} q'_k \mathbf{1}_B(y_k), \quad q'_k = \frac{q_k}{\alpha_m}. \end{aligned}$$

Denote by $Y^m, Y_1^m, \ldots$ i.i.d. rv's with distribution G_m. It follows immediately from the well-known binomial formula for convolutions of mixtures that

$$p_n \equiv \mathbf{P}\{Y_1 + \cdots + Y_n > 0\} \geq \alpha_m^n G_m^{*n}(0, +\infty) = \alpha_m^n \mathbf{P}\{Y_1^m + \cdots + Y_n^m > 0\}.$$

Lemma A.3.3 applies to G_m, so

$$\begin{aligned} l^+ &= \liminf_{n\to\infty} \frac{1}{n} \ln \frac{1}{p_n} \leq -\ln \alpha_m + \Lambda^{(m)}, \\ \Lambda^{(m)} &= \sup_{\tau \geq 0} \left[-\ln \mathbf{E} \exp\{\tau Y^m\}\right]. \end{aligned}$$

For large m the Laplace transform of the component G_m in the mixture approximates that of the "complete" distribution, and its "weight" is large: as $m \to \infty$,

$$\alpha_m \psi_m(t) \nearrow \psi(t), \quad \alpha_m \to 1.$$

For this reason

$$\lim_{m\to\infty} \inf_{t\geq 0} \psi_m(t) \geq \inf_{t \geq 0} \psi(t), \quad \lim_{m\to\infty} \Lambda^{(m)} \leq \Lambda_Y = \sup_{t\geq 0}\left[-\psi(t)\right],$$

and, consequently, for all discrete rv's Y

$$l^+ = \limsup_{n\to\infty} \frac{1}{n} \ln 1/p_n \leq \Lambda_Y. \;\bigcirc \tag{A.3.13}$$

Proof of Proposition A.3.1. To prove the proposition in full generality, it remains to consider the situation where the distribution of Y is not discrete, but condition (A.3.2) is satisfied.

Replace the original rv Y by the discrete one $Y_\varepsilon = \varepsilon\left[\frac{1}{\varepsilon}Y\right]$, using the usual roundoff scheme. By the construction $Y \geq Y_\varepsilon$, so

$$p_n \geq p_{\varepsilon n} = \mathbf{P}\{Y_{\varepsilon 1} + \cdots + Y_{\varepsilon n} > 0\}.$$

By Lemma A.3.4

$$\limsup_{n\to\infty} \frac{1}{n} \ln 1/p_{\varepsilon n} \le \sup_{t\ge 0} [-\ln \mathbf{E} \exp\{tY_\varepsilon\}] \le \sup_{t\ge 0} [-\ln \psi(t) + \varepsilon t]$$

for each $\varepsilon > 0$ (the last of these inequalities makes use of the fact that Y and Y_ε differ by ε at most).

By (A.3.2) the function ψ is strictly convex and attains its lower bound at some finite value of the argument $t_0 \ge 0$. If $\varepsilon > 0$ is small, the upper bound

$$\Lambda_\varepsilon = \sup_{t\ge 0} [-\ln \psi(t) + \varepsilon t]$$

is attained in a small neighbourhood of t_0, and $\lim_{\varepsilon\to 0} [\Lambda_\varepsilon] = \Lambda_Y$.

Hence

$$\limsup_{n\to\infty} \frac{1}{n} \ln 1/p_n \le \Lambda_Y .$$

Condition (A.3.2) includes no assumptions about the nature of the distribution, so the assertion of the proposition follows from the above, the bound of Lemma A.3.2 for the case where (A.3.2) is violated, and the lower bound of Lemma A.3.1. ○

A.3.2 Deviation Function and Information

Consider a distribution F on the real line satisfying the summability condition

$$\forall t \in \mathbf{R} \ \ \psi(t) = \int_{\mathbf{R}} e^{tx} F(dx) < \infty.$$

Proposition A.3.2 *The deviation function* $\Lambda(a) = \sup_{t\ge 0} [at - \ln \psi(t)]$ *satisfies the inequality*

$$\forall\, a \in \mathbf{R} \ \ \Lambda(a) \ge \hat{I}(a)\,,$$

where the right-hand side is the minimal relative information over the class of distributions with expectation a*:*

$$\hat{I}(a) = \inf_{\mathcal{G}_a} \{I_F(G)\}, \quad \mathcal{G}_a = \left\{G : \textstyle\int_{\mathbf{R}} x G(dx) = a\right\},$$

$$I_F(G) = \begin{cases} \displaystyle\int_{\mathbf{R}} \frac{dG}{dF}(x) \ln \frac{dG}{dF}(x) F(dx), & G \ll F, \\ +\infty, & G \not\ll F, \end{cases}$$

Note that both the deviation function and the minimal relative information are defined for distributions in any Banach space (cf. formula (4.2.7), and by Lemma 4.3.1 the converse inequality holds in this general situation: $\hat{I}(a) \ge \Lambda(a)$. Hence the proposition actually implies that the deviation function and the minimal information are equal: $\hat{I}(a) = \Lambda(a)$.

Proof of Proposition A.3.2. Consider first the situation where the upper bound in the definition of the deviation function is attained. In this case the necessary condition for the existence of an extremum yields the equation

$$a = \int_{\mathbf{R}} x F_t^*(dx), \quad F_t^*(dx) = e^{tx} F(dx)/\psi(t),$$

for the extremal value of $t \in \mathbf{R}$. This equation shows that if the extremal value of the parameter t is used to define F_t^*, the Cramér conjugate of F, the expectation of this new distribution is a. Since F_t^* is evidently absolutely continuous with respect to F and its density is $e^{tx}/\psi(t)$, one can write down the relation

$$\Lambda(a) \; = \; I_F\left(F_t^*\right) \; \geq \; \hat{I}(a).$$

Consider next the situation where the upper bound in the definition of the deviation function is not attained. To apply the result already obtained, introduce the mixture of distributions

$$F_\varepsilon = (1-\varepsilon)F + \varepsilon\Phi, \quad \varepsilon \in (0,\, 1),$$

where Φ is the standard normal distribution on $\mathbf{R}$. Its Laplace transform admits the evident bound

$$\ln \varphi_\varepsilon(t) \; = \; \ln\left(\int_{\mathbf{R}} e^{tx} F_\varepsilon(dx)\right) \; \geq \; \tfrac{1}{2}t^2 + \ln \varepsilon.$$

Consequently, the deviation function Λ_ε of this distribution is a *bona fide* maximum which is attained at some point of the real line. By the part of the proposition already proved, this new deviation function equals the minimal information with respect to the distribution F_ε: $\Lambda_\varepsilon(a) \; = \; \hat{I}_\varepsilon(a)$.

The Laplace transform of the mixture satisfies the inequality $\varphi_\varepsilon(t) \geq (1-\varepsilon)\psi(t)$ for all $t \in \mathbf{R}$. Consequently the deviation functions are related through the inequality

$$\Lambda(a) \; = \; \sup_{t\in\mathbf{R}} \left[at - \ln \psi(t)\right] \; \geq \; \Lambda_\varepsilon(a) + \ln(1-\varepsilon),$$

so

$$\Lambda(a) \; \geq \; l \; \stackrel{def}{=} \; \limsup_{\varepsilon\to 0} \Lambda_\varepsilon(a).$$

It remains to evaluate from below the left-hand side of the last inequality.

If $l \; = \; +\infty$, then also $\Lambda(a) \; = \; +\infty$, and the assertion of the proposition is true.

If $l < +\infty$, then one can select a sequence $\varepsilon[j] \searrow 0$ in such a way that for the mixtures $F_{\varepsilon[j]}$ there exist distributions $G_{\varepsilon[j]} \in \mathcal{G}_a$ which satisfy the relation $\lim_{j\to\infty} I_{F_{\varepsilon[j]}}\left(G_{\varepsilon[j]}\right) \; = \; l$.

By the construction of the mixtures, the sequence $F_{\varepsilon[j]}$ is tight, and also the sequence $G_{\varepsilon[j]}$ is tight by Lemma 4.2.3. Hence it is possible to select an infinitesimal subsequence $\varepsilon'[k] = \varepsilon[j_k]$ so that

$$F_{\varepsilon'[k]} \stackrel{weak}{\longrightarrow} F, \quad G_{\varepsilon'[k]} \stackrel{weak}{\longrightarrow} \bar{G}, \quad \bar{G} \in \mathcal{G}_a.$$

One can use a special case of Lemma 4.2.4 to verify that the mean values of the distributions considered also converge (cf. the proof of Lemma 4.2.5):

$$a = \int_{\mathbf{R}} x G_{\varepsilon'(k)}(dx) \rightarrow \int_{\mathbf{R}} x \bar{G}(dx).$$

Thus the limit distribution $\bar{G}$ also has expectation a, so by Lemma 4.2.2

$$l = \limsup_{k \to \infty} I_{F_{\varepsilon'[k]}} \left(G_{\varepsilon'[k]}\right) \geq \hat{I}(a).$$

The proof of the proposition is completed. ○

Appendix B

Bibliographic Commentary

Chapter 1

Section 1.1. Basic properties of Gaussian multidimensional distributions are discussed in detail, e.g., in [*An]. The characterization theorems are demonstrated in [*Fe].

Lemma 1.1.1 and Lemma 1.1.2 were proved in [Ts75].

The proof of Slepian's inequality [Sle62] reproduces the argument of [Ch63]. For detailed exposition of the topic and survey see [*LedTa]. Some related results can be found in [Fer74, JoPP83, Kh93, SamT93].

For two sequences of Gaussian rv's with zero expectations and same variances whose covariances satisfy the restriction $\mathbf{E}\xi^{(j)}\xi^{(l)} \leq \mathbf{E}\eta^{(j)}\eta^{(l)}$, $j \neq l$, the inequality

$$\mathbf{E}\left[\max_{j=\overline{1,n}} \xi^{(j)}\right]^p \geq \mathbf{E}\left[\max_{j=\overline{1,n}} \eta^{(j)}\right]^p, \quad p = 1,\ 2.$$

was established in [Su71] for $p = 1$ and [Chev76] for $p = 1,\ 2$.

Section 1.2. An exhaustive account of modern applications of isoperimetric inequalities in probability theory can be found in [*LedTa]. Here the exposition follows [SuTs74].

A proof of the isoperimetric property of the sphere can be found in [Schm43] and [Ti63] or [*BurZ80].

Section 1.3. The theorems and proofs included in this section are adaptations from [SuTs74]. Note that the proof of Lemma 1.3.4 is basically a version of the argument used to demonstrate the Neyman-Pearson inequality.

Theorem 1.3.4 appeared in [LaSh70] for $\mathbf{P}\{\eta \in H\} > \frac{1}{2}$.

Section 1.4. Detailed exposition of the theory of convex sets (in particular, of polytopes) can be found, e.g., in [*BoF34]. For information on surface integrals consult [*Th72].

The special coordinates of subsection 1.4.1 were suggested in [Ts75].

Section 1.5. Most material included is from [Ts75].

Chapter 2

A systematic exposition of methods based on the isoperimetric inequality can be found in the monograph [*LedTa], as well as a survey of work in this field. Some publications of this direction are [SuTs74, Bor74, Bor75, Eh84, Gor88].

Geometric techniques in the study of Gaussian distributions were developed, e.g., in [Chev76, Fer76, MiP87, Ta87].

Much research in this direction is summarized in [*Su76, *Pis89] and earlier in [Pis86].

Relations of geometry of a Banach space to the properties of distributions of norms of Gaussian random vectors were considered in [ByR88, ByR89, HofSD79, FrT80, Pau82, Pa84, Lin91]. Related results can also be found in [SmL89, By89], see also [Li89, Ry86, RyZ89].

Smoothness of dependence of the Gaussian measure of a ball upon the position of its center was studied in [Lin89, Pau90].

For the squared norm of a Gaussian RV with values in a Hilbert space, many formulae can be derived from the known expression of its CF or Laplace transform. Results of this kind are included in Sections 2.3–2.4.

A systematic treatment of Gaussian processes can be found in [*LedTa]. Some additionsl references are [Li82a, Li82b, Li82c, DaL84, Dm80, Dm81], and, for vector-valued Gaussian processes, [Fer90a, Fer90b, Fer90c].

Methods based on entropy estimates were developed in [Du67, Su69, Su71, Ta88b]. A related topic is the study of continuity of Gaussian processes [Mat84]. See also [*Su76, *Pis89].

For methods exploiting more specific properties of Gaussian processes, e.g., stationarity, see, e.g., [Berm71, Pit81, PitP81, Di81, Eh82, Fer81, We89] and [Li90,Sy79,We80], [Yl65,Yl68], as well as [Ta94, Ji93]. Numerous result in the theory of Gaussian processes are covered by the survey of [Pit82].

The origin of arguments included in Chapter 2 is as follows.

Section 2.1. Theorem 2.1.1 was demonstrated in [Fer70] and Theorem 2.1.2 in [Fer71]. The proofs provided in the section reproduce the arguments of these original papers. Further developments can be found in [Fer75].

The assertion of Theorem 2.1.1 was independently obtained in [LaSh70] by methods of geometric nature. Techniques of this kind were developed in [Bor74, Bor75] and [Su69, Su71, SuTs74, Ts75]. See also Section 2.2. It was shown independently in [Sk70] by a method based on considerations from the theory of continuous-time stochastic processes that for a Gaussian RV in a Banach space $\mathbf{E}e^{\alpha|\eta|} < \infty$. A later development of Skorokhod's method later allowed to use it to derive the original results result of Fernique [Na81].

A proof of exponential summability of the squared norm can be found in [*Ku75].

For related results see also [Chev76, Di81].

Section 2.2. The section is based on [Ts75].

Section 2.3. The principal term of asymptotics for the probability that a Gaussian RV lands outside a large sphere centered at its expectation was first derived in [Zo61]. Later the result was obtained in [Hoe64] by the direct method reproduced in the last subsection. The proof of Theorem 2.3.1 here is a modified version of the original proof of [Zo61]. Theorem 2.3.2 is taken from part II of [Yu91b] (see bibliography to Chapter 6).

Section 2.4. This section is based on [Sy74] (see also [Sy73]). For recent developments of the method, see [LiL93].

Section 2.5. For some additional results, see also [MayZ93, Ost77, Pa84, Pau82, Pau90].

Chapter 3

Section 3.1. Theorem 3.1.1 and its proof reproduce a theorem from [*Sk64].

Lemma 3.1.1, which goes back to [Hof74], and Theorem 3.1.2 are taken from [Na82]. The exposition differs in details. Note that the proof of Theorem 3.1.2 includes features similar to those from the proof of Theorem 2.2.1 reproduced above.

An exposition of techniques based on Lemma 3.1.1 can be found in [*LedTa].

Theorem 3.1.3 follows [Ac81], and Theorem 3.1.4 is an adaptation of a result of [Ac80].

Section 3.2. Lemma 3.2.1 and its proof are reproduced from [*Ka68] (see also [*Bu80]). The argument is a repetition of one originally given in the case of real random variables. As well as Lemma 3.2.2, it seems to have become a part of probabilistic folklore (cf., e.g., [*Lo63]). A systematic exposition of symmetrization methods for sums of random vectors is given in [*LedTa].

Theorems 3.2.1-3.2.2 follow [Na82] and [Ac81].

Section 3.3. The inequality of Proposition 3.3.1 was proved in 1912 (see [Bernα]) and later generalized to what is now known as martingales in [Bernβ]. The exposition here follows [*Be].

Theorems 3.3.1-3.3.3 and their proofs are taken from [PiS85]. They improve an earlier result of [Yu74].

First exponential inequalities for vector and, in particular, Hilbert-space valued variables were derived in [Prα68]. The original technique of obtaining such estimates in finite dimensional Euclidean space was developed later in [Prβ68a, Prβ68b, Prβ68c, Prβ73, Zo68] using Bessel functions of imaginary argument [Prβ68d]. For a survey of early results see [Prα72].

Theorem 3.3.4 and its proof follow [PiS85]. A cruder form of the bound of Theorem 3.3.4 was derived in [Yu70] and later improved to a more complicated bound with the "correct" order of decay (see [Yu76]). A different approach to "correct" estimate is developed in Chapter 6 (see Theorem 6.2.1 and related bibliography).

Related results can be found in [Na83] and a later article [Na87]. See also [Fu90].

Section 3.5. The bounds of the section originate in results of [FuN71] for the real line. Some inequalities of this type were included in [Yu76]. The versions included in the section follow [NaPi77].

Exposition of results pertaining to infinite-dimensional sums of subsection 3.5.3 is based upon [Pi80]. The Burkholder inequalities that it uses were published in [Burk73].

Theorem 3.5.3 follows [Pi78].

For more results in the field see also [AcK83, Burk81, Co82, GiMZ78, GiZ83, Hof72, Kan78], [Ku78]–[Mat91a], [Pi81a]–[Pi93], [Prα83, Zi81].

Chapter 4

The theory of large deviations owes much to [DoV] where it was developed in a setting much more general than that of this chapter.

Exposition of the topic can be found in [*Az, *Var84, *Str84] or [*DeS].

Most of the results included in Chapter 4 can be considered as distant offspring of [Cher52] (see Section A.3 of Appendix A and the corresponding bibliography).

Section 4.1. The exposition follows mainly [BoroM78] and through this paper also [BaZ79] and [GrOR79]. For finite-dimensional RV's the properties of DF (4.1.2) and its relation to LD asymptotics were systematically treated in, e.g., [Boro62, Boro65]. See also [Sl88, Ac84].

Sections 4.2–4.3. The propositions included in these sections are adaptations of the corresponding results from [DoV].

Section 4.5–4.6. Exposition is mainly based upon [BoroM78].

The construction of finite-dimensional approximations of Lemma 4.6.5 follows [Che90].

Chapter 5

It seems natural to discuss first the most important omission of this chapter — it does not include any final results on convergence rates in the infinite-dimensional central limit theorem (CLT) or estimates of finer approximations to laws of normalized sums. For this reason, the results of the chapter are, in a way, preliminary, which explains why all its propositions are labelled "lemma" rather than "theorem".

The reason for not treating convergence rates for the CLT or finer Edgeworth-type approximations is twofold. First, there is a number of monographs and surveys which make the need of one more systematic exposition hardly marked (see, e.g., [*ArG80, *PauR87]). Second, an exposition of convergence rate estimates at the state-of-the-art level would inevitably inflate the chapter to the size of a full-blown book, as each step forward in this direction demands considerable effort.

Standard references on CLT and its refinements in one and several dimensions are the books [*BhRR76, *IbL65, *Lo63, *Pe72].

Much later work on convergence rates in finite dimensions and early infinite-dimensional research is summarized in [*Sa81].

In infinite dimensional CLT setting, the first convergence rate estimates are those of [Ka65, KaV69]. Some finite-dimensional estimates with explicit dependence on the number of dimensions that could, under additional restrictions, produce bounds for the infinite-dimensional case, were obtained, e.g., in [Bahr67a, Bahr67b, Saz68] (bibliography to Ch.1-2) and [VaK68].

It took fairly long to realize that nontrivial convergence rate estimates can be obtained for images of sums' distributions under regular transforms to one or several dimensions. Once this happened, the progress accelerated considerably. The results that had most influence are, probably, the convergence rate estimate from [KK79] — the first one of order $\mathcal{O}(1/n^{\alpha})$, $\alpha > 0$, and that of [Gö79] — the first one with the "correct" exponent $\alpha = \frac{1}{2}$. The symmetrization inequality established in [Gö79] became a standard tool for evaluation of convergence rates (see, e.g., [NaCh86, Yu82, Yu83, Za85c, Za88, Za89b, ZaSU88a, ZaSU88b]. (This inequality is reproduced below in Section 5.7).

A more special problem that attracted much attention is to estimate the accuracy of normal approximation to the probability that a normalized sum of independent RV's lands in a ball of Hilbert space. The aspect of this problem that excited considerable interest dependence of approximation error on the spectrum of covariance operator. The ultimate result in this direction was obtained in [SaUlZa91] (this paper originally appeared in preprint form in 1989), after the result was announced in [ZaSU88b]; a similar assertion was independently announced later in [Na89]). See also [Se89].

Much recent work on the CLT convergence rates and asymptotic expansions is summarized in [BeGöPaRa90].

For estimates of error of finer Edgeworth-type approximations for specific mapping consult, e.g., [NaCh85, Za85a, Za85b, Gö89, NaCh89b, SazUl93].

Sections 5.1-5.2. are based on [Yu91a, Yu94b].

The approximations considered are versions of the classical Edgeworth and Bergström expansions. The latter are the usual intermediate step in construction of asymptotic formulae like (5.2.6). They seem to appear first in [Berg51]; in vector case, they were treated in [NaCh89a].

On the real line or in finite dimensions these have long become standard (see, e.g., [*Fe, *BhRR76]). The infinite-dimensional case has attracted considerable attention lately. It is discussed in numerous research publications (see, e.g., [Gö81, Gö89], [NaCh85, NaCh89a, NaCh89b], [SazUl93]. and surveys [BeGöPaRa90, BeGöZit92]).

Expansions for means of smooth functionals were earlier derived, e.g., in [Be84a] or, in a more general situation, [Be86a].

The new elements of the exposition below are its method and some formulae for terms of expansion that employ the moments of "complex" RV's. The remainder terms in expansions are written out explicitly, but no attempt is made to apply them to rederive known estimates for the approximation error. The "real" form of the expansions considered can be found, e.g., in [BeGöPaRa90].

The method employed below is a version of the classical operator approach of J.W.Lindeberg reflected, e.g., in the classical book [*Fe].

Both LLN and CLT can be treated as manifestations of the fundamental (or elementary) relation between differentiation and translations by very small vectors. In finite dimensions [*Fe, *BhRR76] it proves convenient to exploit this relation through the method of characteristic functions. In infinite dimensions, the more elementary operator approach seems more practical.

The mixture of standard tensor notation (see, e.g., [*vdW]) and "mechanical" (∇, $\langle a, \nabla\rangle$, etc.) was chosen to emphasize the relation between differentiation and translations. It is, probably, eclectic, but seems expressive.

Section 5.3. Subsection 5.3.1 follows [Yu94b].

The remaining part follows [Yu91b, Pt.1] (see references to Chapter 6). It is an adaptation of the standard smoothing technique (see, e.g., [*BhRR76]).

The consideration of mappings to several dimensions does not introduce any difficulties that are new in comparison to those met in the case of smooth mappings to the real line. An alternative approach is to construct asymptotic expansions introducing generalized derivatives of translations of Gaussian measure (see [Be82] in references to Ch. 1–2).

Sections 5.4–5.6. The exposition of two first sections follows [Yu94b]. The calculations reproduced in them are a part of the "CLT folklore".

Section 5.7. The symmetrization inequality of subsection 5.7.1 first appeared in [Gö79] in a somewhat different form, as also the inequality of Lemma 5.7.2. Lemma 5.7.1 is its adaptation used in [Yu82], in conjunction with the technique of mixtures (5.7.7)–(5.7.8).

Theorem 5.7.1 is borrowed from [SazUl93] and Theorem 5.7.2 from [Zit88] (see also [BeGöZit92, Zit89] for related results).

To derive a better-than-Gaussian approximation with an error estimate using the method of Fourier, one needs nontrivial bounds on the CF very far from the origin. The "individual" Cramér-type condition (5.7.10) formulated in terms of a single summand is, unfortunately, very difficult to verify (so far, it is uncertain whether it holds in the classical statistical problems that motivated research on the CLT in Hilbert space). A more realistic "collective" restriction on the squared sum's CF is stated in [BeGöZit92] in terms of the distribution of ζ_n.

The Cramér condition on CF's in one or several dimensions is essentially an assumption that the distribution has an absolutely continuous component. This latter can, in a way, be approximated by a mixture of distributions with "small" compact supports. An attempt to formulate a similar property of a RV and use it to derive nontrivial bounds on CF's of the squared norm of the sum of its copies in Hilbert-space setting was undertaken in [Yur94a].

Chapter 6

Calculations of "fine" LD asymptotics originate in [Cr38] where many tools that later became customary were introduced. A comprehensive survey of results for the real line is contained in [*Pe72] (see also [*IbL65]). The case of sums in finite-dimensional Euclidean spaces was considered by a number of authors (see, e.g., [Os75]) whose tools were basically those developed to treat one-dimensional laws.

The first calculation of infinite-dimensional LD asymptotics in the range of moderately large deviations was performed in [Os78a]. The problem considered was approximation of the probability that a normalized sum of i.i.d. RV's with values in a Hilbert space lands outside a large sphere surrounding its expectation, $\mathbf{P}\{|\zeta_n| \geq r\}$. The formula obtained applied to the whole range of moderately large deviations. It included a form of the "Cramér series" admitting a characterization in terms of its extremal properties. The conditions were all in terms of moments of individual summand's distribution. The version of Cramér's condition it used was more restrictive than (3.6.2). The method used to treat the case when the covariance operator has multiple principal eigenvalue, i.e., the extremum of the DF is not unique, was essentially similar to the "focusing" of this chapter. L. V. Osipov obtained his asymptotic formula before the appearance of most modern refinements of the CLT, and this made the proof fairly complicated. (The most complete exposition of the proof occupies some hundred typewritten pages in [Os78b]).

For a fairly long time the publication [Os78a] remained isolated — probably, because most improvements of the infinite-dimensional CLT appeared after L. V. Osipov's death.

In the range where the Cramér series does not influence asymptotics, the problem was solved under natural restrictions in [Be86a] which provided a bound on the relative error (announced in [Be86b]). The paper [Be86a] is an exception from the rule that calculations of LD asymptotics should have formidable length. Later refinements can be found in [Ra88, Za89a, BeR90].

An estimate $\mathcal{O}(r/\sqrt{n})$ for the relative error of Osipov's approximation was obtained in [Yu91b] under Cramér's condition (3.6.2) and nondegeneracy condition (6.0.1). The method is essentially that of Section 6.3, and the "focusing" is that of subsection 6.2.1.

In the domain of large deviations proper, fine LD asymptotics was obtained in [Bo86, Bo87]. The problem considered there is basically that of calculating $\mathbf{E}\exp\left\{n\psi\left(\frac{1}{n}S_{1,n}[\xi]\right)\right\}$ under the assumption that $\exp\{t|\xi|\} < \infty$ for all t. To derive the asymptotic formulae, Cramér's transform was used to shift the expectation to the point where the difference of ψ and DF is largest. The remaining problem is to calculate an integral with respect to an almost-Gaussian measure. The integrand contains a rapidly decaying exponential, so the problem is solved by a combination of the central limit theorem with the Laplace method. The restrictions of this work were later relaxed in [Chy93].

The method of [Bo86, Bo87, Chy93] is rather similar in essence to that developed in Chapter 6, although many details are quite different. Thus, analysis of the critical manifold is rather more complicated.

In [Ac92] similar analysis was carried out for the range of moderately large deviations under assumption that the Laplace transform of summand's norm distribution is finite for all values of argument. The quantity calculated was $M[\varphi, r_n]$ of (6.0.6) It was assumed that there is only one point where $\varphi - \Lambda_0$ attains maximum, and that φ grows no faster than a linear function of the norm. The resulting formula (with η of (6.0.2)) is

$$M[\psi, r_n] = \exp\left\{r^2 \sup_x \left[\varphi(x) - \Lambda_0(x)\right]\right\}[1 + o(1)]\mathbf{E}\exp\left\{\ \tfrac{1}{2}\langle\eta, \nabla\rangle^2 \varphi(x^*)\ \right\}.$$

The motivation for considering probability (6.0.3) and not the expectation of (6.0.6) in Chapter 6 is the desire to keep possibly close to the classical theorem of [Cr38] and to avoid as much as possible the additional difficulties arising in the analysis of the set of critical points. This was done at the price of postulating a number of conditions that would be available automatically in the setting of (6.0.6).

Sections 6.1–6.2. The exposition is largely an adaptation of techniques of [Yu91b] to a more general setting.

The "focusing device" of subsection 6.2.1 is an improved version of that from [Yu91b] and [Yu91]. Subsection 6.2.2 follows [Yu91].

Appendix A

Section A.1 One can find a sufficiently complete exposition of the theory of surface integrals in [*Th72]. A more detailed exposition of the properties of surface area and related concepts see in [*BurZ80].

Section A.2 The terminology of the section follows the common usage (see, e.g., [*Kr]).

The reminder of measure theory follows mainly [*GikS77] and [*GikS71]. A more complete exposition of the topic can be found, e.g., in [*VaTC85]. An exposition of measure theory in infinite dimensions focusing on its probabilistic aspects is provided also in the more special monographs [*Bu80] and [*Mu79].

The fundamental Proposition A.2.1 is known as the Prokhorov criterion. It had been proved in [PrY56]. The overview of facts related to constructing a linear measurable space using cylindric sets follows the important survey [FoS76]. One can find Proposition A.2.6 in [*GikS71]. Proposition A.2.7 is known as the Minlos–Sazonov theorem (see [Saz58] and [Min59], also [*GikS71]). Further investigations of the possibility to describe in topological terms the conditions which make it possible to construct a countably additive measure in an infinite-dimensional space starting from its characteristic functional are reflected in the monograph [*Mu79].

Section A.3 The proof of Proposition A.3.1 is taken from [Cher52].

Basic Abbreviations and Notation

Abbreviations

a.s.	almost surely
CF	characteristic function or functional
CLPF	convex piecewise-linear function
DF	distribution function or deviation functional
i.i.d.	independent identically distributed
LD	large deviation
rv	real random variable
RV	random vector

Notation

$*$	convolution of measures
$\mathbf{1}_A(u) = \begin{cases} 1, & u \in A, \\ 0, & u \notin A \end{cases}$	indicator of set A
$a_{\pm} = \max[0, \pm a]$	positive and negative parts
card (A)	cardinality of a finite set
$\mathrm{ch}\, a = \frac{1}{2}(e^a + e^{-a})$	hyperbolic cosine
$\mathrm{cov}\,(\xi, \eta) = \mathbf{E}\xi\eta - \mathbf{E}\xi\mathbf{E}\eta$	covariance of ξ, η
$\mathbf{D}\xi = \mathrm{cov}(\xi, \xi)$	variance of ξ
$\det(M)$	determinant of M
$\mathbf{E}\xi = \int_{\Omega} \xi(\omega)\mathbf{P}\,\{d\omega\}$	expectation of rv ξ
$\mathbf{E}\,\{\xi; A\} = \int_A \xi(\omega)\mathbf{P}\,\{d\omega\}$	mean value of rv ξ over the event A
$\mathcal{F}_J = \sigma\,(\xi_j, j \in J)$	σ-algebra generated by RV's, (3.1.1)
$\mathcal{F}_{\leq k} = \mathcal{F}_{\{1,\ldots,k\}}$	"past" of a sequence of RV's, (3.1.1)
$\mathcal{F}_{>k} = \mathcal{F}_{\{k+1,\ldots\}}$	"future" of a sequence of RV's, (3.1.1)
$\mathcal{I}$	the identity operator
I_d	the identity operator in $\mathbf{R}^d$
$\mathcal{L}\,(\xi)$	distribution of rv ξ
$\mathbf{P}\,\{A\}$	probability of event A
$\mathcal{R}\,[z]$	the resolvent of the covariance operator
$S_{1,n}\,[\xi] = \xi_1 + \cdots + \xi_n$	sum of independent RV's
$S_k^l = \lvert S_{k,l} \rvert$	norm of sum (3.1.2)
$S_k^{l*}, S_{k*}^l, S_{k*}^{l*}$	maximal sums (3.1.3)
$\mathrm{tr}(A)$	trace of a matrix or operator
$\bigvee_l^m \xi = \max_{j=\overline{l,m}} \lvert \xi_j \rvert$	maximal summand (3.1.6)
ξ^r	RV truncated at level r (3.7.1)
$\sum_{1\,\uparrow}^{n\,m} [A]$	ordered sum of an array (3.5.9)
$\varphi(u) = \exp\{-u^2/2\}/\sqrt{2\pi}$	the standard normal density
$\Phi(x) = \int_{-\infty}^{x} \varphi(u)dy,\ \Phi^{-1}$	the standard normal DF and its inverse

Bibliography

[*] **BOOKS**

[*An] Anderson T. W. *An Introduction to Multivariate Statistical Analysis.* 2nd ed. J. Wiley & Sons, New York., 1984.

[*ArG80] Araujo A., Giné E. *The Central Limit Theorem for the Real and Banach Valued Random Variables.* J. Wiley & Sons, New York, 1980.

[*Az] Azencott R. *Grandes déviations et applications.* Lect. Notes Math., v.774, Springer-Verlag, 1980.

[*Be] Bernstein S. N. *Theory of Probability,* 4th ed. Gostekhizdat, Moscow–Leningrad, 1946. (Russian)

[*BhRR76] Bhattacharya R. N. and Ranga Rao R. *Normal Approximation and Asymptotic Expansions.* J. Wiley & Sons, New York, 1976.

[*Bi67] Billingsley P. *Convergence of Probability Measures.* J.Wiley & Sons, New York, 1967.

[*BoF34] Bonnesen T., Fenchel W. *Theorie der konvexen Körper.* Verlag J. Springer, Berlin, 1934.

[*Bu80] Buldygin V. V. *Convergence of Random Elements in Topological Spaces.* Naukova Dumka, Kiev, 1980. (Russian)

[*BurZ80] Burago D. M. and Zalgaller V. A. *Geometric Inequalities.* Springer-Verlag, Berlin Heidelberg, 1988 [Russian edn.: Nauka, Leningrad, 1980].

[*DeS] Deuschel J.-D. and Stroock D. W. *Large Deviations.* Pure and Appl. Math., v.137. Academic Press, Boston, 1989.

[*Fed77] Fedoryuk M. V. *The Saddle-Point Method.* Nauka, Moscow, 1977. (Russian)

[*Fe] Feller W. *An Introduction to the Probability Theory and Its Applications,* v.2, J.Wiley & Sons, New York, 1977.

[*GikS77] Gikhman I. I. and Skorokhod A. V. *Introduction to the Theory of Random Processes*, 2nd ed., Nauka, Moscow, 1977. (Russian) (English translation of 1st edn.: Scripta Technica, W. B. Saunders, Philadelphia, 1969.)

[*GikS71] Gikhman I. I. and Skorokhod A. V. *The Theory of Stochastic Processes*, v. 1 / Grundlehren der mathematischen Wissenschaften, 210. Springer-Verlag, Berlin Heidelberg New York, 1980. (Russian ed.: Nauka, Moscow, 1971.)

[*Hof77] Hoffmann-Jørgensen J. *Probability in B-Spaces.* Preprint/Aarhus Univ. Lect. Notes ser.; N48, Aarhus, 1977.

[*IbL65] Ibragimov I. A. and Linnik Yu. V. *Independent and Stationary Sequences of Random Variables*, Wolters-Noordhoff, Groningen, 1971 (Russian edn.:Nauka, Moscow, 1965.)

[*IbR70] Ibragimov I. A. and Rozanov Yu. A. *Gaussian Random Processes.* Springer-Verlag, Berlin New York, 1978. (Russian edn.: Nauka, Moscow, 1970.)

[*Ka68] Kahane J.-P. *Some Random Series of Functions.* Heath Math. Monographs. Cambridge Univ. Press, 1968; 2nd ed. 1985.

[*KolF] Kolmogorov A. N. and Fomin S. V. *Elements of Theory of Functions and Functional Analysis*, 3rd ed. Nauka, Moscow, 1972. (Russian)

[*Kr] Krein S. G., ed. *Functional Analysis: a Handbook*, 2nd ed., Nauka, Moscow, 1972. (Russian)

[*Ku75] Kuo H. S. *Gaussian Measures in Banach Spaces.* Lect. Notes Math., v.463, Springer-Verlag, 1975.

[*LedTa] Ledoux M. and Talagrand M. *Probability in Banach Spaces /Ergebnisse der Mathematik*, 3. Folge, Bd. 23, Springer-Verlag, Berlin Heidelberg, 1991.

[*Leh] Lehmann E. L. *Testing Statistical Hypotheses.* 2nd. ed., J. Wiley & Sons, New York, 1986.

[*Lev] Lévy P. *Problèmes concrets d'analyse fonctionnelle.* Gauthier-Villars, Paris, 1937.

[*Lo63] Loève M. *Probability Theory.* Van Nostrand, New York, 1963.

[*Mu79] Mushtari D. H. *Probabilities and Topologies in Banach Spaces.* University of Kazan, Kazan, 1979. (Manuscript deposited in VINITI, 1979, N 4264-Ya79 DEP; Russian)

[*PauR87] Paulauskas V. J. and Račkauskas A. *Approximation Theory in the Central Limit Theorem. Exact Results in Banach Spaces* Mathematics and Its Appl., Soviet ser., v.32. Kluwer Academic Publishers, Dordrecht, 1989. (Translated from Russian: *The Accuracy of Approximation in the Central Limit Theorem in Banach Spaces.* Mokslas, Vilnius, 1987.)

[*Pe72] Petrov V. V. Sums of Independent Random Variables. Springer-Verlag, Berlin Heidelberg, 1975 (Russian ed.: Nauka, Moscow, 1972.)

[*Pis89] Pisier G. *The Volume of Convex Bodies and Banach Space Geometry.* Cambridge Tracts in Math., v.94, Cambridge Univ. Press, Cambridge, 1989.

[*Sa81] Sazonov V. V. *Normal Approximation — Some Recent Advances.* Lect. Notes Math., Springer-Verlag, v.879, 1981.

[*Sch71] Schaefer C. *Topological Vector Spaces.* Springer-Verlag, New York, 1971.

[*Sk64] Skorokhod A. V. *Stochastic Processes with Independent Increments.* Clearing House of Federal and Technical Information, Wright-Patterson Air Base, US Government Publication, 1964. (Russian edn.: Nauka, Moscow, 1964.) (Russian)

[*Str84] Stroock D. W. *An Introduction to the Theory of Large Deviations.* Springer-Verlag, New York, 1984.

[*Su76] Sudakov V. N. *Geometric Problems in the Theory of Infinite-Dimensional Probability Distributions.*, Trudy Mat. Inst. Steklov, v.141, Nauka, Moscow–Leningrad, 1976. (Russian)

[*Th72] Thorpe J. A. *Elementary Topics in Differential Geometry.* Springer-Verlag, New York, 1972. (Translation into Russian: Mir, Moscow, 1982.)

[*Va71] Vakhania N. N. *Probability Distributions on Linear Spaces.* North-Holland, Amsterdam, 1981 (Russian edn.: Metsniereba, Tbilissi, 1971.)

[*VaTC85] Vakhania N. N., Tarieladze V. I., and Chobanian S. A. *Probability Distributions on Banach Spaces,* Reidel, Dordrecht, 1987 (Russian edn.: Nauka, Moscow, 1985.)

[*Var84] Varadhan S. R. S. *Large Deviations and Applications.* CBMS–NSF Regional Conference Series in Applied Mathematics, v.46. SIAM, Philadelphia, 1984.

[*vdW] van der Waerden B. L. *Algebra,* v. 1–2, 7th ed., Ungar, New York, 1970.

[ ] ARTICLES

[ ] Ch. 1–2

[Bahr67a] von Bahr B. Multi-dimensional integral limit theorems. *Arkiv Mat.*, 1967, v.7, N1, pp.71-88.

[Bahr67b] von Bahr B. Multi-dimensional integral limit theorems for large deviations. *Arkiv Mat.*, 1967, v.7, N1, pp.89-99.

[Be82] Bentkus V. J. Analiticity of Gaussian measures. *Theory Probab. Appl.*, 1982, v.27, N1, pp.155–161.

[Ber75] Beran R. Tail probabilities of noncentral quadratic forms. *Ann. Statist.*, 1975, v.3, N4, pp.969-974.

[Berm71] Berman S. M. Excursions above high level for stationary Gaussian processes. *Pacific J. Math.*, 1971, v.36, N1, pp.63-79.

[Bl73] Blöndal P. H. *Explizite Abschätzungen des Fehlers in mehrdimensionalen zentralen Grenzwertsatz*, Dissertation. Köln, 1973.

[Bor74] Borell C. Convex measures on locally convex spaces. *Arkiv Mat.*, 1974, v.12, N2, pp.239-252.

[Bor75] Borell C. The Brunn-Minkowski inequality in Gauss space. *Invent. Math.*, 1975, v.30, N2, pp.207-216.

[Bor77] Borell C. A note on Gaussian measures which agree on small balls. *Ann. Inst. H. Poincaré*, 1977, ser.B, t.13, N3, pp.231-238.

[Bor81] Borell C. A Gaussian correlation inequality for certain bodies in R^n. *Math. Ann.*, 1981, v.256, N4, pp.569-573.

[By89] Byczkowski T. *On the density of* 0*-concave seminorms on vector spaces.* Preprint, Wroclaw Technical University, Inst. Math., Wroclaw, 1989.

[ByR88] Byczkowski T. and Ryznar M. Series of independent vector valued random variables and absolute continuity of seminorms. *Math. Scand.*, 1988, v.62, N1, pp.59-74.

[ByR89] Byczkowski T. and Ryznar M. *Smoothness of the distribution of the norm in uniformly convex Banach spaces.* Preprint, Wroclaw Technical University, Inst. Math. and Case Western Reserve Univ., 1989.

[Ch63] Chartres B. A. A geometric proof of a theorem due to D. Slepian. *SIAM Review*, 1963, v.5, N4, pp.335-341.

[Chev76] Chevet S. Processus gaussiens et volumes mixtes. *Z. Wahrscheinlichkeitstheorie und verw. Geb.*, 1976, v.36, N1, pp.47-65.

[CuNg93] Cutland N. and Siu-Ah Ng. The Wiener sphere and Wiener measure. *Ann. Probab.*, 1993, v.21, N1, pp.1-13.

[DaL84] Davydov Yu. A. and Lifshits M. A. Fibering methods in some probabilistic problems. J. Sov. Math., 1985, v.31, pp.2796–2858. (Translated from *Itogi Nauki i Tekh.*, Ser. Veroyatn. Mat. Stat. Teor. Kibern., v.22., VINITI, Moscow, 1984, pp.61-201.)

[Di81] Diebolt J. Sur la loi du maximum de certains processus gaussiens sur le tore. *Ann. Inst. H. Poincaré*, 1981, ser.B, t.17, N2, pp.165-179.

[Dm80] Dmitrovskii V. A. On the distribution of maximum and local properties of pre-Gaussian random fields. *Theory Probab. Appl.*, 1980, v.25, N1. (Russ. orig.: pp.204-205.)

[Dm81] Dmitrovskii V. A. On the distribution of the maximum and local properties of realizations of pre-Gaussian fields. *Theory Probab. Math. Stat.*, v.25, 1982, pp.169-180.

[Du67] Dudley R. M. The sizes of compact subsets of Hilbert spaces and continuity of Gaussian processes. *J. Funct. Anal.*, 1967, v.1, N3, pp.290-330.

[Eh82] Ehrhard A. Sur la densité du maximum d'une fonction aléatoire gaussienne. *Lect. Notes. Math.*, 1982, v.920, Springer-Verlag, pp.581-601.

[Eh84] Ehrhard A. Inégalité isopérimetriques et integrales de Dirichlet gaussiennes. *Ann. Scientif. Ec. Norm. Super.*, 1984, v.17, N2, pp.317-332.

[Fer70] Fernique X. Integrabilité de vecteurs gaussiens. *C. R. Acad. Sci. Paris*, 1970, Ser.A, t.270, N25, pp.1698-1699.

[Fer71] Fernique X. Regularité de processus gaussiens. *Invent. Math.*, 1971, v.12, N4, pp.304-320.

[Fer74] Fernique X. Minoration des fonctions aléatoires gaussiennes. *Ann. Inst. Fourier*, 1974, t.24, N2, pp.61-66.

[Fer75] Fernique X. Regularité de trajectoires des fonctions aléatoires gaussiennes. *Lect. Notes Math.*, 1975, v.480, Springer-Verlag, pp.1-96.

[Fer76] Fernique X. Corps convexes et processus gaussiens de petit rang. *Z. Wahrscheinlichkeitstheorie verw. Geb.*, 1976, v.35, N4, pp.349-353.

[Fer81] Fernique X. *Continuité de trajectoires des ecarts aléatoires, applications aux fonctions aléatoires.* Preprint, Inst. de Recherche Mat. Avancée, Univ. L. Pasteur, Strasbourg, 1981.

[Fer90a] Fernique X. Fonctions aléatoires à valeurs dans les espaces lusiniens. *Deux papiers sur la régularité des fonctions aléatoires.* Preprint 416/P-232: Inst. de Recherche Math. Avancée, Univ. L. Pasteur, Strasbourg, 1990.

[Fer90b] Fernique X. Sur la régularité de certaines fonctions aléatoires d'Ornstein–Uhlenbeck. *Deux papiers sur la régularité des fonctions aléatoires.* Preprint 416/P-232: Inst. de Recherche Math. Avancée, Univ. L. Pasteur, Strasbourg, 1990.

[Fer90c] Fernique X. Régularité des fonctions aléatoires gaussiennes à valeurs vectorielles. *Ann. Probab.*, 1990, v.18, N4, pp.1739-1745.

[FrT80] Fremlin D. H. and Talagrand M. A Gaussian measure on l^∞. *Ann. Probab.*, 1980, v.8, N6, pp.1192-1193.

[Gor88] Gordon Y. Gaussian processes and almost spherical sections of convex bodies. *Ann. Probab.*, 1988, v.16, N1, pp.180-188.

[Hoe64] Hoeffding W. On a theorem of V. M. Zolotarev. *Theory Probab. Appl.*, 1964, v.9, N1, pp.89–91.

[HofSD79] Hoffmann-Jørgensen J., Shepp L. A., and Dudley R. M. On the lower tail of Gaussian seminorms. *Ann. Probab.*, 1979, v.7, N2, pp.319-342.

[Ib79] Ibragimov I. A. On the probability that a Gaussian vector with values in a Hilbert space lands in a sphere of small radius. *Zapiski Nauchn. Seminarov LOMI*, 1979, v.85, pp.75-93. (Russian)

[Ji93] Jiayang Sun. Tail probabilities of the maxima of Gaussian random fields. *Ann. Probab.*, 1993, v.21, N1, pp.34-71.

[JoPP83] Joag-dev K., Perlman M. D., and Pitt L. D. Association of normal random variables and Slepian's inequality. *Ann. Probab.*, 1983, v.11, N2, pp.451-455.

[Ka86] Kahane J. P. Une inégalité du type de Slepian et Gordon sur les processus Gaussiens. *Israel J. Math.*, 1986, v.55, pp.109-110.

[Kal70] Kallianpur G. Zero-one laws for Gaussian processes. *Trans. Amer. Math. Soc.*, 1970, v.149, pp.199-211.

[Kh93] Khaoulani B. A vectorial Slepian type inequality. Applications. *Proc. Amer. Math. Soc.*, 1993, v.118, N1, pp.95-102.

[LaSh70] Landau H. J. and Shepp L. A. On the supremum of a Gaussian process. *Sankhyā*, 1970, ser.A, v.32, N4, pp.369-378.

[LeT88a] Ledoux M. and Talagrand M. Un critère sur les petites boules dans le théorème limite central. *Probab. Theory Related Fields*, 1988, v.77, N1, pp.29-41.

[Li82a] Lifshits M. A. Absolute continuity of functionals of the "supremum" type for Gaussian processes. *Zapiski Nauchn. Seminar. Leningr. Otd. Mat. Inst. Steklova*, 1982, v.119, pp.154-166. (Russian)

[Li82b] Lifshits M. A. The fiber method and its applications to the study of functionals of stochastic processes. *Theory Probab. Appl.*, 1982, v.27, N1, pp.69–83.

[Li82c] Lifshits M. A. On the absolute continuity of distributions of functionals of stochastic processes. *Theory Probab. Appl.*, 1982, v.27, N3, pp.600–607.

[Li89] Lifshits M. A. On the norm distribution of Gaussian and other stable vectors. In: *Probability Theory and Mathematical Statistics*, (Vilnius 1989, Conf. Proc.), Mokslas, Vilnius, 1990, pp.97-104.

[Li90] Lifshits M. A. Calculation of the exact asymptotics of some Gaussian large deviations. *Zapiski Nauchn. Seminar. Leningr. Otd. Mat. Inst. Steklova*, 1990, v.184, pp.189-199, 323-324. (Russian)

[LiL93] Li W. V. and Linde W. Small ball problems for non-centered Gaussian measures. *Probab. and Math. Statist.*, 1993, v.14, N2, 231–251.

[Lin89] Linde W. Gaussian measure of translated balls in a Banach space. *Theory Probab. Appl.*, 1989, v.34, N2, pp.307-317.

[Lin91] Linde W. Gaussian measures of large balls in l_p. *Ann. Probab.*, 1991, v.19, N3, pp.1264-1279.

[Mat84] Matsak I. K. The Fernique condition and Gaussian processes. *Ukrainian Math. J.*, 1984, v.36, N2, pp.239-249.

[MayZ93] Mayer-Wolf E. and Zeitouni O. The probability of small Gaussian ellipsoids and associated conditional moments. *Ann. Probab.*, 1993, v.21, N1, pp.14-24.

[MiP87] Milman V. D. and Pisier G. Gaussian processes and mixed volumes. *Ann. Probab.*, 1987, v.15, N1, pp.292-304.

[Na81] Nagaev S. V. On the probabilities of large deviations for a Gaussian distribution in a Banach space. *Izvestia AN Uzbek. SSR*, ser. fiz.-matem. nauk, 1981, N5, pp.18-21. (Russian)

[Ost77] Ostrovskii E. I. Covariance operators and certain bounds for Gaussian vectors in Banach spaces. *Doklady AN SSSR*, 1977, v.236, N3, pp.541-543. (Russian)

[Ost94] Ostrovskii E. I. An exponential bound in the law of the iterated logarithm in Banach space, *Matem. Zametki*, 1994, v.56, N5, 98–107. (Russian)

[Pa84] Pap D. On the distribution of the norm of a Gaussian vector in Banach space. *Litovsk. Matem. Sbornik*, 1984, v.24, N1, pp.140-144. (Russian)

[Pau82] Paulauskas V. J. On the distribution density of the norm of a Gaussian vector in a Banach space. *Sov. Math., Dokl.*, 1982, v.26, N6, pp.499–500.

[Pau90] Paulauskas V. A note on Gaussian measure of balls in Banach space. *Theory Probab. Appl.*, 1990, v.35, N4, pp.802-805.

[Pis86] Pisier G. Probabilistic methods in geometry of Banach spaces. *Lect. Notes. Math.*, 1986, v.1206, Springer-Verlag, pp.167-241.

[PiS85] Pinelis I. F. and Sakhanenko A. I. Remarks on inequalities for probabilities of large deviations. *Theory Probab. Appl.*, 1985, v.30, N1, pp.143–148.

[Pit81] Piterbarg V. I. Comparison of distribution functions of maxima of Gaussian processes. *Theory Probab. Appl.*, 1981, v.26, N4, pp.687–705.

[Pit82] Piterbarg V. I. Gaussian stochastic processes. *J. Soviet. Math.*, 1983, v.23, pp.2599–2626. (Translated from: *Itogi Nauki i Tekh.* Ser. Teor. Veroyatn. Mat. Stat. Teor. Kibern., v.19, VINITI, Moscow, 1982, pp.155–199.)

[PitP81] Piterbarg V. I. and Prisyazhniuk V. P. Exact asymptotic behavior of the probability of a large span of a stationary Gaussian process. *Theory Probab. Appl.*, 1981, v.26, N3, pp.468–484.

[Pitt77] Pitt L. D. A Gaussian correlation inequality for symmetric convex sets. *Ann. Probab.*, 1977, v.5, N3, pp.470-474.

[Pitt82] Pitt L. D. Positively correlated normal variables are associated. *Ann. Probab.*, 1982, v.10, N3, pp.496-499.

[Po78a] Ponomarenko L. S. Inequalities for distributions of quadratic forms in normal random variables. *Theory Probab. Appl.*, 1978, v.23, N3, pp.652–655.

[Po78b] Ponomarenko L. S. An inequality for distributions of quadratic forms of normally distributed random variables. *Theory Probab. Appl.*, 1978, v.23, N3, pp.656–659.

[Pr84] Prisyazhnyuk V. P. *A Study of Asymptotic Behavior of Large Deviation Probabilities of Gaussian Nonstationary Processes and Fields.* Moscow University, Moscow, 1984. (PhD Thesis Abstract) (Russian)

[RRa60] Ranga Rao R. *Some Problems in Probability Theory*, Ph. D. Thesis. Calcutta University, Calcutta, 1960.

[Ry86] Ryznar M. Asymptotic behavior of stable measures near the origin. *Ann. Probab.*, 1986, v.14, N1, pp.287–298.

[RyZ89] Ryznar M. and Zak T. *The measure of a translated ball in uniformly convex spaces.* Preprint: Wroclaw Technical University, Inst. of Math., 1989.

[SamT93] Samorodnitsky G. and Taqqu M. S. Stochastic monotonicity and Slepian-type inequalities for infinitely divisible and stable random vectors. *Ann. Probab.*, 1993, v.21, N1, pp.143-160.

[Saz68] Sazonov V. V. On the multi-dimensional central limit theorem. *Sankhyā*, 1968, ser. A., v.30, N2, pp.369-378.

[Schm43] Schmidt E. Beweis der isoperimetrischen Eigenschaft der Kugel in hyperbolischen und sphärischen Raum jeder Dimensionszahl. *Math. Z.*, 1943, Bd.49, N5, pp.1-109.

[Sk70] Skorokhod A. V. A remark on Gaussian measures in a Banach space. *Theory Probab. Appl.*, 1970, v.15, N3, pp.508–509.

[Sle62] Slepian D. The one-sided barrier problem for Gaussian noise. *Bell Syst. Tech. J.*, 1962, v.41, N2, pp.463-501.

[SmL89] Smorodina N. V. and Lifschitz A. M. On the distribution of the norm of a stable vector. *Theory Probab. Appl.*, 1989, v.34, N2. (Russ. orig.: pp.304-313.)

[Su69] Sudakov V. N. Gaussian measures, Cauchy measures and ε-entropy. *Soviet Math. Dokl.*, 1969, v.10, pp. 310–313.

[Su71] Sudakov V. N. Gaussian random processes and measures of solid angles in Hilbert space. *Soviet Math. Dokl.*, 1971, v.12, pp.412–415.

[SuTs74] Sudakov V. N. and Tsirel'son B. S. Extremal properties of half-spaces for spherically-invariant measures. *J. Soviet Math.*, 1978, v.9, pp.9–18. (Translated from: *Zapiski Nauchn. Seminarov LOMI*, 1974, v.41, pp.14-24.)

[Sy73] Sytaya G. N. On certain local representations for the Gaussian measure in Hilbert space. In: *Vilnius Internat. Conf. on Probab. and Statist.; Abstracts*, v.2. Vilnius, 1973, pp.267-268. (Russian)

[Sy74] Sytaya G. N. On certain asymptotic representations for the Gaussian measure in Hilbert space. *Teoriya Sluchain. Protsessov*, N2, Kiev, 1974, pp.93-104. (Russian)

[Sy79] Sytaya G. N. On the problem of asymptotic behavior of the Wiener measure of small spheres in uniform metric. *Analiticheskie Metody Teorii Veroyatnostey*, Kiev, 1979, pp.96-98.

[Ta87] Talagrand M. Regularity of Gaussian processes. *Acta Math.*, 1987, v.159, N1-2, pp.99-149.

[Ta88a] Talagrand M. Small tails for the supremum of a Gaussian process. *Ann. Inst. H. Poincaré*, 1988, ser.B, t.24, N2, pp.307-315.

[Ta88b] Talagrand M. The structure of sign-invariant G-B sets of certain Gaussian measures. *Ann. Probab.*, 1988, v.16, N1, pp.172-179.

[Ta94] Talagrand M. Sharper bounds for Gaussian and empirical processes. *Ann. Probab.*, 1994, v.22, N1, pp.28-76.

[Ti63] Tippe J. Zur isoperimetrische Eigenschaft der Kugel in Riemannschen Räumen konstanter positiver Krümmung. *Math. Ann.*, 1963, Bd.152, N2, pp.120-148.

[Ts75] Tsirelson B.S. The density of the distribution of the maximum of a Gaussian process. *Theory Probab. Appl.*, 1975, v.20, pp.847–856.

[We80] Weber M. Analyse asymptotique des processus gaussiens stationnaires. *Ann. Inst. H. Poincaré*, 1980, ser.B, v.16, N2, pp.117-176.

[We89] Weber M. The supremum of Gaussian processes with a constant variance. *Probab. Theory Related Fields*, 1989, v.81,N4, pp.585-591.

[Yl65] Ylvisaker D. The expected number of zeroes of a stationary Gaussian process. *Ann. Math. Statist.*, 1965, v.36, N3, pp.1043-1046.

[Yl68] Ylvisaker D. A note on the absence of tangencies in Gaussian sample paths. *Ann. Math. Statist.*, 1968, v.39, N1, pp.261-262.

[Zo61] Zolotarev V. M. Concerning a certain probability problem. *Theory Probab. Appl.*, 1961, v.6, N2, pp.201–203.

[Zo84] Zolotarev V. M. On asymptotics of a Gaussian measure in l_2. *Problemi Ustoichivosti Stokhasticheskikh Modelei* /Seminar Proc., VNIISI, Moscow, 1984, pp.54-58. (Russian)

[] **Ch. 3**

[Ac80] de Acosta A. Strong exponential integrability of sums of independent B-valued random vectors. *Probab. Math. Statist.*, 1980, v.1, N2, pp.133-150.

[Ac81] de Acosta A. Inequalities for B-valued random vectors with applications to the strong law of large numbers. *Ann. Probab.*, 1981, v.9,N1, pp.157-161.

[AcK83] de Acosta A. and Kuelbs J. Limit theorems for moving averages of independent random variables. *Z. Wahrscheinlichkeitstheor. Verw. Geb.*, 1983, v.64, N1, pp.67-123.

[Ber75] Beran R. Tail probabilities of noncentral quadratic forms. *Ann. Statist.*, 1975, v.3, N4, pp.969-974.

[Bernα] Bernstein S. N. On a modification of the Chebyshev inequality and the error of the Laplace formula. In: *Collected Works*, v.4, pp.71-79. Nauka, Moscow, 1964. (Russian)

[Bernβ] Bernstein S. N. On certain modifications of the Chebyshev inequality. In: *Collected Works*, v.4, pp.331-333. Nauka, Moscow, 1964. (Russian)

[Burk73] Burkholder D. L. Distribution function inequalities for martingales. *Ann. Probab.*, 1973, v.1, N1, pp.19-42.

[Burk81] Burkholder D. L. A geometrical characterization of Banach spaces in which martingale difference sequences are unconditional. *Ann. Probab.*, 1981, v.9, N6, pp.997-1011.

[Co82] Cox D. C. Note on a martingale inequality of Pisier. *Math. Proc. Cambridge Phil. Soc.*, 1982, v.92, N1, pp.163-165.

[Fu90] Fukuda R. Exponential summability of sub-Gaussian vectors. *Probab. Theory Related Fields*, 1990, v.85, N4, pp.505-521.

[FuN71] Fuk D. Kh. and Nagaev S. V. Probability inequalities for sums of independent random variables. *Theory Probab. Appl.*, 1971, v.16, N4, pp.643–660.

[GiMZ78] Giné E., Mandrekar V., and Zinn J. On sums of independent random variables with values in L_p, $2 \leq p \leq \infty$. *Lect. Notes Math.*, 1978, v.709, Springer-Verlag, pp.111-124.

[GiZ83] Giné E. and Zinn J. Central limit theorems and weak laws of large numbers in certain Banach spaces. *Z. Wahrscheinlichkeitstheor. Verw. Geb.*, 1983, v.62, N3, pp.323-354.

[Hof72] Hoffmann-Jørgensen J. *Sums of Independent Banach Space-Valued Random Variables*. Preprint/Ser. Mat. Inst. Aarhus Univ., v.15, Aarhus, 1972/1973.

[Hof74] Hoffmann-Jørgensen J. Sums of independent Banach space-valued random variables. *Studia Math.*, 1974, v.52., N3, pp.159-186.

[HofP76] Hoffmann-Jørgensen J. and Pisier G. The law of large numbers and the central limit theorem in Banach spaces. *Ann. Probab.*, 1976, v.4, N4, pp.587-599.

[Ja75] Jain N. C. Tail probabilities for sums of independent Banach space values random variables. *Z. Wahrscheinlichkeitstheorie verw. Geb.*, 1975, v.33, N3, pp.155-166.

[JaM75] Jain N. C. and Marcus M. B. Integrability of infinite sums of independent vector-valued random variables. *Trans. Amer. Math. Soc.*, 1975, v.212, pp.1-35.

[Kan78] Kanter M. Probability inequalities for convex sets. *J. Multivar. Anal.*, 1978, v.8, pp.222-236.

[Kol29] Kolmogoroff A. N. Über das Gesetz des iterierten Logarithmus. *Math. Ann.*, 1929, v.101, pp.126-135.

[Ku78] Kuelbs J. Some exponential moments of sums of independent random variables. *Trans. Amer. Math. Soc.*, 1978, v.240, pp.145-162.

[Ku80] Kuelbs J. The law of the iterated logarithm for Banach space valued random variables. *Lect. Notes Math.*, 1980, v.860, Springer-Verlag, pp.268-278.

[KuZ83] Kuelbs J. and Zinn J. Some results on LIL behavior. *Ann. Probab.*, 1983, v.11, N3, pp.506-557.

[LeT88b] Ledoux M. and Talagrand M. Characterization of the law of the iterated logarithm in Banach spaces. *Ann. Probab.*, 1988, v.16, N3, pp.1242-1264.

[Mat91a] Matsak J. K. An asymptotic estimate for sums of independent random variables in a Banach space. *Ukrainian Math. J.*, 1991, v.43, N2, pp.239-242.

[Na82] Nagaev S. V. Probability inequalities for sums of independent random variables with values in a Banach space. In: *Trudy Inst. Matematiki SO AN SSSR* (Proc. Inst. Math. of Siber. Div. of USSR Acad. Sc.), 1982, v.2, pp.159-167. (Russian)

[Na83] Nagaev S. V. On probabilities of large deviations in Banach spaces. *Matem. Zametki*, 1983, v.34, N2, pp.309-313. (Russian)

[Na87] Probability inequalities for sums of independent random variables with values in a Banach space. *Sib. Math. J.*, 1987, v.28, N4, pp.652–664.

[NaPi77] Nagaev S. V. and Pinelis I. F. Some inequalities for distributions of sums of independent random variables. *Theory Probab. Appl.*, 1977, v.22, N2, pp. 254–263.

[Pi78] Pinelis I. F. On the distribution of sums of independent random variables with values in a Banach space. *Theory Probab. Appl.*, 1978, v.23, N3, pp. 608–615.

[Pi80] Pinelis I. F. Estimates of moments of infinite-dimensional martingales. *Math. Notes*, 1980, v.27, pp.459–462.

[Pi81a] Pinelis I. F. Some limit theorems for large deviations of infinite dimensional random walks. *Theory Probab. Appl.*, 1981, v.26, N3. (Russ. orig.: pp.645-646.)

[Pi81b] Pinelis I. F. On a bound for probabilities of large deviations. *Predel'nie Teoremi i Statisticheskie Vivodi*, Tashkent, 1981, pp. 160-165. (Russian)

[Pi84] Pinelis I. F. Some probabilistic extremal problems in Hilbert space. *Theory Probab. Appl.*, 1984, v.29, N2 (Russ. orig.: pp.408-409.)

[Pi93] Pinelis I. *Optimum bounds for the distributions of martingales in Banach spaces.* Preprint, Michigan Technological University, 1993.

[PiS85] Pinelis I. F. and Sakhanenko A. I. Remarks on inequalities for the probabilities of large deviations. *Theory Probab. Appl.*, 1985, v.30, N1, pp. 143–148.

[Pis75a] Pisier G. Martingales with values in uniformly convex spaces. *Israel J. Math.*, 1975, v.20, N3-4, pp.326-350.

[Pis75b] Pisier G. *Le théorème de la limite centrale et la loi du logarithme itéré dans les espaces de Banach.* Preprint: Ecole Polytechnique, Seminaire Maurey–Schwartz, 1973, exposés 3-4, Paris, 1975.

[Prα68] Prokhorov Yu. V. An extension of S.N. Bernstein's inequalities to multidimensional distributions. *Theory Probab. Appl.*, 1968, v.13, N2, pp.260–267.

[Prα72] Prokhorov Yu. V. Multidimensional distributions: inequalities and limit theorems. In: *Itogi Nauki i Tekh.* Ser. Teor. Veroyatn. Mat. Stat. Teor. Kibern., v.10. VINITI, Moscow, 1972. pp.5-25. (Russian)

[Prα83] Prokhorov Yu. V. On sums of random vectors with values in Hilbert space. *Theory Probab. Appl.*, 1983, v.28, N2, pp.375–378.

[Prβ68a] Prokhorov A. V. On multidimensional analogues of S. N. Bernstein's inequalities. *Theory Probab. Appl.*, 1968, v.13, N2, pp.268–280.

[Prβ68b] Prokhorov A. V. On a probability inequality. *Matem. Zametki*, 1968, v.3, N6, pp.737-744. (Russian)

[Prβ68c] Prokhorov A. V. S. N. Bernstein's inequalities in multidimensional case. *Theory Probab. Appl.*, 1968, v.13, N3, pp.438–447. (Russ. orig.: pp.462-470.)

[Prβ68d] Prokhorov A. V. Inequalities for Bessel functions of a purely imaginary argument. *Theory Probab. Appl.*, 1968, v.13, N3, pp.496–501.

[Prβ73] Prokhorov A. V. On sums of random vectors. *Theory Probab. Appl.*, 1973, v.18, N1, pp.186–187.

[Sa84] Sakhanenko A. I. On Lévy–Kolmogorov inequalities for random variables with values in a Banach space. *Theory Probab. Appl.*, 1984, v.29, N4, pp.830–836.

[Yu70] Yurinsky V. V. On an infinite-dimensional version of S. N. Bernstein's inequalities. *Theory Probab. Appl.*, 1970, v.15, N1, pp.108–109.

[Yu74] Yurinsky V. V. Exponential bounds for large deviations. *Theory Probab. Appl.*, 1974, v.19, N1, pp.154-155.

[Yu76] Yurinsky V. V. Exponential inequalities for sums of random vectors. *J. Multivar. Anal.*, 1976, v.6, N4, pp.473-499.

[Zi81] Zinn J. Inequalities in Banach spaces with applications to limit theorems in probability. A survey. *Lect. Notes Math.*, 1981, v. 860, Springer-Verlag, pp.324-329.

[Zo68] Zolotarev V. M. Some remarks on multidimensional Bernstein–Kolmogorov–type inequalities. *Theory Probab. Appl.*, 1968, v.13, N2, pp.281–286.

[ ] **Ch. 4**

[Ac84] de Acosta A. On large deviations of sums of independent random vectors. Lect. Notes Math., Springer-Verlag, 1985

[BaZ79] Bahadur R. R. and Zabell S. L. Large deviations of the sample mean in general vector spaces. *Ann. Probab.*, 1979, v.7, N4, pp.587-621.

[BaxJ67] Baxter J. R. and Jain N. C. A comparison principle for large deviations. *Proc. Amer. Math. Soc.*, 1988, v.103, N4, pp.1235-1240.

[Boro62] Borovkov A.A. New limit theorems in boundary problems for sums of independent summands. *Sibirsk. Matem. Zhurn.*, 1962, v.3, N5, pp.645-695. (Russian)

[Boro65] Borovkov A.A. Analysis of large deviations in boundary problems with arbitrary boundaries. *Sibirsk. Matem. Zhurn.*, 1965, v.5, N2, pp.253-289. (Russian)

[BoroM78] Borovkov A.A. and Mogulskii A.A. On large deviation probabilities in topological spaces. I–II. *Sibirsk. Matem. Zhurn.*, 1978, v.19, N5, pp.988-1004 and 1980, v.21, N5, pp.12-26. (Russian)

[Che90] Chen Xia. Probabilities of moderate deviations for independent random vectors. *Appl. Probab. and Statist.*, (Chinese)

[DoV] Donsker M. D. and Varadhan S. R. S. Asymptotic evaluation of certain Markov process expectations for large time. *Comm. Pure and Appl. Math.*, Pt.I — 1975, v.28, N1, pp.1-47; Pt.II —1975, v.28, N2, pp.297-301; Pt.III — 1976, v.29, N4, pp.389-461.

[GrOR79] Groenboom P., Oosterhoff J., and Ruymgart F. H. Large deviations theorems for empirical probability measures. *Ann. Probab.*, 1979, v.7, N4, pp.553-586.

[Sl88] Slaby M. On the upper bound for large deviations of sums of i.i.d. random vectors. *Ann. Probab.*, 1988, v.16, N3, pp.978-990.

[ ] **Ch. 5**

[Be84a] Bentkus V. Asymptotic expansions in the CLT in Hilbert space. *Lithuanian Math. J.*, 1984, v.24, p.210-225.

[Be84b] Bentkus V. Asymptotic expansions for distributions of sums of independent random elements of a Hilbert space. *Lithuanian Math. J.*, 1984, v.24, p.305-319.

[Be86c] Bentkus V. Asymptotic expansions for moments in the CLT in Banach spaces. *Lithuanian Math. J.*, 1986, v.26, p.10-26.

[BeGöPaRa90] Bentkus V., Götze F., Paulauskas V, and Račkauskas A. *The Accuracy of Gaussian Approximation in Banach Spaces.* Preprint 90-100, Sonderforschungsbereich 343 "Diskrete Strukturen in der Mathematik", Universität Bielefeld, 1990.

[BeGöZit92] Bentkus V., Götze F. and Zitikis R. *Asymptotic expansions in the integral and local limit theorems in Banach spaces with applications to ω-statistics.* Preprint 92-043, Sonderforschungsbereich 343 "Diskrete Strukturen in der Mathematik", Universität Bielefeld, 1992.

[Berg51] Bergström H. On asymptotic expansions of probability functions. *Skand. Aktuarietidskrift*, v.1-2, 1951, 1-34.

[Bo95] Borisov I. S. Bounds for characteristic functions of additive functionals of order statistics, *Siberian Adv. Math.*, 1995, v.5, N1, pp.1–15.

[Gö79] Götze F. Asymptotic expansions for bivariate von Mises functionals. *Z. Wahrscheinlichkeitstheorie verw. Geb.*, 1979, v. 50, 333-355.

[Gö81] Götze F. On Edgeworth expansions in Banach spaces. *Ann. Probab.*, 1981, v.9, 852-859.

[Gö89] Götze F. Edgeworth expansions in functional limit theorems. *Ann. Probab.*, 1989, v.17, 1602-1634.

[Ka65] Kandelaki N. P. On limit theorem in Hilbert space. In: *Trudy Vychisl. Centra Akad. Nauk. Gruzin. SSR*, 1965, N11, pp.46-55. (Russian)

[KaV69] Kandelaki N. P. and Vakhania N. N. On the estimate of the rate of convergence in central limit theorem in Hilbert space. In: *Trudy Vychisl. Centra Akad. Nauk Gruzin. SSR*, 1969, N9, pp.150-160.

[KK79] Kuelbs J. and Kurtz T. Berry–Esseén estimates in Hilbert space and an application to the law of the iterated logarithm. *Ann. Probab.*, 1974, v.2, pp.387-407.

[Na89] Nagaev S. V. On the new approach for the investigation of the distribution of the norm of a random variable in the Hilbert space. In: *5th Vilnius Conf. on Probab. Th. and Math. Statist.*, Abstracts, Vilnius, 1989, pp.77-78.

[NaCh85] Nagaev S. V. and Chebotarev V. I. Asymptotic expansions for the distribution of the sum of i.i.d. Hilbert-space valued r.v. *Probability Theory and Mathematical Statistics*, Vilnius 1985, (Proc. of Internat. Conf.), v.2. *VNU Science Press*, Utrecht, The Netherlands, 1987.

[NaCh86] Nagaev S.V. and Chebotarev V. I. A refinement of the error estimate of the normal approximation in a Hilbert space. *Siberian Math. J.*, 1986, p.434-450.

[NaCh89a] Nagaev S. V. and Chebotarev V. I. On asymptotic expansions of the Bergström type in Hilbert space. *Asymptotic Analysis of Distributions of Stochastic Processes*, Proc. Inst. Math. of Siber. Branch. of USSR Acad. Sc., v.13, 1989, 66-77.

[NaCh89b] Nagaev S. V. and Cheboterev V. I. *On Edgeworth expansion in Hilbert space.* Preprint, Institute of Applied Mathematics /USSR Academy of Sciences, Far-Eastern Branch., 1989, p.1-62.

[Ra81] Račkauskas A. On proximity in uniform metric of sums of independent Hilbert space valued random variables. *Litovsk. Matem. Sbornik*, 1981, v.21, N3, pp.83-90. (Russian)

[SazUl93] Sazonov V.V. and Ulyanov V.V. *Asymptotic Expansions in Hilbert Space.* Research Reports, Dept. of Theoretical Statistics, Institute of Math., Univ. of Aarhus, N 265, 1993.

[SaUlZa91] Sazonov V. V., Ulyanov V. V., and Zalesskii B. A. A precise estimate of the rate of convergence in the CLT in Hilbert space. *Math. USSR. Sbornik*, 1991, v.68, pp.453-482.

[Se84] Senatov V.V. On the orders of magnitude in estimates for the convergence rate in the central limit theorem in Hilbert space. *Problemi Ustoichivosti Stokhasticheskikh Modelei* /Seminar Proc., VNIISI, Moscow, 1984, pp.128-135.

[Se89] Senatov V. V. On the estimate of the rate of convergence in the central limit theorem in HIlbert space. In: *Lect. Notes Math.*, 1989, v.1412, Springer-Verlag, pp.309-327.

[VaK68] Vakhania N. N. and Kandelaki N. P. On an estimate of convergence rate in the multidimensional central limit theorem. *Soobshch. AN Gruz. SSR*, 1968, v.50, N2, pp.273-276. (Russian)

[Yu82] Yurinsky V. V. On the accuracy of normal approximation of the probability of hitting a ball. *Theory Probab. Appl.*, 1982, v.27, N2, pp.280–289.

[Yu83] Yurinsky V. V. Error of normal approximation. *Siberian Math. J.*, 1983, v.24, pp.977-987.

[Yu91a] Yurinsky V. V. Edgeworth-type expansions for the statistics of samples with independent vector observations. *Proc. 6th Soviet-Japanese Symposium on Probability Theory*, August 1991, Kiev.

[Yur94a] Yurinsky V. V. Characteristic functions "too far" from the origin. *Siberian Adv. Math.*, 1994, v.4, N2, p.136-150.

[Yu94b] Yurinsky V. V. Asymptotic expansions for vector-valued sums. *Siberian Adv. Math.*, 1994, v.4, N4, p.118-150.

[Za85a] Zalesskii B.A. *Edgeworth Expansions for a Class of Sets in Hilbert Space.* Preprint, Institute of Mathematics /Academy of Sciences of the Byelorussian SSR, N5 (214), February 1985. (Russian)

[Za85b] Zalesskii B.A. *Edgeworth Expansions in Banach Spaces.* Preprint, Institute of Mathematics /Academy of Sciences of the Byelorussian SSR, N24 (233), October 1985. (Russian)

[Za85c] Zalesskii B. A. On the convergence rate in the central limit theorem on a class of sets in Hilbert space. *Theory Probab. Appl.*, 1985, v.30, N4, pp.702-711.

[Za88] Zalesskii B. A. On the accuracy of normal approximation in Banach spaces. *Theory Probab. Appl.*, 1988, v.33, N2, pp.239–247.

[Za89b] Zalesskii B. A. Accuracy of Gaussian approximation in Banach spaces. *Theory Probab. Appl.*, 1989, v.34, N4. (Russ. orig.: pp.815-817.)

[ZaSU88a] Zalesskii B. A., Sazonov V. V., and Ulyanov V. V. Normal approximation in Hilbert space. *Theory Probab. Appl.*, 1988, v.33, NN3-4. (Russ. orig.: v.33, N3, 225-245 and N4, pp.508-521.)

[ZaSU88b] Zalesskii B. A., Sazonov V. V., and Ulyanov V. V. The correct estimate for the precision of normal approximation in Hilbert space. *Theory Probab. Appl.*, 1988, v.33, N4. (Russ. orig.: pp.753-754.)

[Zit88] Zitikis R. Asymptotic expansions in the local limit theorem for ω_n^2-statistics. *Litovsk. Matem. Sb*, 1988, v.28, p.461-474.

[Zit89] Zitikis R. Asymptotic expansions for the derivatives of the Andersen–Darling statistics distribution function. *Litovsk. Mat. Sb.*, 1989, v.29, 35-53.

[ ] **Ch. 6**

[Ac92] de Acosta A. Moderate deviations and associated Laplace approximations for sums of independent random vectors. *Trans. Amer. Math. Soc.*, 1992, v.329, N1, pp.357-375.

[Be86a] Bentkus V. J. Large deviations in Banach spaces. *Theory Probab. Appl.*, 1986, v.31, N4, pp.627–632.

[Be86b] Bentkus V. J. On large deviations in Banach spaces. In: *First World Congress of Bernoulli Society, Abstracts.*, p.757. Nauka, Moscow, 1986.

[BeR90] Bentkus V. J. and Račkauskas A. J. On probabilities of large deviations in Banach spaces. *Probab. Theory Related Fields*, 1990, v.86, pp.131-154.

[Bo86] Bolthausen E. Laplace approximation for sums of independent random vectors. *Probab. Theory Related Fields*, 1986, v.72, pp.305-318.

[Bo87] Bolthausen E. Laplace approximation for sums of independent random vectors. Part II. Degenerate maxima and manifolds of maxima. *Probab. Theory Related Fields*, 1987, v.76, pp.167/206.

[Chy93] Chyonobu T. On a precise Laplace-type asymptotic formula for sums of independent random vectors. In: *Asymptotic Problems of Probability Theory: Wiener Functionals and Asymptotics.* Eds. Elworthy K.D., Ikeda N. /Pitman Research Notes in Mathematics, v.284. Longman Scientific and Technical, Burnt Mill, Harlow, Essex, 1993, pp.122-135.

[Cr38] Cramér H. Sur un nouveau théorème-limite de la théorie des probabilités. *Actual. Sci. et Industr., Paris*, 1938, N736.

[Os75] Osipov L. V. Multidimensional limit theorems for large deviations. *Theory Probab. Appl.*, 1975, v.20, N1, pp.38–56.

[Os78a] Osipov L. V. On large deviation probabilities for sums of independent random vectors. *Theory Probab. Appl.*, 1978, v.23, N3, pp.490–506.

[Os78b] Osipov L. V. *Probabilities of Large Deviations for Sums of Independent Random Vectors.* DSc Thesis. Leningrad University, Leningrad, 1978. (Russian)

[Ra88] Račkauskas A. Probabilities of large deviations in Linnik zones in Hilbert space. *Litovsk. Matem. Sbornik*, 1988, v.28, N3, pp.520-533. (Russian)

[Yu91] Yurinsky V.V. Exponential estimates of Bernstein type. *Siber. Math. J.*, 1991, v.32, pp.716-724.

[Yu91b] Yurinsky V. V. On the asymptotics of large deviations in Hilbert space. I–III. *Theory Probab. Appl.*, I: 1991, v.36, N1, pp.99–114; II: N3, pp.548–554; III: 1992, v.37, N2, pp.261–268.

[Za89a] Zalesskii B. A. Probabilities of large deviations in a Hilbert space. *Theory Probab. Appl.*, 1989, v.34, N4. (Russian original: pp.650-655.)

[ ] **App. A**

[Cher52] Chernoff H. A measure of asymptotic efficiency for tests of a hypothesis based on sums of observations. *Ann. Math. Statist.*, 1952, v.23, N4, pp.493-507.

[FoS76] Fomin S. V. and Smolianov O. G. Measures on topological linear spaces. *Uspekhi Matem, Nauk*, 1976, v.31, N4(190), pp.3-56. (Russian; Engl. transl. in *Russian Math. Surveys*)

[Min59] Minlos R. A. Generalized stochastic processes and their extension to measures. *Trudy Moskovskogo Matem. Obshchestva*, 1959, v.8, pp.497-518. (Russian)

[PrY56] Prokhorov Yu. V. Convergence of random processes and limit theorems in probability theory. *Theory Probab. Appl.*, 1956, v. 1, N 2, pp.157–222.

[Saz58] Sazonov V. V. A remark on characteristic functionals. *Theory Probab. Appl.*, 1958, v.3, N2, pp.188–191.

Index

Vol. 1526: J. Azéma, P. A. Meyer, M. Yor (Eds.), Séminaire de Probabilités XXVI. X, 633 pages. 1992.

Vol. 1527: M. I. Freidlin, J.-F. Le Gall, Ecole d'Eté de Probabilités de Saint-Flour XX – 1990. Editor: P. L. Hennequin. VIII, 244 pages. 1992.

Vol. 1528: G. Isac, Complementarity Problems. VI, 297 pages. 1992.

Vol. 1529: J. van Neerven, The Adjoint of a Semigroup of Linear Operators. X, 195 pages. 1992.

Vol. 1530: J. G. Heywood, K. Masuda, R. Rautmann, S. A. Solonnikov (Eds.), The Navier-Stokes Equations II – Theory and Numerical Methods. IX, 322 pages. 1992.

Vol. 1531: M. Stoer, Design of Survivable Networks. IV, 206 pages. 1992.

Vol. 1532: J. F. Colombeau, Multiplication of Distributions. X, 184 pages. 1992.

Vol. 1533: P. Jipsen, H. Rose, Varieties of Lattices. X, 162 pages. 1992.

Vol. 1534: C. Greither, Cyclic Galois Extensions of Commutative Rings. X, 145 pages. 1992.

Vol. 1535: A. B. Evans, Orthomorphism Graphs of Groups. VIII, 114 pages. 1992.

Vol. 1536: M. K. Kwong, A. Zettl, Norm Inequalities for Derivatives and Differences. VII, 150 pages. 1992.

Vol. 1537: P. Fitzpatrick, M. Martelli, J. Mawhin, R. Nussbaum, Topological Methods for Ordinary Differential Equations. Montecatini Terme, 1991. Editors: M. Furi, P. Zecca. VII, 218 pages. 1993.

Vol. 1538: P.-A. Meyer, Quantum Probability for Probabilists. X, 287 pages. 1993.

Vol. 1539: M. Coornaert, A. Papadopoulos, Symbolic Dynamics and Hyperbolic Groups. VIII, 138 pages. 1993.

Vol. 1540: H. Komatsu (Ed.), Functional Analysis and Related Topics, 1991. Proceedings. XXI, 413 pages. 1993.

Vol. 1541: D. A. Dawson, B. Maisonneuve, J. Spencer, Ecole d'Eté de Probabilités de Saint-Flour XXI - 1991. Editor: P. L. Hennequin. VIII, 356 pages. 1993.

Vol. 1542: J.Fröhlich, Th.Kerler, Quantum Groups, Quantum Categories and Quantum Field Theory. VII, 431 pages. 1993.

Vol. 1543: A. L. Dontchev, T. Zolezzi, Well-Posed Optimization Problems. XII, 421 pages. 1993.

Vol. 1544: M.Schürmann, White Noise on Bialgebras. VII, 146 pages. 1993.

Vol. 1545: J. Morgan, K. O'Grady, Differential Topology of Complex Surfaces. VIII, 224 pages. 1993.

Vol. 1546: V. V. Kalashnikov, V. M. Zolotarev (Eds.), Stability Problems for Stochastic Models. Proceedings, 1991. VIII, 229 pages. 1993.

Vol. 1547: P. Harmand, D. Werner, W. Werner, M-ideals in Banach Spaces and Banach Algebras. VIII, 387 pages. 1993.

Vol. 1548: T. Urabe, Dynkin Graphs and Quadrilateral Singularities. VI, 233 pages. 1993.

Vol. 1549: G. Vainikko, Multidimensional Weakly Singular Integral Equations. XI, 159 pages. 1993.

Vol. 1550: A. A. Gonchar, E. B. Saff (Eds.), Methods of Approximation Theory in Complex Analysis and Mathematical Physics IV, 222 pages, 1993.

Vol. 1551: L. Arkeryd, P. L. Lions, P.A. Markowich, S.R. S. Varadhan. Nonequilibrium Problems in Many-Particle Systems. Montecatini, 1992. Editors: C. Cercignani, M. Pulvirenti. VII, 158 pages 1993.

Vol. 1552: J. Hilgert, K.-H. Neeb, Lie Semigroups and their Applications. XII, 315 pages. 1993.

Vol. 1553: J.-L- Colliot-Thélène, J. Kato, P. Vojta. Arithmetic Algebraic Geometry. Trento, 1991. Editor: E. Ballico. VII, 223 pages. 1993.

Vol. 1554: A. K. Lenstra, H. W. Lenstra, Jr. (Eds.), The Development of the Number Field Sieve. VIII, 131 pages. 1993.

Vol. 1555: O. Liess, Conical Refraction and Higher Microlocalization. X, 389 pages. 1993.

Vol. 1556: S. B. Kuksin, Nearly Integrable Infinite-Dimensional Hamiltonian Systems. XXVII, 101 pages. 1993.

Vol. 1557: J. Azéma, P. A. Meyer, M. Yor (Eds.), Séminaire de Probabilités XXVII. VI, 327 pages. 1993.

Vol. 1558: T. J. Bridges, J. E. Furter, Singularity Theory and Equivariant Symplectic Maps. VI, 226 pages. 1993.

Vol. 1559: V. G. Sprindžuk, Classical Diophantine Equations. XII, 228 pages. 1993.

Vol. 1560: T. Bartsch, Topological Methods for Variational Problems with Symmetries. X, 152 pages. 1993.

Vol. 1561: I. S. Molchanov, Limit Theorems for Unions of Random Closed Sets. X, 157 pages. 1993.

Vol. 1562: G. Harder, Eisensteinkohomologie und die Konstruktion gemischter Motive. XX, 184 pages. 1993.

Vol. 1563: E. Fabes, M. Fukushima, L. Gross, C. Kenig, M. Röckner, D. W. Stroock, Dirichlet Forms. Varenna, 1992. Editors: G. Dell'Antonio, U. Mosco. VII, 245 pages. 1993.

Vol. 1564: J. Jorgenson, S. Lang, Basic Analysis of Regularized Series and Products. IX, 122 pages. 1993.

Vol. 1565: L. Boutet de Monvel, C. De Concini, C. Procesi, P. Schapira, M. Vergne. D-modules, Representation Theory, and Quantum Groups. Venezia, 1992. Editors: G. Zampieri, A. D'Agnolo. VII, 217 pages. 1993.

Vol. 1566: B. Edixhoven, J.-H. Evertse (Eds.), Diophantine Approximation and Abelian Varieties. XIII, 127 pages. 1993.

Vol. 1567: R. L. Dobrushin, S. Kusuoka, Statistical Mechanics and Fractals. VII, 98 pages. 1993.

Vol. 1568: F. Weisz, Martingale Hardy Spaces and their Application in Fourier Analysis. VIII, 217 pages. 1994.

Vol. 1569: V. Totik, Weighted Approximation with Varying Weight. VI, 117 pages. 1994.

Vol. 1570: R. deLaubenfels, Existence Families, Functional Calculi and Evolution Equations. XV, 234 pages. 1994.

Vol. 1571: S. Yu. Pilyugin, The Space of Dynamical Systems with the C^0-Topology. X, 188 pages. 1994.

Vol. 1572: L. Göttsche, Hilbert Schemes of Zero-Dimensional Subschemes of Smooth Varieties. IX, 196 pages. 1994.

Vol. 1573: V. P. Havin, N. K. Nikolski (Eds.), Linear and Complex Analysis – Problem Book 3 – Part I. XXII, 489 pages. 1994.

Vol. 1574: V. P. Havin, N. K. Nikolski (Eds.), Linear and Complex Analysis – Problem Book 3 – Part II. XXII, 507 pages. 1994.

Vol. 1575: M. Mitrea, Clifford Wavelets, Singular Integrals, and Hardy Spaces. XI, 116 pages. 1994.

Vol. 1576: K. Kitahara, Spaces of Approximating Functions with Haar-Like Conditions. X, 110 pages. 1994.

Vol. 1577: N. Obata, White Noise Calculus and Fock Space. X, 183 pages. 1994.

Vol. 1578: J. Bernstein, V. Lunts, Equivariant Sheaves and Functors. V, 139 pages. 1994.

Vol. 1579: N. Kazamaki, Continuous Exponential Martingales and *BMO*. VII, 91 pages. 1994.

Vol. 1580: M. Milman, Extrapolation and Optimal Decompositions with Applications to Analysis. XI, 161 pages. 1994.

Vol. 1581: D. Bakry, R. D. Gill, S. A. Molchanov, Lectures on Probability Theory. Editor: P. Bernard. VIII, 420 pages. 1994.

Vol. 1582: W. Balser, From Divergent Power Series to Analytic Functions. X, 108 pages. 1994.

Vol. 1583: J. Azéma, P. A. Meyer, M. Yor (Eds.), Séminaire de Probabilités XXVIII. VI, 334 pages. 1994.

Vol. 1584: M. Brokate, N. Kenmochi, I. Müller, J. F. Rodriguez, C. Verdi, Phase Transitions and Hysteresis. Montecatini Terme, 1993. Editor: A. Visintin. VII, 291 pages. 1994.

Vol. 1585: G. Frey (Ed.), On Artin's Conjecture for Odd 2-dimensional Representations. VIII, 148 pages. 1994.

Vol. 1586: R. Nillsen, Difference Spaces and Invariant Linear Forms. XII, 186 pages. 1994.

Vol. 1587: N. Xi, Representations of Affine Hecke Algebras. VIII, 137 pages. 1994.

Vol. 1588: C. Scheiderer, Real and Étale Cohomology. XXIV, 273 pages. 1994.

Vol. 1589: J. Bellissard, M. Degli Esposti, G. Forni, S. Graffi, S. Isola, J. N. Mather, Transition to Chaos in Classical and Quantum Mechanics. Montecatini Terme, 1991. Editor: S. Graffi. VII, 192 pages. 1994.

Vol. 1590: P. M. Soardi, Potential Theory on Infinite Networks. VIII, 187 pages. 1994.

Vol. 1591: M. Abate, G. Patrizio, Finsler Metrics – A Global Approach. IX, 180 pages. 1994.

Vol. 1592: K. W. Breitung, Asymptotic Approximations for Probability Integrals. IX, 146 pages. 1994.

Vol. 1593: J. Jorgenson & S. Lang, D. Goldfeld, Explicit Formulas for Regularized Products and Series. VIII, 154 pages. 1994.

Vol. 1594: M. Green, J. Murre, C. Voisin, Algebraic Cycles and Hodge Theory. Torino, 1993. Editors: A. Albano, F. Bardelli. VII, 275 pages. 1994.

Vol. 1595: R.D.M. Accola, Topics in the Theory of Riemann Surfaces. IX, 105 pages. 1994.

Vol. 1596: L. Heindorf, L. B. Shapiro, Nearly Projective Boolean Algebras. X, 202 pages. 1994.

Vol. 1597: B. Herzog, Kodaira-Spencer Maps in Local Algebra. XVII, 176 pages. 1994.

Vol. 1598: J. Berndt, F. Tricerri, L. Vanhecke, Generalized Heisenberg Groups and Damek-Ricci Harmonic Spaces. VIII, 125 pages. 1995.

Vol. 1599: K. Johannson, Topology and Combinatorics of 3-Manifolds. XVIII, 446 pages. 1995.

Vol. 1600: W. Narkiewicz, Polynomial Mappings. VII, 130 pages. 1995.

Vol. 1601: A. Pott, Finite Geometry and Character Theory. VII, 181 pages. 1995.

Vol. 1602: J. Winkelmann, The Classification of Three-dimensional Homogeneous Complex Manifolds. XI, 230 pages. 1995.

Vol. 1603: V. Ene, Real Functions – Current Topics. XIII, 310 pages. 1995.

Vol. 1604: A. Huber, Mixed Motives and their Realization in Derived Categories. XV, 207 pages. 1995.

Vol. 1605: L. B. Wahlbin, Superconvergence in Galerkin Finite Element Methods. XI, 166 pages. 1995.

Vol. 1606: P.-D. Liu, M. Qian, Smooth Ergodic Theory of Random Dynamical Systems. XI, 221 pages. 1995.

Vol. 1607: G. Schwarz, Hodge Decomposition – A Method for Solving Boundary Value Problems. VII, 155 pages. 1995.

Vol. 1608: P. Biane, R. Durrett, Lectures on Probability Theory. VII, 210 pages. 1995.

Vol. 1609: L. Arnold, C. Jones, K. Mischaikow, G. Raugel, Dynamical Systems. Montecatini Terme, 1994. Editor: R. Johnson. VIII, 329 pages. 1995.

Vol. 1610: A. S. Üstünel, An Introduction to Analysis on Wiener Space. X, 95 pages. 1995.

Vol. 1611: N. Knarr, Translation Planes. VI, 112 pages. 1995.

Vol. 1612: W. Kühnel, Tight Polyhedral Submanifolds and Tight Triangulations. VII, 122 pages. 1995.

Vol. 1613: J. Azéma, M. Emery, P. A. Meyer, M. Yor (Eds.), Séminaire de Probabilités XXIX. VI, 326 pages. 1995.

Vol. 1614: A. Koshelev, Regularity Problem for Quasilinear Elliptic and Parabolic Systems. XXI, 255 pages. 1995.

Vol. 1615: D. B. Massey, Lê Cycles and Hypersurface Singularities. XI, 131 pages. 1995.

Vol. 1616: I. Moerdijk, Classifying Spaces and Classifying Topoi. VII, 94 pages. 1995.

Vol. 1617: Vadim Yurinsky, Sums and Gaussian Vectors. XI, 305 pages. 1995.

Vol. 1538 – Second Edition: P.-A. Meyer, Quantum Probability for Probabilists. X, 312 pages. 1995.